TROUBLED WATERS

TROUBLED WATERS

Steamboat Disasters, River Improvements, and American Public Policy, 1821–1860

PAUL F. PASKOFF

Louisiana State University Press
Baton Rouge

Published by Louisiana State University Press
Copyright © 2007 by Louisiana State University Press
All rights reserved
Manufactured in the United States of America
First printing

Designer: Laura Roubique Gleason
Typeface: Whitman
Typesetter: J. Jarrett Engineering, Inc.
Printer and binder: Thomson-Shore, Inc.

Library of Congress Cataloging-in-Publication Data

Paskoff, Paul F.
 Troubled waters : steamboat disasters, river improvements, and American public policy, 1821–1860 / Paul F. Paskoff.
     p. cm.
 Includes bibliographical references and index.
 ISBN 978-0-8071-3268-5 (cloth : alk. paper)
 1. Mississippi River—Navigation—History—19th century. 2. Inland navigation—Middle West—Safety measures—History—19th century. 3. River engineering—Government policy—United States—History—19th century. 4. River engineering—United States—Finance—History—19th century. 5. Steamboat disasters—United States—Prevention—History—19th century. 6. United States—Politics and government—1815–1861. 7. Mississippi River Valley—History—19th century. I. Title.
 HE630.M6P37 2007
 363.12′3—dc22

2007001976

Portions of this book appeared previously, in slightly different form, as "Hazard Removal on the Western Rivers as a Problem of Public Policy, 1821–1860," *Louisiana History* 40 (1999): 261–82.

The paper in this book meets the guidelines for permanence and durability of the Committee on Production Guidelines for Book Longevity of the Council on Library Resources.∞

For Beth

CONTENTS

List of Illustrations / ix

Preface / xv

Introduction / 1

PART I

1.
Troubled Waters / 11

2.
Politics before Polk / 40

3.
Politics: Polk and Post-Polk / 64

PART II

4.
Ways and Means / 109

5.
Factors of Destruction / 138

6.
The Success of Public Policy / 164

Epilogue / 189

Appendices / 201

Notes / 251

Works Cited / 291

Index / 313

ILLUSTRATIONS

MAPS

1. Major Navigable Rivers and Competing Railroad Lines, c. 1860 / 10
2. A Bad Stretch of River: The Southern End of the "Graveyard" / 158

PLATES

1. Cairo, Illinois / 13
2. The Packet *Belfast* / 28
3. Cover of *Lloyd's Steamboat and Railroad Directory* / 31
4. The Sinking of the *Shepherdess* / 32
5. Drawing of Snag Boat from Patent Application / 135
6. The Packet *Jacob Strader* / 151

FIGURES

1. Steamboats in Operation and Steamboats Lost to Natural Hazards on Western Rivers, 1821–1860 / 39
2. Steamboat Tonnage in Operation and Tonnage Lost to Natural Hazards on Western Rivers, 1821–1860 / 39
3. Memorials for Improvements and Water-Related Appropriations: 19th–36th Congresses, 1825–1861 / 76
4. Federal Government's Receipts, Expenditures, and Budget Balance, 1821–1860 / 112
5. Receipts from Public Land Sales as a Percentage of Total Federal Revenues, 1821–1860 / 113
6. State-Specific, Miscellaneous, and Total Appropriations for River and Harbor Improvements, 1821–1860 / 116
7. Total Appropriations for and Net Expenditures on Western River Improvements, 1821–1860 / 117

ILLUSTRATIONS

8. Total Appropriations for Rivers and Harbors and Appropriations for Lighthouses, Buoys, and Markers, 1821–1860 / 120

9. Tonnage Engaged in the Coasting Trade and Appropriations for Light Stations, Marker Buoys, and Beacons, 1821–1860 / 120

10. Net Tonnage Cleared from U.S. Ports and Appropriations for Light Stations, Marker Buoys, and Beacons, 1821–1860 / 121

11. Total Federal Spending and the Military's Share, 1821–1860 / 125

12. Net Expenditures on Western River Improvements as a Percentage of Net Federal Expenditures, 1821–1860 / 126

13. Shares of Total Federal Expenditures Going to the Army and Navy and to Western River Improvements, 1821–1860 / 127

14. Public Land Sales and Town Formation in Four States along the Western Rivers, 1821–1860: IN, MO, OH, and MS (MS to 1848) / 131

15. Public Land Sales, Number of Towns Formed, and Miles of Railroad Track Laid: Ohio, 1821–1860 / 132

16. Public Land Sales, Number of Towns Formed, and Miles of Railroad Track Laid: Indiana, 1821–1860 / 133

17. Net Federal Expenditures on Western River Improvements and Number of Towns Formed in Five Affected States: IN, KY, MO, OH, and MS, 1821–1860 / 133

18. Convergence of the Ratio of Depth to Breadth of Steamboat Hulls across Tonnage Classes, 1818–1880 / 141

19. Convergence of the Ratio of Depth to Length of Steamboat Hulls across Tonnage Classes, 1818–1880 / 141

20. Convergence of the Ratio of Hull Depth to Hull Breadth across Tonnage Classes of Western River Steam Packets, 1842–1860 / 142

21. Convergence of the Ratio of Hull Depth to Hull Length across Tonnage Classes of Western River Steam Packets, 1842–1860 / 142

22. Relationship between Vessel Tonnage and Hull Length of Steamboat Packets on the Western Rivers, 1831–1861 / 144

23. Relationship between Tonnage and Hull Area of Steamboat Packets on the Western Rivers, 1831–1861 / 145

24. Steam Packets on the Western Rivers: Rate of Loss Due to Snags by Ratio of Hull Depth to Hull Length, 1844–1860 / 146

ILLUSTRATIONS

25. Units of Weight per Horsepower of Different Forms of Motive Power: Steamboats, Locomotives, and Gasoline Tractors, Selected Years / 148
26. Average Cylinder Inside Diameter (c.i.d.) and Average Stroke of Engines of Western River Steam Packets, 1843–1860 / 149
27. Average Total Cylinder Volume of Engines of Steam Packets on the Western Rivers, 1843–1860 / 149
28. Average Total Engine Cylinder Volume per Vessel Ton of Steamboat Packets on the Western Rivers, 1844–1860 / 153
29. Snagged Tonnage as Percentage of Operating Tonnage by Engine Volume per Vessel Ton of Western River Packets, 1843–1860 / 154
30. Interest Rates in Pennsylvania, outside Philadelphia, 1834–1860 / 155
31. Time Series of Engine Volume per Ton of Vessel Displacement and the Interest Rate in Pennsylvania, outside Philadelphia, 1844–1860 / 156
32. Engine Volume per Ton of Displacement of Western River Steamboat Packets by the Interest Rate in Pennsylvania, outside Philadelphia, 1843–1860 / 157
33. State-Specific Appropriations for River and Harbor Improvements and the 1850 Population of the States, 1821–1860 / 170
34. Total Appropriations for River and Harbor Improvements and Number of Wrecks Due to Natural Causes, 1821–1860 / 170
35. Tons Lost to Snags as a Percentage of Tons in Operation on the Western Rivers, 1821–1860 / 172
36. Tons Lost to Snags as a Percentage of Tons Lost to All Natural Hazards on Western Rivers, 1821–1860 / 172
37. Loss Rate Due to Snags and Federal Expenditures on Western River Improvements, 1821–1860 / 173
38. Mean Size of Steamboats in Operation and Mean Size of Steamboats Lost to Snags on the Western Rivers, 1821–1860 / 174
39. Average Age in Years of Western River Steamboats, 1821–1860 / 176
40. Average Longevity of Side-wheeler and Stern-wheeler Steamboats on the Western Rivers, 1832–1860 / 177
41. Cumulative Percentages of Steamboats on Western Rivers Not Yet Wrecked, by Vessel Age, 1831–1860 / 179
42. Weighted Index of Input Prices for Mississippi River Steamboats on the Louisville to New Orleans Run, 1821–1860 / 180

ILLUSTRATIONS

43. Weighted Index of Output Prices for Mississippi River Steamboats on the Louisville to New Orleans Run, 1821–1860 / 181

44. Index of Total Factor Productivity for Mississippi River Steamboats on the Louisville to New Orleans Run, 1821–1860 / 181

45. Weighted Index of Output Prices of Steamboats and Mean Size of Steamboats on the Western Rivers, 1821–1860 / 182

46. Index of Total Factor Productivity and Mean Size of Mississippi River Steamboats, 1821–1860 / 183

47. Weighted Index of Output Prices (freight rates) and Loss Rate of Tonnage Due to Snags, 1821–1860 / 184

48. Index of Steamboat Productivity (Pi/Po) and Loss Rate of Tonnage Due to Snags, 1821–1860 / 185

49. Tons Actually Lost to Snags and Projected Tons Lost to Snags, 1821–1860 / 187

50. Spending on Rivers and Harbors as a Percentage of Total Federal Spending, Exclusive of Debt-Servicing, 1821–1882 / 192

C.1. Net Expenditures for River and Harbor Improvements in Current and Constant Dollars, 1821–1860, Using Cincinnati, New Orleans, and Warren & Pearson Price Indexes / 212

C.2. Differences between Nominal and Adjusted Net Expenditures for River and Harbor Improvements, 1821–1860, Using Warren & Pearson, Cincinnati, and New Orleans Deflators / 212

C.3. Net Expenditures for River and Harbor Improvements, 1821–1860: Nominal and Adjusted for Inflation and Deflation / 213

TABLES

1. Centers of Construction of Steamboat Packets Plying the Western Rivers, 1840–1860 / 25

2. Number of Steamboat Arrivals and Tons of Freight Arriving at New Orleans by River from the Interior, 1840–1860 / 29

3. Positions and Distances from Mouth of the Missouri (downward) to Natchez / 37

4. Internal Improvements Legislation and Presidential Administrations, 1789–1861 / 42

5. Internal Improvements Appropriations in Each Presidential Administration, 1789–1861 / 43

ILLUSTRATIONS

6. Some Measures of the Scale of the Federal Government, 1821–1861 / 110
7. Hull Dimensions and Tonnage of Steam Packets, 1831–1860 / 140
8. Public Land Grants by the Federal Government in Support of Transportation, 1823–1857 / 162
9. Federal Land Grants to Railroads and Railroad Capital Formation, 1847–1857 / 162
10. Average Age in Years of Western River Steamboats, 1832–1860 / 175
11. Average Age of Steam Packets Lost by All Causes on Western Rivers by Five-Year Period and River, 1841–1860 / 178
B.1. Shares of Ohio River Mileage and the 1827 Miscellaneous Appropriation / 207
C.1. Wholesale Price Indices, 1821–1860 / 210
D.1. Number of Wrecks on U.S. Waters, by Cause and Five-Year Period, 1821–1899 / 214
D.2. Tonnage of Wrecks on U.S. Waters, by Cause and Five-Year Period, 1821–1899 / 215
D.3. Side-wheel and Stern-wheel Steam Packets on the Western Rivers, 1841–1860 / 216
D.4. Data on Steam Packet Engines / 217
D.5. Public Land Sales Receipts in Selected States, 1814–1860 / 218
D.6. State-Specific Federal Appropriations and Expenditures for Rivers and Harbors: 1790–1860, 1861–1882, and 1790–1882 / 220
D.7. Specific Federal Appropriations and Expenditures for Rivers and Harbors, 1790–1860: Specified by State / 222
D.8. Appropriations and Net Expenditures for Miscellaneous Improvements of Rivers and Harbors, prior to 1861 / 227
D.9. Appropriations and Net Expenditures for Miscellaneous Improvements of Rivers and Harbors, 1861–1882 / 230
D.10. Appropriations and Expenditures for Lighthouses, Light Stations, Beacons, and Buoys in Each State, 1791–1860 / 233
D.11. Number of Memorials Concerning Internal Improvements Sent by Each State to the 19th–36th Congresses, 1825–1861 / 234
D.12. Maverick Members of the House of Representatives in the 29th–32nd Congresses, 1845–1853 / 236

ILLUSTRATIONS

D.13. Maverick Members of the House of Representatives in the 29th Congress, 1845–1847, on the Issues of the Tariff and Internal Improvements / 243

D.14. Regional Origin of Memorials Concerning Improvements Received by the 19th–36th Congresses, 1825–1861 / 245

D.15. Maverick Voting on Internal Improvements in the House of Representatives in the 29th–32d Congresses, 1845–1853 / 246

D.16. Maverick Voting on Internal Improvements and a Protective Tariff in the House of Representatives in the 29th Congress, 1845–1847 / 247

D.17. Major Sources of Federal Revenue in Each Presidential Administration / 248

D.18. Major Objects of Federal Expenditures in Each Presidential Administration / 249

D.19. Incorporation of Transportation Companies in Ohio, 1821–1860 / 250

PREFACE

The work for this book began several years ago with what seemed at the time to be a fairly straightforward objective in mind. I wanted to establish quantitatively the extent of the federal government's failure to make the country's major rivers safe for steamboat navigation during the decades before the Civil War. Firmly establishing that failure would, I thought, have important implications for historians' understanding of the larger question of whether federal public policy contributed significantly to antebellum economic growth and development. I decided to study the rather narrow subject of river improvements because, unlike other aspects of the internal improvements program, such as roads or canals, any increase in safety on the rivers brought about by the reduction or elimination of natural hazards to navigation could be determined empirically. I imagined that such an empirical determination would not be especially difficult. In the end, however, it proved to be considerably more challenging than I had initially assumed.

A much greater surprise, however, resulted from a quantitative analysis of the data concerning steamboat construction and operation and federal spending on river improvements. Rather than confirming a record of failure, the analysis indicated a federal program that had largely succeeded in making steamboat navigation safer by removing natural hazards on the Mississippi and its major tributaries. That finding necessarily led me to question my initial assumptions concerning the public sector's role in the economy. This book is the result of that inquiry.

My research benefited from the expertise and kindness of librarians and archivists at many repositories, especially the Special Collections and Government Documents departments and the Map Library of the Louisiana State University Libraries in Baton Rouge. Jennifer Cargill, Dean of Libraries, and Faye Philips, Associate Dean, LSU Libraries, and Director of Special Collections, were helpful at every turn. The staff of the Louisiana and Lower Mississippi Valley Collections, particularly Judy Bolton, Gina Costello, and Leah Jewett, were invariably patient before what must have seemed an endless stream of requests by me for documents, volumes, and photographs. This book would not have been possible without their assistance.

The holdings of LSU Libraries' Government Documents collections also provided essential materials, especially the record of congressional debates on the river and harbor improvements program, memorials and petitions to Congress from constituents interested in the program, and data concerning appropriations for and expenditures on improvements. Smitty Bolner, once head of Government Documents and now retired, seemed always to know where to find needed materials, and I benefited from her expertise. The holdings of LSU's Map Library enabled me to trace the changing course of the Mississippi River and its major tributaries, and I am indebted to its curator, John Anderson, for his help. The staff of LSU Libraries' Reference Department, especially Tom Diamond, Head of Reference and Collection Development Services, and one of his graduate student workers, Kirsten Corby, have my thanks for having helped me find my way through a thicket of patent files.

The collections of the library of the Army Corps of Engineers in Washington, D.C., hold manuscript sources that were of considerable value to this study. Similarly, visits to the Corps' Mississippi River research station in Vicksburg and a more than eighty-acre outdoor model of the Lower Mississippi River system in Clinton, Mississippi, gave me a perspective on the river's dynamics that might otherwise have eluded me. I also benefited from the outstanding collections of the Cartographic Research Laboratory of the University of Alabama and from the kind assistance of its Director, Craig Remington. The collections and staffs of the Alabama Department of Archives and History in Montgomery and the Mississippi Department of Archives and History in Jackson were also helpful.

Over the years of research, I have had the help of some first-rate assistants, Guice Giambrone, Erin Murray, Justin Poché, Joel Boussert, and my daughters, Catherine Chang and Martha Welsh, all of whom spent part of their undergraduate student years tracking down sources and entering data. My gratitude to them all is profound. I wish also to thank Ronald Perritt and Benjamin Price, who gave generously of their skill in computer graphics to convert the book's many charts into publishable form, and Clifford "Dupe" Duplechin of the Cartographic Laboratory of LSU's Department of Geography and Anthropology, who produced the map in chapter 1. I am also grateful to Karl Roider, Jane Collins, and Guillermo Ferreyra, each successively Dean of the College of Arts and Sciences while I worked on the book, for providing essential research support.

While writing and revising the manuscript, I benefited from the criticisms and suggestions of several readers. William Cooper, Roger Ekirch, Gaines Foster, and Victor Stater, friends and colleagues, read chapters, often more than once, and their incisive comments improved the book's writing and reasoning. Other friends and colleagues—Carl Brasseaux, Steven Collins, John Henderson, Suzanne Marchand, Karl Roider, Charles Royster, Carol Shiner-Wilson, and Daniel Wilson—patiently listened to descriptions of this or that part of the manuscript and provided valuable advice. A friend and former colleague, Robert Becker, applied his remarkable editorial eye to an early draft of the book and saved me from committing numerous errors. Stanley Engerman read a subsequent draft, and his criticisms and encouragement were particularly helpful. He and James Huston read the manuscript for LSU Press and provided page-by-page, line-by-line criticism, which I took to heart and without which the book would have been considerably weaker. I am also grateful to Rand Dotson, my editor at LSU Press, and MaryKatherine Callaway, the Director of the Press, for their confidence in the value of the manuscript, and to Susan Brady, my copy editor, for her able assistance in turning a manuscript into a book.

I have dedicated this book to my wife, Beth, my friend of more than thirty-five years. She read chapters or listened to me read the entire text aloud and offered gentle but nonetheless incisive criticisms and helpful suggestions. Her criticisms and advice, like the criticisms and advice offered by other readers, made the book stronger, but ultimately the responsibility for any weaknesses it may still have is mine, alone.

TROUBLED WATERS

INTRODUCTION

On July 14, 1814, Robert Fulton's steamboat, the 371-ton *New Orleans,* ran onto a thick, spearlike submerged tree trunk and sank on the Mississippi River off Baton Rouge, Louisiana. Less than three years earlier, the *New Orleans* had demonstrated the practicability of steamboat transportation on the western rivers when it completed its initial run from Pittsburgh to New Orleans, arriving in the Crescent City in January 1812 after a voyage of not quite three months. By virtue of those two trips, its first and its last, the *New Orleans* achieved two distinctions: it was the first steamboat to navigate the Ohio and Mississippi rivers, and it became the first one to fall prey to the natural hazards that infested those and the other rivers of the American interior.[1]

The thing that destroyed the *New Orleans* was colloquially known by its generic name of "snag," and an encounter with one was called a "snagging." During the next several decades, as steam transformed the Mississippi, Ohio, and Missouri river valleys into one of the nation's most dynamic economic regions, snags and other naturally occurring hazards on the rivers claimed a mounting toll of steamboats, cargoes, and lives. During the 1850s, they destroyed 348 steamboats, or 80,043 tons of shipping, well over half the total number and tonnage of steamboats lost to all causes, including fire and explosion, and about 5 percent of all the steamboats in operation each year of that decade. The rate of loss had been even higher in earlier decades.[2] Such losses made the question of what might be done to reduce those hazards one of the most pressing issues of public policy in the history of the United States. One would think that any program dedicated to removing such hazards would enjoy the support of Congress and Americans in all parts of the United States. One would be wrong.

On March 10, 1846, Alabama congressman William Lowndes Yancey, excitable and given to rhetorical excess, threatened his House colleagues that southern states would secede from the Union unless government policy changed. About a week later, Jefferson Davis, his colleague from Mississippi, used more temperate language to warn of the same thing.[3] They were not alarmed over real or imagined dangers to the survival of slavery, as they would be within only a few years. And they were not distressed over protective tariffs, as some southerners had been little more than a decade before. Instead, what exercised them was the use of federal money to improve the great rivers and lakes of the nation's interior. Of course, few if any political matters of consequence arise or fester in isolation from others, and so much was true of the river and harbor improvements policy.

On a fundamental level, some of the intensity of the controversy over improvements derived from southern fears for the future of slavery, though the connection between them was seldom explicitly drawn. And, while slavery ultimately caused the disruption of the Union, the improvements issue carried enough constitutional baggage in its own right that in the decade and a half before the firing on Fort Sumter few contemporaries would have discounted its disruptive potential. At a time when notions of Hamiltonian federalism vied with those of Jeffersonian confederationism, people in and out of Congress professed a willingness to abandon and even to destroy the Union over this matter of public policy.

Why such an apparently worthy object as the improvement of rivers and harbors should have excited such passion is a question worth pondering, as is the related question of whether the federal program to achieve that object succeeded in making the nation's inland waterways and harbors safer for navigation. The river improvements program touched on questions of economic development, constitutional law, partisan politics, and sectional rivalry. As such, it was one of the most volatile issues in national, sectional, and state politics. Like the protective tariff with which it was so often linked by the enemies of both, the program was one of those antebellum tender spots, easily irritated and prone to inspire overheated rhetoric about a national union imperiled, depending upon one's perspective, by a greedy group of iniquitous, nationalizing capitalists or by a coterie of myopic, small-minded, backward-looking parochialists.

The adversarial pairs in the battles over the river improvements program assumed other, sometimes overlapping, forms, as well. As strict constructionists clashed with loose constructionists, Democrats and Whigs (later, Democrats and Republicans) relentlessly contending for political advantage found the improvements issue to be a useful arena for partisanship. Older, more settled parts of the country nervously eyed the younger, rapidly growing West, aware that as its population continued to swell, its political clout and therefore its role in making national policy would grow. Western spokesmen insisted that their region contributed far more to maintaining the federal government than it received in benefits from the national treasury and laid the blame for such inequality at the feet of the Atlantic seaboard states. The politicians of the Upper South, eventually joined by those of the Lower South, studied successive population censuses, nursed their insecurities, and came to see any proposal for river and harbor improvements as a part of a complex but not subtle assault on the ancient prerogatives of the South by those hostile to its interests. It was these considerations that had prompted William Yancey to insist "I fear not the West" and then to make his thinly veiled threat that the South would secede if the West did not relent in its demands for ever larger shares of the nation's treasure.[4] And yet, contrary to the beliefs of many opponents, especially those in the South, the program of federally funded river improvements proceeded without any grand design and was, after the demise of the "American System," a policy implemented and sustained without an underlying ideology.

Instead, federally funded river and harbor improvements had the important but prosaic and pragmatic goal of reducing the risk of movement on the Great Lakes and especially the western rivers. Danger on the rivers came in a variety of forms. Shoals, rapids, ice, rocks, uprooted trees and submerged steamboat wrecks embedded in river beds, and sandbars were the most common naturally occurring perils and accounted for the largest number of steamboat disasters. Boiler explosions, fires, collisions, and human recklessness and incompetence also destroyed a large number of vessels. This dauntingly large array of river hazards required a similarly broad range of efforts to remove or at least ameliorate them.

Collectively, the federal government's river and harbor improvements effort constituted the single-largest and longest-lasting component of an overarching program of internal improvements that may be said to have begun with George Washington's call for a national military road. Controversial in its entirety, as well as in its constituent parts, the public policy of internal improvements was also the political issue of internal improvements, and no part of the internal improvements policy was more fiercely contested or more bitterly resented and defended than the river improvements program. The program was very much a product of a rich mix of partisan, sectional, and fiscal politics, which profoundly influenced its geographical scope and fiscal extent. Ultimately, however, the rivers themselves determined the direction and intensity of the policy. The economic vitality and importance of the Mississippi River valley and that region's rapidly growing population drove the policy. As steamboat traffic increased and losses of steamboats and their cargoes and passengers mounted, clearing the rivers of their hazards became a matter of some urgency.

Even the program's bitterest opponents in Congress never called into question the necessity for improving the rivers. What they objected to were not the ends of the policy but its means. In their view, states and localities along the rivers and private capital should bear the responsibility and the costs of clearing the rivers of their obstructions and dangers. That approach, they insisted, was consistent with the Constitution. Advocates of a federal program rejected that constitutional argument, but they also maintained that the geographical, technical, and fiscal requirements of the work exceeded the capacities of individuals and state and local governments.

In the end, what mattered to the people who lived along the great western rivers and whose livelihood depended on the rivers' steamboat traffic was that something be done to make the rivers safer. An economic and technical problem of considerable magnitude, it nevertheless potentially had a solution. Between 1821, when the program to improve the rivers began, and 1861, the federal government applied money, engineering expertise, and technology in an effort to reduce the naturally occurring hazards on the major rivers of the nation's interior and thereby lower the rate of steamboat losses. Attempts to eliminate these hazards focused on three main lines of attack: surveying and mapping river channels and passes; blasting rock outcroppings and ledges and dredging the rivers and their approaches to remove sandbars; and pulling snags and sunken steamboat and flatboat wrecks from the channels and preemptively removing trees from riverbanks before they could be swept into the channels to form snags.

These improvements were projected and executed within a political, economic, and technological context that determined their pace, timing, and effectiveness. In the political arena, proponents of state governmental supremacy jealously contested any effort to project federal power through a general program of improvements. Another reason for the intensity of the political struggle over the river improvements program was the program's fiscal magnitude, which, apart from outlays for the army and navy, exceeded federal spending in many years for almost every other purpose. Competition between and within the political parties and an increasingly vituperative sectional dispute over an array of policies, including a national bank and the tariff, also complicated the political setting. Had opposition to a federal program of river improvements been driven only by such political disputes, however, funding for the program might well have benefited from the sausage grinder of political compromise. But any such possibility was complicated and probably precluded by a rapidly expanding contest between champions of laissez-faire and advocates of an activist role for government, especially the federal government, in the promotion of economic development.

Many historians have concluded that in this contest the laissez-fairists were more in harmony with the spirit of the age, especially with respect to internal improvements. According to this view, the triumph of laissez-faire capitalism—embodied in the rapid growth of the railroad industry and built on the failings of state government–funded enterprise—was substantially achieved before the Civil War.[5] A corollary to this interpretation is the conclusion that the federal program of internal improvements, especially river and harbor improvements, ultimately failed to promote economic growth and development. This view became fairly well-established by the mid-1970s and represented a departure from "the [then] current orthodoxy that emphasizes significant public involvement in nineteenth-century internal transportation developments which appears to have resulted from preoccupation with canals and railroads."[6] In fact, until that time, most of the literature on the subject did attribute much of the progress in river and lake transportation to investment by states and, especially, the federal government. The displacement of this perspective by one that emphasized the role of laissez-faire capitalism marked the triumph of what has since become a new orthodoxy.

But this new interpretation goes too far. The available evidence offers qualified support for the traditional view that government played an important role in the projection, financing, and construction of the nation's transportation and communication infrastructure. A recent examination of the benefits arising from the federal government's expenditures on antebellum internal improvements, mainly roads, credits the program with opening the frontier and preparing the way for extensive western settlement and development.[7] There is, in fact, a case to be made for the measurable impact of government's role in promoting economic development during the decades before the Civil War, and that case is set forth in this analysis of the federal river improvements program.

The basic argument offered here is that the federal government played an important role in the economic development of the United States before the Civil War by undertaking internal improvements projects, especially projects directed at removing hazards to navigation on the great rivers of the nation's interior, and through extensive public land sales. The primary benefits from these improvements were a reduction and stabilization of the rate of loss of steamboats and an increase in their working life and productivity. Ultimately, increased safety on the rivers facilitated the settling and economic development of the greater Mississippi Valley. These conclusions rest on a detailed analysis of the quantitative record of steamboat wrecks, federal appropriations and expenditures for river improvements, the technology of the steamboats that operated on the western rivers, and public land sales and town formation in the river states.

The success of the federal government's program of river improvements has some important implications for our understanding of antebellum public policy. It poses a challenge to the current orthodoxy concerning the essential role played by the heroic agents of a triumphant laissez-faire capitalism in the evolution of the antebellum political economy. According to this view, the public sector had only a supporting role in the drama of national economic development, and then more often than not as the knave or the fool. Put in these terms, this characterization is admittedly oversimplified, but it is not caricature. However important mobilized private capital and state-level aid were to turnpike building and railroad construction, both were inadequate to the task of clearing hazards to navigation from the western rivers, the prewar era's major arteries of commerce. The notion that an entrepreneurial capitalism, if unfettered by federal interference, might have managed that accomplishment is unrealistic and is refuted by the historical record. Steamboat owners and their shipping customers were keenly interested in safer rivers, but they lacked the means to achieve that result, and they knew it.

Even among exponents of laissez-faire, few purists were to be found, and most, especially those who were railroad promoters, were pragmatic in their approach to government's role in the economy. Where capital investment was concerned, government might provide assistance but must not impose direction. While that approach worked for the railroad industry, it would have failed miserably had it been applied to the clearing of hazards in the great rivers. The difference between railroads and the rivers in this respect was that entrepreneurs were able to realize sustained profits from their capital investments in railroads and from the assistance given them by the federal government and various state governments. No such prospect could be held out for work to remove hazards on the rivers. Moreover, local- and state-level efforts to improve the nation's major rivers seldom succeeded in marshaling the large amounts of money necessary to accomplish that task.

Advocates of a federally funded program of river improvements often argued that the solution to the problem of natural hazards on the major rivers should and had to be national in character because the rivers themselves were national in extent and significance. Their view of the matter would triumph after the Civil War, when a national political consensus about wielding federal power and about the importance of making the rivers safer unleashed millions of federal dollars for that purpose. But even before the war, the federal program to remove natural hazards from the great rivers of the interior, funded much more modestly and sporadically, succeeded in making steamboat transportation safer and more reliable. That success and its stimulative effect on economic development were significant achievements of the antebellum national state.

PART I

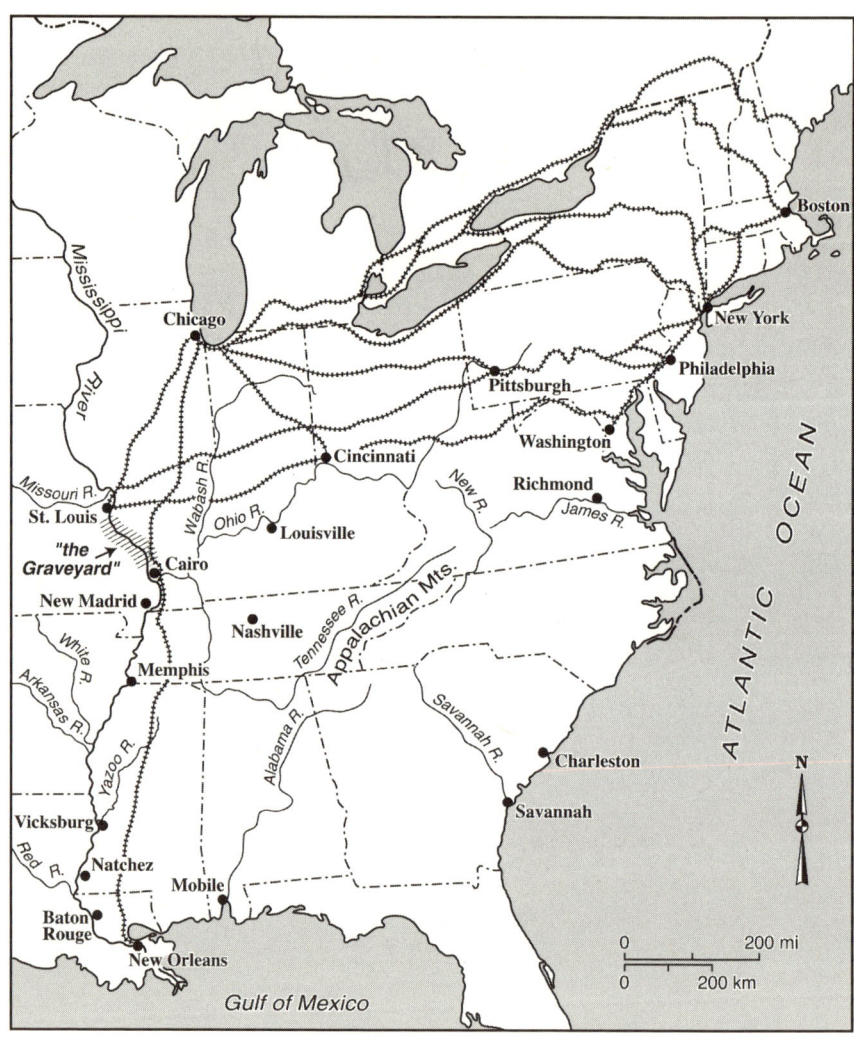

Map 1. Major navigable rivers and competing railroad lines, c. 1860. Map by Clifford Duplechin.

1

TROUBLED WATERS

In 1849, it was possible to travel the navigable length of the Mississippi River, from the Gulf of Mexico to Minnesota's Falls of St. Anthony, a trip usually requiring perhaps two weeks and covering a distance of well over two thousand miles, in a matter of only a few hours. The cities, farms, and woods on the river's banks would slip by, as would the occasional steamboat wreck and sandbar. Such a journey was, of course, a virtual and not an actual one, the product of a painter's art and a mechanic's craft. The "passengers" on such a voyage were really spectators, theatergoers who, for the price of a ticket, could watch roll after roll of one of the great Mississippi River panoramas of the day unwind.[1]

The rolls were seemingly endless expanses of canvas that were stitched together and wound on spindles. Moved by massive mechanisms before seated audiences, they offered viewers in the late 1840s and early 1850s "the illusion of steamboat travel" on the Mississippi.[2] They were the work of gifted landscape painters, notably Leon Pomarède and Henry Lewis, who had contrived their representations of the river as a money-making form of popular entertainment for audiences hungry for sensation and edification. Although the publicity for these panoramic spectacles claimed canvas sizes measured in miles, their actual lengths were probably on the order of a few to several hundred yards, some perhaps as long as thousands of feet and as high as twelve feet. Even so, viewing times of two to three hours were reported by newspaper reporters who attested to the realism and beauty of the work.[3] None of the panoramas survives today, but it is possible to reconstruct what their audiences would have seen.[4]

Viewers of the panoramas could begin their vicarious voyage at the river's mouth on the Gulf of Mexico. After steaming upstream somewhat more than one hundred miles, they would draw abreast of New Orleans, which lay behind its levee on the river's east bank "in the great bend of the Mississippi, . . . in the shape of a crescent, . . . the principal city in the South, and the third commercial mart of the United States."[5] Visible behind the levee was the city's great cathedral, and across the river on the opposite bank was the town of Algiers. The New Orleans riverfront was an impressively busy place as fleets of steamboats, sailing craft, and flatboats took on and put off cargo while moored along the levee.

Continuing upstream past Plaquemine and Donaldsonville to Baton Rouge, Louisiana's capital, they could see the still uncompleted capitol building, the structure that Mark Twain called "that Gothic horror." Proceeding past Bayou Sara and St. Francisville, they would steam onward to Natchez, Mississippi, to gaze at the city's tall cliffs towering above them, and then go on to the steep defiles of Vicksburg. Further up the river, meandering through its vast delta, northward to Memphis, Tennessee, the line of sight would jog to the west, then to the east, presenting the armchair travelers with a remarkable series of different riverscapes.

This stretch of the Mississippi was one of oxbow lakes, the remains of meanders cut off years before. This was where the river had been convulsed by the New Madrid earthquake of 1811, thought to have been one of the most violent seismic events in the history of the Mississippi Valley. According to some accounts, the quake threw up towering swells in the river, which briefly reversed direction and ran northward, obliterating much of its bankland, and was felt as far away as Pittsburgh.[6] Aftershocks persisted for weeks, even months, and left once-fertile fields stained, as they are still, by large, barren patches of sand deposited by the tortured river. A large area of once-elevated land along the river subsided, to be known thereafter as "the sunk country."[7]

Beyond the New Madrid area lay the middle reaches of the Mississippi, where the Ohio joined it at Cairo, Illinois. All but surrounded by the Ohio and Mississippi, Cairo was prone to frequent flooding and was almost always damp, contrary to the shameless claims for its pleasing aspect and healthy climate made by its boosters. One visitor to Cairo who had believed those claims was Charles Dickens. He arrived by steamboat in 1842, en route from Cincinnati via Louisville to St. Louis. Expecting to see one kind of place and finding quite another, he vented his disappointment in a scathing characterization of Cairo, which he called "a hotbed of disease, an ugly sepulchre, a grave uncheered by any gleam of promise; a place without one single quality, in earth or air or water, to commend it: such is this dismal Cairo." The city would remain for him "the detestable morass called Cairo."[8] If somewhat harsh, Dickens nevertheless saw Cairo's vulnerabilities clearly enough. Flooding was not merely possible; it was likely. As a steamboat directory and river guide published in 1856 noted, when there was unusually high water on the Ohio River, as there had been in 1847, "the water . . . almost inundates Cairo."[9] In 1858, a flood put Cairo under water "9 to 16 feet deep throughout the entire town."[10]

Well above the confluence of the Ohio with the Mississippi, spectators would see the junction of the Missouri and the Mississippi, as the Missouri flowed into it from the left, or west. But before they got there, they came upon St. Louis, the most impressive city on the Mississippi above New Orleans. St. Louis was a good place to be at midcentury. It was at about the geometric center of a United States newly enlarged by territory seized from Mexico in the war that had ended about a year earlier. The city's political leaders were proud of the contribution that St. Louis had made to the war effort and had occasion to remind Congress that the city's federal arsenal was "the place whence have been fitted out the military expeditions which have conquered New Mexico and Chihuahua; and the greater portion of the field ammunition and shells used in the reduction of Vera Cruz and the city of Mexico have been manufactured and issued to the army at the St. Louis arsenal."[11] Men and ammunition had traveled from St. Louis down the Mississippi River to New Orleans and, from there, to the battlefields of Mexico.

Cairo, Illinois, not long after Charles Dickens visited the city.
Source: Lloyd, *Lloyd's Steamboat and Railroad Directory, and Disasters on the Western Waters*, 256, Louisiana and Lower Mississippi Valley Collections, Louisiana State University Libraries, Baton Rouge. Courtesy of Louisiana State University Libraries' Special Collections.

Even well before the war, St. Louis had profited from an expanding trade up and down the river and its major tributaries, and, with a population of almost seventy-eight thousand, it was, after Cincinnati, the largest city of the interior.[12] So rapid and extensive was its progress that it had begun to rival New Orleans for the distinction of being the most important city on the Mississippi. Once primarily an entrepôt for the trans-shipment of manufactured goods, plantation cotton, and grain, St. Louis had become a manufacturing, technological, and financial center and, despite Missouri's ferocious attachment to slavery, had come to resemble Cincinnati more than New Orleans. The reasons for St. Louis's growing size and prosperity were rooted in geography and technology and were widely recognized at the time. The *Western Journal and Civilian*, a local monthly, published an article in its January 1848 issue that confidently predicted that, because of the city's strategic location, "she must necessarily ultimately become one of the largest Manufacturing and Commercial marts on the American Continent."[13] Such exuberant confidence about the city's prospects was in the air. That same year, a gazetteer of the inland rivers, *Conclin's New River Guide,* asserted flatly that "[T]here is no town in the western country more favorably situated, as the seat of an immense trade.... It has this obvious advantage over any town on the Ohio, that steam-boats can run between here and New Orleans at the lowest stage of water."[14] That year steamboats landed at St. Louis 3,179 times.[15]

The city's residents understood that they owed their prosperity to the river and particularly to the steamboats that had transformed the Mississippi, Ohio, and Missouri from continental drains into two-way arteries of commerce. In a request to Congress in 1844 for money to improve the western rivers and, in particular, St. Louis's harbor, a citizens' committee noted that "[t]he carrying trade on the Western waters has grown up, through the agency of steam, to an amount and importance and with a rapidity that are incredible to those whose attention has not been particularly turned to the subject."[16] The petitioners also informed Congress that their city, "although the remotest city in the West, and in the interior, has kept pace with the rapid growth of this trade.... In steamboat tonnage she is the third city in the Union—the great commercial marts of the North and South, New York and New Orleans, only exceeding her."[17]

When the mayor and city council of St. Louis petitioned Congress in 1848 and recalled their city's role in the Mexican War, they too hoped to persuade the House and Senate to vote an appropriation to improve the city's harbor on the Mississippi. As the petition of four years earlier had done, the one in 1848 stressed the importance of the river steamboat to the city's prosperity. It pointed out that the river's depth below the city was sufficiently great that "[t]he largest class of boats engaged on the lower Mississippi can ascend to St. Louis, with scarcely any exception, during all stages of the water, and at all seasons of the year—thus St. Louis is likewise made the great depot of the ascending trade for the upper Mississippi and Missouri."[18] The mayor and council were confident that so great was the traffic in and out of the city that the entire country must be "interested in its prosperity and conservation, and any calamity that would befall it would be a national calamity."[19] Such geocentrism was hardly unique to St. Louis. It was, in fact, an essential premise of the argument made by westerners in favor of federal funding to improve the Mississippi, Ohio, and Missouri rivers by eliminating natural hazards to steamboat navigation.

Steamboats had made St. Louis.[20] On the night of May 17, 1849, they almost destroyed it. It was this St. Louis, a city fighting for its life, that the great panoramas presented. Laid out before the spectators was the most arresting sight of the entire virtual voyage, the horrible but thrilling scene of St. Louis ablaze in a great fire, steamboats flaming like torches along the levee as the city's waterfront and much of its interior burned. The fire had begun aboard the packet steamboat *White Cloud*, which had been moored on the levee along the Mississippi River. As the vessel burned, embers fell on the nearby 299-ton *Edward Bates*, another packet, and ignited it as it moved downstream past a long row of steamboats that were nosed into the arc of the riverbank at the levee. Perhaps the flames burned through the hemp steering cables of the *Edward Bates;* perhaps its crew panicked. In any event, the vessel, out of control, bumped into one boat after another, consecutively setting them ablaze.[21]

Firemen from the city's volunteer fire companies devoted their initial efforts to extinguishing the flames aboard the ships in the hope of preventing the fires on the water from spreading to the land. That approach quickly proved futile as sparks and flames jumped from the line of burning steamboats to the wharves, warehouses, and other wooden buildings along the levee. In the flame-lit darkness, the firemen shifted their attention to the burning buildings, attempting to extinguish the multiple fires before they joined together as one great fire. They failed, in part because the intense heat and walls of flame prevented them from getting close enough to the river to draw water from it for their pumps, hoses, and buckets.[22]

Soon the wind off the river swept the flames up the streets leading from the water toward the heart of the city, and along those corridors one wooden building after another caught fire. The fire was now close to being out of control and threatened to destroy much, even most of the city. At that point, St. Louis found a savior in Thomas B. Targee, a captain of the Missouri Company Number 5 volunteer fire company.[23] Targee proposed to create a firebreak by blowing up buildings in the path of the advancing flames, hoping that the fire, deprived of further fuel, would burn itself out, thereby sparing most of the city.[24]

Having explained his plan to the other firemen around him, he ordered kegs of black powder, probably surplus stock left over from the Mexican War and stored at the city's federal arsenal, to be brought to the fire line.[25] He went home briefly to tell his wife what he had done and what he intended to do, and then returned to the fire.[26] When the kegs arrived, Targee began to supervise their placement in buildings located strategically in the fire's path. Once placed, Targee intended to order their detonation by fuse. He was putting one of the last kegs in a drug store close to the fire line when it detonated, killing him instantly as "his body was blown in every direction."[27] His comrades, continuing to work according to his plan, set off the kegs in the other buildings without further mishap. The explosions did just what Targee had hoped they would do, and the fire subsided to a level at which firemen could extinguish it. The buildings in the city's heart had been saved; its riverfront was a ruin; and fifty-seven looters were arrested.[28]

Along the levee, 5,603 tons of shipping—twenty-two steamboats—had been lost, many burned almost to their waterlines.[29] In terms of the number of ships destroyed, it was the single-worst day in the history of the inland navigation of the United States.[30] That it may have begun as the work of an arsonist—another steamboat had burned under suspicious circumstances some weeks earlier at the same berth at which the *White Cloud* had caught fire—of course mattered little to the people of St. Louis, from whom the fire had taken 418 buildings and twenty lives.[31]

And yet, while the night of May 17–18 was terrible and memorable, it marked only a brief interruption of the city's rapid growth, and only for a short time did it receive much attention in the city's press. An article in the *Western Journal and Civilian* that appeared under the heading "Great Fire in St. Louis" is noteworthy more for what it omitted than for what it said. Although it gave the dollar values of destroyed buildings and mercantile inventories and steamboats and cargoes, as well as the amount of each category that was insured, the article asserted that the actual cost of the fire would "probably never be known." And, in any case, that figure had "been greatly exaggerated in many of the public prints."[32] Nothing was said of the number of buildings burned and lives lost. Instead, readers were informed that "though there may be cases of individual loss, yet in the aggregate the assurance and the enhanced value of the ground in the burnt district is quite equal to the value of the buildings destroyed."[33] Thus, even the catastrophe had been transformed into a blessing by puffery and the peculiar alchemy of St. Louis's civic economy.

This boundless optimism precluded pessimism or bad news, even if that required ignoring or distorting facts, such as the extent of the losses suffered in the fire. The *Western Journal* put the value of property lost at below $1 million and assured its readers that "no fire of the same extent, perhaps, ever took place that caused as little real suffering, or where so few individuals were rendered totally destitute."[34] Had an epidemic not swept the city after the fire, the paper said, many of the destroyed buildings would have been rebuilt immediately.[35] Such boosterism characterized every issue of this periodical, but the tone was typical of publications of its sort and time.[36] The playing down of the damage caused by the fire was likely done with an eye to the continued promotion of St. Louis's economy and was aimed especially at the commercial interests of the city and their trading partners in other cities. Despite such denials, the fire had inflicted considerable damage on St. Louis and had cut its telegraph connection with the rest of the country for at least three days, causing considerable alarm as far away as New York, where the *New York Herald* reported: "We have nothing later from St. Louis. The wires are deranged between St. Louis and Louisville."[37] If anything, the rendering of a great city incommunicado with the rest of the nation, even temporarily, could only have heightened the national public's fascination with the fire, both as an urban catastrophe and as a spectacle on a grand, heroic scale.

Reported in the nation's newspapers and periodical press, the conflagration was almost immediately captured on canvas by landscape painters and pen-and-ink illustrators, with the most influential rendering being that by Nathaniel Currier, later of Currier & Ives fame.[38] Newspapers around the country reproduced Currier's lithograph, giving many of their readers a vicarious glimpse of a great fire, the single-greatest danger faced by the residents of any city of the day.[39] By far the most compelling depictions of the great St. Louis fire were those presented in Pomarède's and Lewis's panoramas, or "scenic newsreels," of the Mississippi River.[40] Pomarède's panorama, which was shown that summer, was the first to include the fire.

With memories of the fire still fresh and evidence of its destructiveness still visible, the *St. Louis Weekly Reveille* understandably enough made Pomarède's depiction of the fire the centerpiece of its review of his panorama. First shown was "a long line of steamers wrapped in flames.... In the background, a heavy mass of black clouds, looking more sullen from the contrast, lowers threateningly over the city." The city is in flames. As the canvas moved on, the light from the fire dimmed and then went out as "the darkness of night again shrouds the city. Then follows morning, revealing, magically enough, a totally different scene, a view of 'St. Louis *in ruins*,' wherein a mass of charred wrecks choking the harbor, whole blocks of blackened and tottering walls, groups of citizens surveying their own ruins &c., &c., form the principal features." The reviewer concluded with the observation that "[t]he whole catastrophe ... is truthfully recorded and must reflect credit on the genius and skill of the artist."[41]

Pomarède had beaten his competition and set a standard for spectacle. His success, particularly the impact of his depiction of the fire, convinced Henry Lewis to change the design of his own great Mississippi River panorama. In a letter to his brother in late July 1849, he indicated that he had expected to finish the painting considerably earlier, but that he had decided to incorporate a depiction of the fire that would "show this under 3 effects—namely, the sun setting on St. Louis as it was—the moon rising—and then the first boat taking fire and communicating to all the others and the grand conflagration will finish the spectacle. It will make a grand finale."[42] The critical reception of his massive painting gave Lewis ample evidence that his effort had been successful.

Perhaps a determination to put the fire and its unfortunate associations out of the public's mind accounts for the fact that a review of Lewis's Mississippi River panorama published in October by the boosterist *Western Journal and Civilian* made no mention of the fire. Instead, the reviewer emphasized the work's scope and physical size, which, at a reported 1,325 yards long by three yards high, covered the river's course "from New Orleans to the Falls of St. Anthony" and was the "largest picture, we imagine, that has ever been painted." Also mentioned were the exertions of the artist, himself a resident of St. Louis who, having been a mechanic, became an accomplished and "self-instructed" painter. Of the beauty of the panorama itself, the reviewer asserted that "we can say with truth, that we never enjoyed a higher degree of pleasure from the exhibition of any work of art that we have witnessed."[43]

The initial popularity of the panoramas in cities throughout the country had owed a lot to their novelty and size, and to the theatrical effect of trompe l'oeil.[44] But much of their appeal for the public was likely due to their joint subject matter, the Mississippi River and its steamboats. The Mississippi, of vital interest to those who lived along it, fascinated people back east who perhaps had only a vague idea of where it was. Because of its fundamental economic importance and impressive scale, its boosters delighted in calling it "the American Nile."[45] A similar heroic stature was accorded the steamboat and the railroad, the great innovative transportation technologies of the first half of the nineteenth century. But, of the two, it was the steamboat that, by midcentury, had had the more profoundly important economic and social influence and had secured the more prominent place in the American imagination, where it loomed large. Early in 1850, *Scientific American* carried a brief account of the construction of a small, shallow-draft steamboat designed to navigate the Mississippi River above the Falls of St. Anthony. The story appeared under the heading "Where is the Wilderness?" The magazine's answer was emphatic: "Where the steamboat goes, there the wilderness disappears."[46]

STEAM AND SAFETY

Americans were still attempting to come to terms with the consequences of a steam-engine technology that most did not comprehend, but that permitted the generation of enormous motive power. Their Faustian bargain entailed acknowledging that fires aboard steamboats, while all too common, were at some minimal frequency of occurrence unavoidable. Explosions of steamboat boilers, while less common, were even more horrible and thrilling, as well, especially when seen and felt from a distance or contemplated as an image on canvas. Like fires aboard steamboats, they were also understood to be a reducible risk, but one that was intrinsic to that mode of river transportation and its technology. Nevertheless, such incidents alarmed the public, who demanded that something be done to make steamboat travel safer.

Many advocates of greater steamboat safety were convinced that steamboat owners and operators could not be relied upon to police themselves. They also recognized that state-level regulation was spotty, inconsistent, and necessarily limited in its geographical reach.[47] The situation presented, for all practical purposes, a regulatory vacuum, into which, logically, if somewhat warily, the federal government was twice drawn. In 1838 and again in 1852, Congress passed legislation requiring inspection of steamboat boilers. The act of 1852, more ambitious than its predecessor, also mandated the licensing and regulation of river pilots on vessels carrying passengers.[48]

The aim of the 1852 law was to minimize the perils of steamboat travel arising from either design flaws in the vessels and their engines or the recklessness and cupidity of their human operators. There was every reason to expect that the legislation would succeed in reducing vessel losses due to fires and explosions and so save the lives of people who otherwise would perish on the western rivers. Historian Louis Hunter thought that the regulatory legislation had had a significant and immediate effect, and other historians have agreed, saying that the federal inspection laws passed in 1838 and 1852 to regulate the operation of marine steam engines had significantly reduced the loss of life from fire and explosion aboard the nation's steamboats.[49] Although these assertions are, on their face, plausible enough, they are not entirely borne out by the evidence.

A comparison of data for the five-year periods preceding and following 1852 indicates that during the earlier five-year period, 107 vessels burned and another 28 exploded, while during the five years following the act's passage, 125 burned and 28 exploded. Although the number of vessels in service during the second five-year period was somewhat greater than during the first five-year period (an average of 650 from 1847–51 and an average of 733 from 1853–57), the aggregate rate of loss due to fire and explosion was virtually the same—about 21 percent of the vessels in service—from one period to the other.[50]

The Inspection Act of 1852 looks better when the number of steamboats lost only to explosions is analyzed with an eye to determining whether the act's higher standards for new boiler construction contributed to reducing the rate of loss. Of seventeen steamboats destroyed by explosions during the first five years of the act's enforcement, all but two had been constructed before the act and its standards took effect. Presumably, boats launched after 1852 were equipped with boilers built to the new higher standards, a development that Inspector W. W. Guthrie believed was responsible for a dramatically reduced incidence of explosion: "It is worthy of remark that there has been no explosion or collapse of flue of any boilers manufactured since the passage of the law by Congress of August 30th, 1852, and coming under the reduction of steam pressure. In every instance, the disasters have been from boilers made previous to the passage of that law."[51] If so, then the record of the next five years, 1858–62, would suggest that boilermakers and steamboatmen reverted to their old ways. Of the fifteen steamboats wrecked by exploding boilers, eleven had been built during the first ten years of the operation of the Inspection Act. Nine of those eleven steamboats had been constructed, and presumably equipped with new boilers, during the initial five-year period of the act's enforcement.[52]

That the Board of Supervising Inspectors, set up under the act of 1852, was able to report a dramatic decrease in fatalities involved in steamboat accidents from the first to the second five-year period may have had as much to do with the enforcement of the act's provisions requiring life preservers and lifeboats as with boiler inspections.[53] This is, of course, not to say that such regulatory legislation was without effect, and it may in fact be reasonable to infer from the evidence that the rate of loss due to fire and/or explosion would have been even worse in the absence of such regulation and inspection. Still, the technical limitations of the pressure and temperature gauges, fittings, and other elements of steamboat boiler construction, as well as the human factor, worked against anything like consistent, reliable boiler operation, as many a steamboat captain could have pointed out.[54] The result, until better construction and operating techniques were introduced, well after the Civil War, was a depressingly large toll of ships and lives.

The demand for steamboat inspection and regulation was a response to the carnage on the nation's lakes, coastal waters, and rivers, in the East as well as in the West. Any citizen, in any part of the country, ran the risk of becoming a casualty of a burst boiler aboard a steamboat, whether the vessel plied Long Island Sound, one of the Great Lakes, the Hudson River, or one of the great rivers of the West. In that sense, boiler explosions and fires aboard steamboats were a national problem that excited attention and action from Congress. The problem was, in part, a technical one of requiring that steamboat boilers be better made and equipped with safety gauges and other features, and also a behavioral one of ensuring that engineers, captains, and pilots were properly trained and did not operate their vessels' engines in a dangerous manner. The inspection acts of 1838 and 1852, especially the latter, confronted this dual problem and represented the first concerted effort by the federal government to regulate an aspect of the operation of the market; as such, the legislation was important at the time and also is of great historical significance. But it was the federal government's efforts, over the course of decades, to address the systemic aspect of the problem of steamboat safety—that is, the danger posed by natural hazards to navigation on the rivers—that ultimately had greater success.

A solution to the problem of how to make steamboats safer was, in fact, easier to achieve and more politically attractive to implement than was a solution to the problem of how to make the western rivers on which most steamboats operated less dangerous. In great part, this was true because, as a technical problem and a political matter, the issue of steamboat safety had a national constituency that was vitally interested in its resolution. The far greater and less tractable problem of removing natural hazards to steamboat navigation from the Mississippi, Ohio, and Missouri rivers, while also a technical and political challenge, was of most immediate interest to the people who lived in the states, cities, and towns along those rivers, and to their congressional delegations. Unlike a boiler explosion, which could happen anywhere, the destruction of a steamboat by an encounter with a submerged tree or a steamboat wreck or by grounding on a very recently formed and therefore uncharted sandbar was almost uniquely an occurrence confined to the western rivers. Many congressmen and senators, irrespective of party affiliation, from states east of the Appalachians could more readily understand the urgency of reducing the danger of exploding boilers than of clearing the western rivers of their hazards. For the people west of the Appalachians, especially those along the rivers, this inability or, in some cases, refusal to understand their situation was, to say the least, puzzling, particularly because to their minds the great rivers of the interior and the cities and towns along them and the steamboats that traveled upon them were essential to the economic and political strength of their region and of the nation. Some eastern politicians, notably Daniel Webster, did grasp the dynamism of the West and the importance of its rivers. Speaking at a dinner at Philadelphia in December 1846, he was moved to exclaim: "Why, what a world is there! What rivers and what cities on their banks!—Cincinnati, Louisville, St. Louis, Natchez, New Orleans, and others spring up while we are talking of them, or, indeed, before we begin to speak of them; commercial marts, great places for the exchange of commodities along these rivers, which are, as it were so many inland seas!"[55]

The claim, often made by western river interests in support of their call for federally funded river improvements, that "the annual value of the commerce of the great valley [the Mississippi River valley, broadly defined] . . . [was] more than twice that of the whole foreign commerce of our country" was substantially correct.[56] Steamboats plying the western rivers had turned a public domain into territories and territories into states. The enthusiasm for the steamboat and the acute interest in the rivers among those who lived along them, and among others as well, stemmed in part from this fascination with progress and its agencies and also from a conviction that there was something uniquely American about this form of progress.

Denison Olmsted, professor of natural science and astronomy at Yale and prominent biographer of Eli Whitney, said as much in an article published in the January 1856 issue of the *American Journal of Education*. The article's title, "On the Democratic Tendencies of Science," is a succinct statement of its general thesis and suggests its hortatory tone and the exuberance to which even sophisticated observers were prone when contemplating the high technology of their day. Steam-driven motive power, embodied in the steamboat and the locomotive, according to Olmsted, was having a profound effect on the nation's growth and development, not of its economy alone but also of its aesthetic and political sensibilities: "Not only has there been great progress all over the country within the period of steamboats and railroads, in taste for the embellishments of art and the refinements of civilized life, but the steamboats and railroads have themselves furnished the means of gratifying that taste."[57] Even viewing the steamboat at some remove from the part of the country where it was having its most pronounced and sweeping effects, Olmsted was still able to comprehend the steamboat's capacity to transform the material and cultural life of the nation's vast interior.

The steamboat's stimulating influence touched virtually all the river cities and was an integral and bustling part of everyday life in them. Mail and news moved by packet steamboats that followed regular, posted schedules between the major river ports. Before the telegraph system connected the nation's cities and towns, the steamboat was the fastest means of communication up and down the great western rivers. It was, for example, by steamboat that residents of the river cities and their hinterlands learned of the outbreak of fighting between the United States and Mexico in April 1846. The packet *Peytona* brought the news up the Mississippi from New Orleans to Louisville. From there, another packet, *Yorktown*, carried word of the outbreak of war to Cincinnati, from which still another packet, *Cambria*, steamed up the Ohio to bring the dispatches to Wheeling, Virginia, and to Pittsburgh. At the same time, the *J. M. White*, one of the most celebrated steamboats of the day and famed for its speed records, steamed up the Mississippi from New Orleans to St. Louis with the news.[58]

The keen interest taken by residents of the river cities in the construction and destruction of steamboats and in the vessels' comings and goings was also due to their pragmatic understanding that steamboats on the western rivers had created or at the least had nurtured their market towns and cities. In 1842, a group of citizens of Cincinnati, seeking federal aid to improve the Ohio River and other western rivers, said as much in a petition to Congress: "Of all the elements of the prosperity of the West—of all the causes of its rapid increase in population, its growth in wealth, resources, and improvement, its immense commerce, and gigantic energies, the most efficient has been the navigation by steam."[59]

THE STEAMBOAT ECONOMY

For several cities and towns along the western rivers, steamboats were more than an essential means of transportation. In these places, the vessels were themselves products of a boat-building industry that was an important part of the local economy. As table 1 indicates, most of the steamboat packets that plied the western rivers during the twenty years before the Civil War came from boatyards in six states and, within these, from eleven places.[60] As one would expect, the centers of steamboat construction tended to clump together along the rivers of the nation's interior. This sort of geographical concentration was, and is still, common enough in industrial production and was, no doubt, dictated by a variety of considerations, the most critical of which probably was the availability of the necessary skilled labor. We can arrive at a quantitative estimate of the economic importance of steamboat building by doing some rough-hewn calculations. If $15,000 is used as a conservative estimate of the value of the average steamboat at the time of its construction, it becomes readily apparent that steamboat building was a major industry in a number of the river states and their cities and towns.[61] For example, during the 1850s, the small Monongahela River town of California, Pennsylvania, south of Pittsburgh, built sixty-one steamboats worth just under $1 million. The boatyards of Cincinnati, the "Queen City" of the Ohio River valley, turned out more than $1 million in steamboats during the same decade, while, downriver, Louisville's steamboat production exceeded $700,000.

Table 1
Centers of Construction of Steamboat Packets Plying the Western Rivers, 1840–1860

State	Number of Places Building Steamboats	Number of Steamboats Built
AL	1	1
AR	2	2
IL	2	4
IN	5 (2)	156 (143)
KY	7 (1)	82 (67)
LA	4	4
MN	1	1
MO	2 (1)	19 (18)
MS	1	1
OH	7 (1)	143 (127)
PA	18 (5)	228 (159)
TN	2	4
VA	6 (1)	25 (16)[a]
WI	1	1
Total	59 (11)	671 (530)[a]

Source: Steamboat Packet Database, derived from Way, *Way's Packet Guide*, passim.

Note: for each state listed, the number in parentheses in the first column denotes the number of places, of the total number listed for the state, that were significant steamboat-building centers, while the number in parentheses in the second column indicates the corresponding number of steamboats built at those centers.

[a] Includes one packet built in 1832 at Wheeling, Virginia.

These amounts may not seem especially large, but they should be viewed in the context of the urban economies in which they were generated. Louisville and especially Cincinnati were, of course, major cities with complex, multisector economies, of which steamboat building and related industries were only constituent parts. It is, therefore, hardly surprising that, in 1850, the value of steamboats turned out in each city accounted for only about one-half of 1 percent of the value of the manufacturing output of each city's respective county.[62] In California, Pennsylvania, however, steamboat construction represented almost all of surrounding Washington County's $1.1 million in manufacturing product and almost certainly was the community's major source of nonfarm income and employment.[63]

Significantly, few of the steamboats built in Ohio and Kentucky remained in those states: just over 2 percent of the steamboats built in Ohio stayed there; less than 5 percent of those built in Kentucky stayed within the state.[64] Instead, the boats built in those states plied the waters of other states, especially those of the Lower Mississippi River valley. More than 40 percent of Ohio's steamboat packet construction and 50 percent of Kentucky's ended up in Louisiana.[65] Even so, the number of steamboats on the Ohio River increased throughout the period, with more than 60 percent of them coming from boatyards in Pennsylvania. These yards constructed four of every ten steamboat packets of under two hundred tons displacement, the size that could fit in the locks of the Ohio River's Louisville Canal.[66] So heavy had steamboat traffic along the Ohio River to and from Cincinnati become by 1850 that a guide to the city's business community listed twenty-seven firms whose operations had directly to do with steamboats: six steamboat agents, three steamboat packet company offices, eleven engine builders, two steamboat inspectors, and five other steamboat-related businesses.[67]

Memphis experienced comparable growth in the part of its economy directly tied to the steamboat, and, during the last half of the 1850s, the number of steamboat agents in the city increased from three in 1856 to five in 1859 to seven in 1860.[68] Residents of Tennessee's chief Mississippi River port knew that their city owed its prosperity to its location on the river and to the steamboat and railroad. Moreover, they recognized the role of the steamboat as a driving force in the city's rapidly diversifying economy. In a city directory published in 1849, the compiler hailed the city's citizens for their "foresight, deserving all praise" for understanding "that cities cannot prosper by commerce alone" and for their "strong feeling in favor of manufactures."[69] Singled out for mention were a cotton textile mill, "an extensive Foundry for building steam boat and other machinery," and a steamboat construction yard from which two vessels had been launched that year.[70] A half-dozen years later, an advertisement in another business directory for a foundry that specialized in the building and repairing of steam engines, especially for steamboats, was flanked by two engravings, one of a steamboat and the other of a locomotive.[71]

Steamboat traffic stimulated more than heavy industry in Memphis and other large cities along the major rivers of the Mississippi Valley. Steamboats calling at Memphis, Nashville, and Vicksburg, like those arriving at New Orleans, St. Louis, and Cincinnati, were catered to by all manner of businesses.

These included hardware and cutlery dealers, suppliers of furniture, commission merchants, grocers, ship chandlers, and suppliers of ice, some of which, advertised to be "pure Spring Lake Illinois Ice," had itself come down the Mississippi by steamboat.[72]

The importance of the steamboat to the economic life of the cities and towns along the rivers grew with almost every passing year as the size of the fleet of steamboats and the value of freight carried by them increased. An indication of the pace and magnitude of this growth is provided in table 2, which traces the expansion of the volume and value of steamboat traffic coming into New Orleans from the rest of the Mississippi River system during the last two decades before the Civil War. Although the greater part of this traffic—between 40 and 50 percent—during the 1840s and 1850s originated in other stretches of the Mississippi and the Ohio and Red rivers, a significant proportion—15 to 20 percent—came from the towns, plantations, and farms along the bayous of the Crescent City's hinterland.[73] There was also substantial steamboat tonnage at work on the middle and upper reaches of the Mississippi River, energizing the economies of river port cities such as St. Louis and Minneapolis. There was, in fact, considerable specialization by steamboat owners with respect to routes, and many steamboats were built for particular trades, such as those between Pittsburgh and Cincinnati, Cincinnati and St. Louis, Louisville and St. Louis, New Orleans and St. Louis, and New Orleans and Louisville.[74] Specialization of this sort made good sense, both in terms of increasing the productivity of river commerce on steamboats and reducing the risk of operating on the rivers.[75] In the case of river navigation, familiarity did not breed contempt; it increased a steamboat's chances of survival on waters plagued by natural hazards.

NATURAL HAZARDS

Disaster on the western rivers could take any number of forms (see tables D.1 and D.2). A steamboat could come to grief by grounding on a sandbar in a channel; by foundering in a changing current or rapid; by snagging; by colliding with another boat or running onto the submerged wreck of a steamboat; by becoming imprisoned and then crushed by ice; by catching fire and burning to the waterline, as happened to so many boats along the St. Louis levee that terrible night in May 1849; and by an exploding boiler. Although fire and explosion were the most dramatic causes of steamboat wrecks and accounted for the largest number of lives lost on the rivers, such disasters accounted for a relatively small share of the total number of wrecks. Most steamboats wrecked on the western rivers fell victim to the rivers' natural hazards, especially the snag.

Snag was a generic term for an uprooted tree that had been swept into a river, only to lodge its root-end obliquely into the river's bed, while its crown or upper trunk broke or lay just below the surface of the water. Strictly speaking, a snag was such a tree that was pointing upstream, while a tree that pointed downstream was called a "sawyer," after the sawing motion it described as the current caused it to bob. In addition to snags and sawyers, rivers could be obstructed by "planters," that is, trees left standing in a channel newly cut by the river.[76] Snags (the term will be used here to denote sawyers and planters, as well) lay in wait for steamboats and other river craft, such as flatboats and an occasional

The 751-ton packet *Belfast* had an eventful six years on the Mississippi. Built in 1854 for the Memphis to New Orleans route, the *Belfast* collided with and sank another steamboat in February 1857, grounded itself in March 1857, and then sank and was a total loss in November 1860. (The original collector of the photograph drew the arrow and wrote the word "Eclipse" to identify another, much smaller steamboat berthed next to *Belfast*. See Way, *Way's Packet Directory*, entry 1690, 139.)

Sources: E. B. and N. Philip Norman Collection, Mss. 1084, Louisiana and Lower Mississippi Valley Collections, Louisiana State University Libraries. Courtesy of Louisiana State University Libraries' Special Collections. Information about the unhappy history of this packet is from Way, *Way's Packet Directory*, entry 0478, 40.

sailing ship. The outcome of an encounter between a ship and a snag was often catastrophic, and snaggings were the leading cause of steamboat disasters on western waters. As the fate of the *New Orleans* indicates, snags were a problem from the outset of steamboat navigation on the western rivers.[77] So frequent were snaggings that a widely circulated guide to steamboats and steamboat di-

Table 2
Number of Steamboat Arrivals and Tons of Freight Arriving at New Orleans by River from the Interior, 1840–1860

Year[a]	Steamboat Arrivals	Tons of Freight
1840–41	2,187	557,500
1841–42	2,132	566,500
1842–43	2,324	782,600
1843–44	2,570	652,000
1844–45	2,530	868,000
1845–46	2,770	971,700
1846–47	3,024[b]	937,600
1847–48	2,977[c]	1,025,900
1848–49	2,873	1,009,900
1849–50	2,784	886,000
1850–51	2,918	1,058,200
1851–52	2,778	1,160,000
1852–53	3,252	1,328,800
1853–54	3,076	1,286,300
1854–55	2,763	1,247,200
1855–56	2,956	1,500,200
1856–57	2,745	1,431,800
1857–58	3,264	1,572,700
1858–59	3,259	1,803,400
1859–60	3,566	2,187,560

Source: U.S. Congress, House, Department of the Treasury, *Report on the Internal Commerce of the United States 1887*, William F. Switzler, part 2 of *Report on the Commerce and Navigation of the United States*, 50th Cong., 1st sess., 1888, H. Exec. Doc. 6, serial 2552, table, "Steam-boats Arrived at New Orleans, Together with Tons of Freight Received by River," 221.

[a] Years end on August 31.

[b] The figure of 4,024 given in the source is a misprint and has been corrected here by using the figure supplied in Hunter, *Steamboats on the Western Rivers*, table 2.

[c] The figure of 2,917 given in the source is a misprint and has been corrected here by using the figure supplied in Hunter, *Steamboats on the Western Rivers*, table 2.

sasters used a print of a snagging, rather than one of an exploding boiler, as the illustration for its cover.[78]

Essentially an ambush, a snagging was almost always a complete surprise to a steamboat's crew and passengers and often occurred at night or at other times of reduced visibility. Although not as grisly in its effects as an explosion, a snagging carried its own special horrors, of which the drowning of scores of people could be one.[79] The effects could be catastrophic, as they were when the *Shepherdess* was snagged on the Mississippi River on a January night in 1844. The *Shepherdess* had been built less than two years earlier in a boatyard at Ripley, Ohio, for the Cincinnati–St. Louis trade and, at 133 tons, was a relatively small vessel for that route. The snagging of the *Shepherdess* occurred just below St. Louis "at 11 o'clock, in a dark and stormy night . . . just above the mouth of Cahokia creek," a stream that entered from the east out of Illinois.[80] Ironically enough, the steamboat had only a short while earlier negotiated the single-most dangerous section of the Mississippi, called the "Graveyard," and, relieved at their deliverance, the passengers had gone to bed.[81] They were soon jarred awake when the *Shepherdess* ran into the snag, which must have been huge. Almost immediately, wintry cold river water began to pour into the steamboat's hull, which had been ripped open by the snag, and, within three minutes, a substantial part of the vessel's superstructure was underwater.

Some passengers, seeking to escape the doomed steamboat either jumped or fell into the water, and many of these drowned or died of hypothermia. Others, more fortunate, climbed up to the hurricane deck, the third and highest of a western river steamboat's three decks, and were able to stay above the water as the boat continued to sink. Soon, however, many of those who had made their way to the hurricane deck were flung into the river when the sinking *Shepherdess*, now without power, drifted downstream onto another snag. This encounter finished off the boat, as "the hull and cabin parted; the former sunk and lodged on a bar . . . while the cabin floated down to the point of the bar . . . where it lodged and became stationary."[82] Thanks to the efforts of rescuers from other steamboats and smaller craft on the river, most of the passengers and crew aboard the *Shepherdess* were saved. The number of people lost that terrible night varied with the different accounts of the snagging from forty to "not less than seventy" to "between sixty and one hundred."[83]

Snags infested the Mississippi and Ohio rivers, as well as the lower reaches of the Missouri and a host of other, smaller rivers, such as the Tennessee and

Cover of *Lloyd's Steamboat and Railroad Directory, and Disasters on the Western Rivers*, depicting the snagging of a steamboat.

Source: Lloyd, *Lloyd's Steamboat and Railroad Directory, and Disasters on the Western Waters*, Louisiana and Lower Mississippi Valley Collections, Louisiana State University Libraries, Baton Rouge. Courtesy of Louisiana State University Libraries' Special Collections.

Sinking of the *Shepherdess* by a snag on the Mississippi River below St. Louis in 1844.
Source: Lloyd, *Lloyd's Steamboat and Railroad Directory, and Disasters on the Western Waters*, 165, Louisiana and Lower Mississippi Valley Collections, Louisiana State University Libraries, Baton Rouge. Courtesy of Louisiana State University Libraries' Special Collections.

the Yazoo.[84] The danger they presented to river navigation was recognized and justly feared by all who shipped or traveled on the western rivers. Describing her voyage down the Mississippi to New Orleans on the 846-ton steam packet *Autocrat*, "one of those 'floating steam-palaces'" that plied the St. Louis–New Orleans trade, the English traveler Emmeline Stuart-Wortley recalled a series of encounters with snags, two of which were damaging, though not ruinously destructive.[85] Of the first of these, which occurred some distance below Memphis, she noted that, "though we survived it . . . I assure you the shock might give one a faint idea of being blown up." The snag pierced the left paddle wheel so that it could not turn until the vessel was brought to a stop and the crew removed the snag.[86] The second notable snagging was by far the more serious of the two and occurred at night while she and her fellow passengers slept. The snag pierced the side of the steamboat's hull with such force that the sleepers "were suddenly woke up by an immense stunning shock, and the steamer stopped immediately, quivering, so to say, in every nerve of her huge body." Fortunately, the hull remained watertight and, not long after coming to a stop, "on went the powerful steamer again, plunging through the thick darkness with the great blunt arrow that had struck her so sorely, fast in her poor wounded side; but, this time, it had just missed the wheel." Soon after starting up again, the steamboat was snagged again, this time on the other side of the hull, but again not in a paddle wheel.[87] And that snagging was not the end of the ordeal.

Even when an encounter with a snag did no damage to a steamboat, the experience of those aboard was still a harrowing one because they knew that another encounter was quite possible and that its results could be catastrophic. Throughout the night, as the *Autocrat* moved downriver toward Natchez, it passed through a particularly bad stretch of water and "suffered a long succession of bumps and thumps (as well as her passengers) from a whole series of snags.... They would not let one repose for a quarter of an hour together in peace. The vessel went, jarring and jumping along in as disagreeable a manner as it is possible to imagine; very much as if she was playing at leap-frog, or hopping on one paddle for a wager."[88] Charles Dickens had had a similar experience on a trip up the Mississippi in that stretch of the river between Cairo and St. Louis. The trip took two days as his steamboat struck "constantly against the floating timber, or [stopped] . . . to avoid those more dangerous obstacles, the snags, or sawyers."[89] The steamboat's progress upriver was slowed at night by the frequent necessity of its captain having to stop the boat's engine in response to a forward lookout's ringing of the alarm bell, which signaled that a "great impediment" lay ahead. It was an unnerving experience and "always in the night this bell has work to do, and, after every ring, there comes a blow which renders it no easy matter to remain in bed."[90]

So feared were snags that other steamboat mishaps, groundings on bars, for example, were sometimes initially mistaken for snaggings. Frances Trollope reported that such an incident had occurred during her voyage in 1828 from New Orleans to Memphis aboard the *Belvidere*, a 158-ton steamboat.[91] As the steamboat proceeded upstream, "a sudden and violent shock startled us frightfully. 'It is a sawyer!' said one. 'It is a snag!' cried another. 'We are aground!' exclaimed the captain." After almost two days, another, much larger steamboat came along and pulled the grounded *Belvidere* free of the bar.[92]

The frequency with which groundings and snaggings occurred gave the Mississippi River its reputation for treachery and danger. For Dickens, the river was "a slimy monster hideous to behold"[93] and "this foul stream."[94] He did not care for much of what he saw and experienced during his stay in the United States, but he reserved some of his most stinging sarcasms for the Mississippi River. It seemed almost a living thing, a primitive, malevolent creature that could be written about but not understood: "But what words shall describe the Mississippi, great father of waters, who (praise be to Heaven!) has no young children like him?"[95] Incidents of the sort described by Dickens and by Frances Trollope and Emmeline Stuart-Wortley were not uncommon and could have happened at any number of places in the channel of the Mississippi River. Several stretches of the river were worse, however, and some were far worse, than others. These places were the ones targeted for special attention by the U.S. Army's Bureau of Topographical Engineers when the federal government conducted its campaign to remove snags, rocks, bars, and other obstructions from the Mississippi, Ohio, Missouri, and other western rivers.

There had been some efforts made by various states to clear the rivers within or along their respective jurisdictions, though most of the interest in internal improvements by state governments focused on turnpikes, canals, and railroads.[96] The waterways that received attention by a state were, almost without exception, those rivers and creeks lying entirely or mainly within its territory. For example, North Carolina's legislature subsidized work to improve navigation on the Cape Fear, Roanoke, Tar, and Neuse rivers and, in northwest Louisiana and east Texas, from the mid-1840s until the late 1850s, efforts to clear each state's respective parts of Cypress Bayou west of Shreveport were underwritten by private and state funding. Similar efforts, financed by local, parish, and state authorities before 1820 and again during the 1830s, targeted obstructions to navigation on the bayous of south Louisiana.[97] Any success resulting from these small-scale undertakings proved to be of limited value and short duration. But, even more ambitious projects were not notably successful.

In 1826, the Illinois state legislature considered, and ultimately rejected, proposals to fund the removal of hazards on the Little Wabash River, Big Bay Creek, and the Cache River.[98] Ten years later, however, during the 1836–37 session of the Illinois legislature, support from politicians across party lines, including the Whig Abraham Lincoln and the Democrat Stephen A. Douglas, propelled an ambitious internal improvements program through to passage. The legislation provided for $400,000 in funding to improve five rivers.[99] The economic crisis that began in 1839, first signaled by the Panic of 1837, put an end to such ambitions and left the state treasury of Illinois deeply in the red.[100] Illinois was not alone in its embarrassments, as the finances of other states, Tennessee and Indiana among them, crumbled under the strain of the severe economic contraction.[101]

Even before the crisis of 1837, efforts by states to improve their rivers frequently ran into a wall of political obstacles, the sources of which more often than not lay within the borders of those states. Intrastate sectionalism in Tennessee and other southern states resulted in political gridlocks that forestalled attempts to provide adequate funding for improvements.[102] In other states, Illinois for one, where the political resources were mobilized for projecting river improvements, the state's financial resources were inadequate to pay for the work.[103] The desirability of improving a river of interest to two states occasionally prompted their legislatures and governors to negotiate a cooperative effort by them to fund the improvements. Illinois and Indiana undertook such an arrangement in 1833 when each agreed to appropriate $12,000 to improve the Wabash River, which formed the border between them and emptied into the Ohio River.[104] Almost a decade later, in December 1842, as the depression that eventually had followed the Panic of 1837 continued, the State of Indiana asked the United States Congress to fund the improvements to the Wabash River, citing "the financial condition of this and the adjoining states" as a key reason for requesting federal funds.[105] The financial difficulties that had frustrated the efforts by various states to improve their rivers underscored for westerners the need for a federally funded program of river improvements. They considered the need for such a program to be particularly pressing with respect to the removal of natural hazards on the larger, interstate rivers, which were beyond the capacity of any one state or group of states to improve.

In March 1854, Secretary of War Jefferson Davis forwarded a report to Congress concerning the work done on the rivers during the preceding twelve months by his department's engineers. They had given the greatest attention to the most hazardous stretches of the Mississippi's channel, removing snags and cutting up logs lying on bars and shoals. They also cut down trees growing along the river's banks in a proactive effort to avoid the formation of snags, which likely would occur were a future flood to undercut the banks on which the trees grew.[106] The report detailed the areas on which they had concentrated their efforts in the form of "a list of the more formidable and dangerous passes."[107] The list is reproduced in table 3 and gives a good idea of what the army's topographical engineers were up against. Of the 1,000 miles of river covered in the list, beginning where the Missouri River flows into the Mississippi just above St. Louis and ending at Natchez, Mississippi, almost 750 miles, three-fourths of the total distance, were classified as "Snaggy and dangerous," "Dangerous," or "Very Dangerous."

The engineers cleared thousands of hazards from the Mississippi and hundreds more from the Missouri, Ohio, Arkansas, and Illinois rivers in 1853. They worked over stretches of the river on and along which they and their colleagues had labored in years past because snags often recurred in areas that had been infested with snags and had then been cleared. And yet, as figures 1 and 2 illustrate, despite all these efforts, the number and tonnage of steamboats lost to snags and other natural hazards on the western rivers rose inexorably from one decade to another.[108] The more recondite reasons for the persistent increase in losses due to natural hazards are explored in detail later. For now, it is enough to point out that the rise in the number and tonnage of steamboat losses more or less paced the rise in the number and tonnage of steamboats operating on the western rivers, suggesting that, at the very least, the army's engineers managed to prevent conditions on the rivers from getting any worse. The notable exceptions to this generalization are the sharp upward spikes in losses in two sets of consecutive years, 1841–42 and 1851–52.

The explanation for these anomalies likely has to do with the amount of precipitation in the watershed of each of the rivers, especially that of the Mississippi, and the water levels in the river channels.[109] During the second year of each pair of years, water levels in the Mississippi's channel were unusually low, a condition that was remarked upon at the time.[110] The river's channel was most dangerous during periods of low water because it was then that there was the least distance between the underside of a steamboat's hull and the river's snags, bars, and other hazards. There was, in fact a functional relationship between the effects of a succession of years of higher than normal precipitation in the watershed and the consequences of those effects during subsequent years of low water levels in the rivers. Greater precipitation could result in freshets and floods, as happened in 1851, with the result that an unusually large number of

Table 3
Positions and Distances from Mouth of the Missouri (downward) to Natchez

Designation of Passes	Character of Passes	Intermediate Distances	Total Distances
Sawyer's Bend	Snaggy and dangerous	10	10
There to Turkey Island	Snags and scattering	60	70
Turkey Island Bend	Very dangerous	—	—
Grand Tower	Very snaggy	15	85
Tower Island	Occasional snags	5	90
Cairo	Snaggy and dangerous	110	200
Bend at Island No. 10	Dangerous	60	260
Island No. 18	Very dangerous	55	315
Island No. 21	Dangerous	5	320
Bend at Island No. 25	Dangerous	15	335
Islands No. 26 and 27	Dangerous	10	345
Head of Island No. 30	Dangerous	10	355
Bends at Plum Point No. 33	Very dangerous	10	365
Bends at Island No. 34	Dangerous	10	375
Bends at Island No. 35	Dangerous	10	385
Bends at Island No. 37	Very dangerous	10	395
Devil's Elbow[a]	Dangerous	10	405
Bend at Brandywine Bar	Dangerous	10	415
Paddy's Hen and Chickens	Dangerous	10[b]	435[b]
President Island	Dangerous	10	445
Cow Island	Dangerous	10	455
Buck Island	Dangerous	20	475
Commerce Island	Dangerous	10	485
Council Bend[a]	Dangerous	10	495
Grand Cut-off	Dangerous	10	505
Walnut Bend	Dangerous	10	515
Ship Island	Dangerous	5	520
St. Francis' Island	Dangerous	5	525
Helena Island No. 60	Dangerous	10	535
Montezuma Bar	Very dangerous	5	540
Horse-shoe Cut-off	Very dangerous	10	550
Old Town Bend	Dangerous	10	560
Islands No. 62 and 63	Dangerous	5	565
Island No. 64	Dangerous	10	575
Island No. 65	Dangerous	5	580
Indian Charley's Bend[a]	Dangerous	10	590
Islands No. 67 and 68	Dangerous	10	600

(Table 3, continued)

Designation of Passes	Character of Passes	Intermediate Distances	Total Distances
Island No. 69	Dangerous	10	610
Islands No. 70 and 71	Dangerous	10	620
Chicot Island Bend	Very dangerous	65	685
Kentucky Bend	Very dangerous	30	715
Princeton Island No. 89	Dangerous	10	725
Lara Island[a]	Dangerous	15	740
Bunches' Cut-off	Dangerous	10	750
Island No. 93	Dangerous	5	755
Island No. 95[a]	Dangerous	25	780
Islands No. 96 and 97	Dangerous	10	790
Island No. 98[a]	Very dangerous	5	795
Island No. 100	Dangerous	10	805
Milliken's Bend	Dangerous	15	820
Papaw Island No. 103[a]	Very dangerous	10	830
Natchez	Occasional snags	170	1,000

Source: U.S. Congress, Senate, *Report of the Secretary of War Made in Compliance with a Resolution of the Senate in Relation to the Work Done under the Appropriations of 1852 for the Improvement of Western Rivers and Harbors*, 33rd Cong., 1st sess., 1854, S. Exec. Doc. 51, table, "Positions and Distances from Mouth of the Missouri (downward) to Natchez," 4.

[a] Passes no longer exist because of the Mississippi River's continuous sculpting of its channel, and because of federal construction, dredging, and other work on the river in the last quarter of the nineteenth century and, especially, throughout the twentieth century.

[b] Either the intermediate distance of 10 miles given in the source from the Bend at Brandywine Bar and Paddy's Hen and Chickens should be 20 miles, or the total distance traveled down to Paddy's Hen and Chickens should be 425 miles, 10 miles less than in the source.

trees were uprooted and washed into the river channels where many of them became lodged as snags.[111] The upshot was that, during a subsequent year or two of low water, the river channels were more snaggy and therefore more dangerous and claimed an unusually large number of steamboats.

Even under the best of conditions, the work to clear the rivers of their many hazards to navigation presented an engineering and technical problem of considerable magnitude. And, because this work of improving the rivers exceeded the fiscal capacities of the states and entailed the spending of unprecedentedly large sums of money from the federal treasury in the face of determined opposition within Congress, the removal of hazards on the western rivers was also a political problem. Until the sectional collision over the territorial extension of slavery occluded all other matters before the American people, the issue of river improvements would be one of the most bitter and disruptive political controversies in the history of the United States.

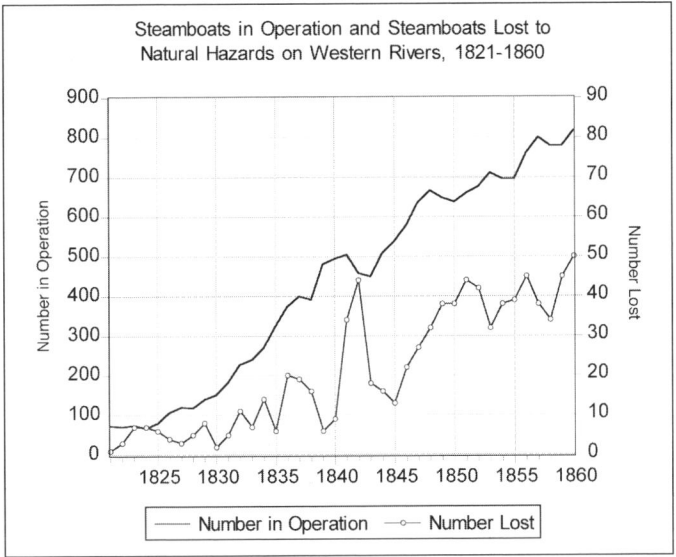

Figure 1. *Sources:* See notes to text.

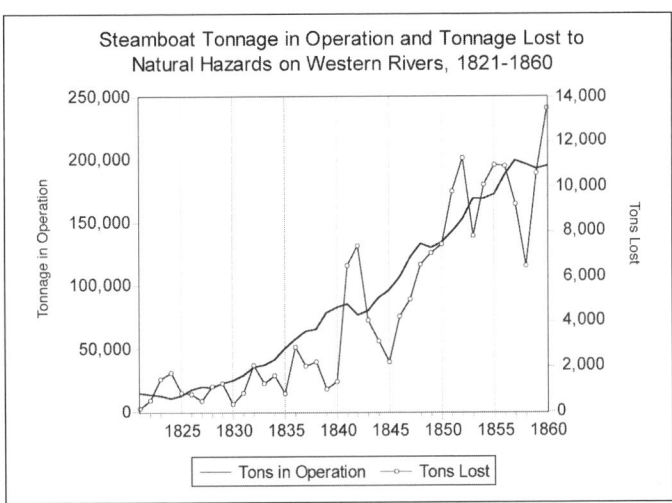

Figure 2. *Sources:* See notes to text.

2

POLITICS BEFORE POLK

As was true of virtually everything that Congress did, decisions to appropriate federal funds for the clearing of navigational hazards from the Mississippi River or the construction of a breakwater for Cleveland on Lake Erie were intensely political. Vote trading, buying and selling, and appeals to principles and to national, state, and local interests entered into the consideration of the location, timing, and level of support of various projects. The avowed purpose of this admixture of high and low politics was, depending upon with whom one spoke, either an appropriation of funds for projects that would advance the national welfare, even as they necessarily served local public and private interests, or the cynical and reckless overturning of the constitutional basis of the American republic. These competing interests and points of view did not emerge fully formed at one particular moment but instead developed gradually as Congress repeatedly took up the issue of internal improvements during the last four decades before the Civil War.

Although the internal improvements program that fired the imagination and indignation of so many nineteenth-century Americans took shape in James Monroe's second term, it had really begun much earlier, during Washington's presidency. Then, the sole object of the program was the construction of lighthouses, buoys, and beacons on the main points of the Atlantic coastline and at the entrances to ocean harbors. All told, seventeen bills for these purposes arrived on George Washington's desk, and he signed every one of them.[1] His successor, John Adams, was even more active in that line, approving eleven such measures in his four years in office. It was, however, in Thomas Jefferson's administration that the program struck out on its first, albeit modest, departure from its rather limited original purpose.

Jefferson more than matched Washington's support for lighthouses and the like, but in a significant change from what had gone before, he also signed into law two bills to build roads and one to support the construction of a canal. The more important of the road bills, signed in 1806, was for the Cumberland

Road; the other, also passed that year, appropriated money to build a road from Georgia to Louisiana.² Although Jefferson's innovative policy was a significant first step toward the realization of Washington's dream of a national highway, it was his successor, James Madison, who put support for roads onto an entirely new plane. Jefferson's two bills became Madison's eleven, and his $48,400 for roads became Madison's $698,800, and this despite the extraordinary expense of the War of 1812 (see tables 4 and 5).

The end of Madison's second term in March 1817 marked the end of whatever innocence and general support the internal improvements program had once enjoyed. On March 3, the day before he left office, Madison vetoed the so-called "Bonus Bill." Supporters of the bill had proposed it as a novel means by which to provide regular, recurring federal funding for road and canal building as part of what in later years would come to be called a "general system" of internal improvements.³ They had hoped to use the one-time payment of the $1.5 million charter fee or bonus paid to the Treasury to establish the Second Bank of the United States, as well as a recurring annual dividend of $650,000 on bank stock, to finance the plan; hence the name "Bonus Bill."⁴ Madison's veto of the bill reflected his doubts about a loose construction of the Constitution and was a blow to the bill's chief architect, John C. Calhoun, who, then still in his neo-Hamiltonian incarnation, had counted on his plan to promote the economic development of the South and West in order to enable those regions to keep pace with a rapidly developing Northeast.⁵

James Monroe shared Madison's view of the proper construction of the Constitution and, in May 1822, successfully vetoed an improvements measure, this one intended to permit the levying of tolls—essentially users' fees—along the Cumberland Road to defray the cost of its improvement.⁶ Friends of the bill may well have scratched their heads in frustrated puzzlement, thinking that, by providing for tolls, they had done what was necessary to avoid incurring the opposition of the president, who, more than four years earlier, had warned the House and the Senate not to test his resolve to prevent unconstitutional impositions on the Treasury for the purpose of funding internal improvements. In vetoing the Cumberland Road bill, Monroe deliberately set out to make a constitutional point, which he developed in a lengthy essay that laid out his sense of how internal improvements should be executed. Notwithstanding its length and complexity, his thinking emerged clearly enough for the program's supporters and opponents to discern it.⁷ He had not so much imposed an explicit constitutional

Table 4
Internal Improvements Legislation and Presidential Administrations, 1789–1861

President	Period	No. Years	Number of Signed Bills for					Number of Vetoed Bills for				All Vetoes
			Lights[a]	Roads	Canals	R&H[b]	All Bills	Lights[a]	Roads	Canals	R&H[b]	
Washington	1789–1797	8	17	0	0	0	17	0	0	0	0	0
Adams, John	1797–1801	4	11	0	0	0	11	0	0	0	0	0
Jefferson	1801–1809	8	18	2	1	0	21	0	0	0	0	0
Madison	1809–1817	8	13	11	0	0	24	0	—1—		0	1[c]
Monroe	1817–1825	8	14	18	0	5	37	0	1	0	0	1
Adams, J. Q.	1825–1829	4	1	18	10	12	41	0	0	0	0	0
Jackson	1829–1837	8	13	32	6	15	66	1	2	1	2–3	6–7[d]
Van Buren	1837–1841	4	5	7	1	6	19	0	0	0	0	0
Harrison/Tyler	1841–1845	4	3	4	1	11	19	0	0	0	1	1
Polk	1845–1849	4	7	1	1	5	14	0	0	0	2	2
Taylor/Fillmore	1849–1853	4	8	4	1	2	15	0	0	0	0	0
Pierce	1853–1857	4	6	19	0	5	30	0	0	0	6	6
Buchanan	1857–1861	4	4	8	0	0	12	0	0	0	2	2

Source: Derived from E. C. Nelson, "Presidential Influence on the Policy of Internal Improvements," *Iowa Journal of History and Politics* 4 (1906), appendices A and B.

[a] Lighthouses, buoys, and beacons.
[b] Rivers and Harbors.
[c] The vetoed bill was both a road and a canal bill.
[d] Includes a veto of a bill to distribute the Treasury's surplus to the states.

Table 5
Internal Improvements Appropriations in Each Presidential Administration, 1789–1861
(in current dollars)

Administration	Period	Lighthouses	Canals	Roads	Rivers and Harbors	Total
Washington	1789–1797	320,297	0	0	0	320,297
Adams, John	1797–1801	234,052	0	0	0	234,052
Jefferson	1801–1809	862,307	25,000	48,400	0	935,707
Madison	1809–1817	801,982	0	698,800	0	1,500,782
Monroe	1817–1825	734,780	0	1,342,905	123,150	2,200,835
Adams, J. Q.	1825–1829	0	1,564,085[a]	1,278,937	1,208,894	4,051,916
Jackson	1829–1837	3,034,492	473,400[a]	4,972,546	4,984,943	13,465,381
Van Buren	1837–1841	1,573,935	[a]	601,300	1,456,817	3,632,052
Harrison/Tyler	1841–1845	1,100,901	[a]	58,800	932,105	2,091,806
Polk	1845–1849	2,881,311	[a]	30,000	29,231	2,940,542
Taylor/Fillmore	1849–1853	2,989,301	[a]	97,118	2,101,255	5,187,674
Pierce	1853–1857	3,510,029	0	1,391,590	931,000	5,832,619
Buchanan	1857–1861	2,181,986	0	504,247	0	2,686,233
Totals	1789–1861	20,225,373	2,062,485	11,024,643	11,767,395	45,079,896

Source: Derived from E. C. Nelson, "Presidential Influence on the Policy of Internal Improvements," appendix A.

[a] Includes grants of land and, in the case of the John Quincy Adams administration, the purchase of the stock of canal companies. In neither case is a cash value assigned in the source. Where no dollar figure is given, only grants of land came from the federal government to the states.

criterion as established a utilitarian standard that improvements bills had to meet: Did a proposed measure transcend local interests and advance the larger national interest, that is, promote the general welfare? If it did, and if the suggested improvement met with the approval of the states directly affected by it, Monroe suggested, it would win his endorsement.

His veto in 1822 notwithstanding, it was during Monroe's two terms that the improvements program began to assume the magnitude and scope called for by its friends in earlier years. During his first term, especially before the Panic of 1819 and its immediate aftermath, two large appropriations were made for the Cumberland Road that, along with other bills for that project passed by the end of 1820, brought the total for roads to more than $1 million.[8] By the time he left office in March 1825, appropriations for roads during the preceding eight years amounted to more than $1,342,000, almost double the amount authorized during Madison's presidency. More significant in terms of how the internal improvements program was to develop were the appropriations made in 1823 and 1824 for surveys of and improvements to the great western rivers, especially the Ohio and the Mississippi. With the passage of these measures, the essential features of the improvements program were in place that would make it a centerpiece of political controversy for the next four decades.

Monroe had done more than simply sign road and river bills into law. His veto of the Cumberland Road bill in 1822 had laid out the criteria that improvements measures had to satisfy in order to gain his approval. Friends of an ambitious system of internal improvements consisting of road and canal projects had the opportunity to put this other Monroe doctrine to the test in 1824, when the General Survey Bill passed both houses of Congress on April 30. Monroe immediately signed it. The act authorized the expenditure of $20,000 to employ War Department engineers to survey road and canal routes to link the seaboard states with the rapidly growing new states of the trans-Appalachian West.[9]

The General Survey Bill was very much the handiwork of Monroe's secretary of war, John C. Calhoun. Although Calhoun had reluctantly given ground before James Madison's veto of the Bonus Bill, a measure that had also been of Calhoun's devising, he had not abandoned his view that the language of the Constitution implied a range of powers to be used to further the national interest, including the chartering of a national bank and a federally funded system of general, as opposed to local, internal improvements. This was the thrust of the General Survey Bill that had passed Congress and won Monroe's approval.

Considering the subjects and participants in the political antagonisms of the decades that followed, there is no little irony in the fact that Calhoun's greatest congressional ally in the cause of the General Survey was Henry Clay, who was already touting his "American System."

Such positions and posturings had, of course, a decided partisan and personal political cast. Both Calhoun and Clay wanted the presidency in 1824, and both played to a western gallery. But to note the political aspect to the General Survey is not to call into question its profound economic and demographic rationales. Between 1815 and 1821, Indiana, Mississippi, Illinois, Alabama, and Missouri—all states beyond the Appalachian Mountains and all watered by the great western rivers—had been admitted to the Union (Maine had also come in, in 1820). As territories in 1810, their combined population had been just under 100,000; in 1820, it exceeded 470,000; by 1830, in part because of transportation improvements, their combined population would exceed 1 million.[10] At decennial net rates of increase of 377 percent from 1810 to 1820 and 130 percent from 1820 to 1830, these trans-Appalachian western states were growing faster than almost any other part of the country.

This population growth and the dramatic increases in agricultural and manufacturing output that accompanied it gave these states and their northwestern and southwestern neighbors great economic importance. Their growing and pivotal political importance was only dimly reflected in the combined size of their delegations in the House of Representatives, which went from seven seats in 1820 to thirteen in 1830.[11] Their votes in 1824, in an alliance with the Middle Atlantic states with which they traded, were sufficient to pass the General Survey over strong opposition from New England and the seaboard South.

By 1824, then, the foundation of the internal improvements program had been laid, which is by no means to say that it was secure. Contributing to its somewhat precarious position were the complexities and confusion that attended that year's presidential election. Each of the major candidates at the beginning of the year—John Quincy Adams of Massachusetts, William Crawford of Georgia, Calhoun of South Carolina, Clay of Kentucky, and Andrew Jackson of Tennessee—jockeyed for political advantage over his rivals, which meant attempts at coalition building and pandering to special interests. In this respect and in most others, as well, the history of this election is too familiar to require a detailed retelling here.

The presidential election campaign of 1824 shattered the illusion of the permanency of the "Era of Good Feelings" and the health of one contender, Crawford; deflected the ambitions of another, Calhoun; and forged an alliance between two others, Adams and Clay, to defeat Jackson. Their alleged bargain—the White House for Adams and the State Department for Clay—gave the election, which ended up in the House of Representatives, to Adams on February 9, 1825. Andrew Jackson's presidential campaign of 1828 probably began that day.

THE AMERICAN SYSTEM

After Adams's and Clay's victory in the House (it was as much Clay's as it was Adams's), they began to plan the implementation of what by now was *their* American System. The key to defraying its expenditures was to be the protective tariff that, over the four years of Adams's presidency, would provide 87 percent of the Treasury's average annual receipts.[12] A reduced reliance on public land sales to generate revenue—5.1 percent of total revenue compared to 7.8 percent under Monroe—signified the new direction being charted by Adams.[13] Rather than actively encourage the country's *extensive* growth by releasing more public land acreage for sale, he and Clay promoted its *intensive* growth by developing its transportation infrastructure.

In the light of what came after, when Jackson was president, the most fateful aspect of the American System, and the source of so many of its difficulties, was its reliance on the tariff. As enemies of a federal program of improvements would assert, and not without reason, improvements and protection were two sides of the same federal coin. Without the large expenditures called for by the proposed federal program of internal improvements, they insisted, the government's profound dependence on the tariff for revenue—a dependency that had only been exceeded when tariffs generated 92 percent of federal revenue during Jefferson's two terms—would not be necessary.[14] And, turning that unholy relationship on its head, they also pointed out that, were it not for the revenues generated by the tariff, the federal government could not presume to impose upon the states by undertaking a general program of road and canal construction and river and harbor improvements.

Opponents of Adams and Clay and their American System invoked these arguments during Adams's tenure in office as he and an agreeable congressional majority projected and undertook one segment after another of the national road and involved the federal government as a promoter of and stockholder in various intrastate canal companies. There is no little irony in the fact that the improvements program that, under Adams, evoked such hostility was a far more modest affair than it would become under Jackson. Although Jackson and Adams differed profoundly in their readings of the Constitution, there would, in fact, turn out to be only two real *operational* differences in their respective approaches to the question of improvements: Jackson would oppose federal subscriptions to canal company stock offerings while Adams had supported them, and Adams had relied on tariff revenues to fund the program while Jackson would use public land sales to do that.

On May 19 and 24, 1828, Adams signed road bills appropriating funds for continued work on the Cumberland Road, some other roads entirely within the state of Michigan, and a military road in Maine. He also approved on May 24 the purchase by the federal government of stock in the Chesapeake and Ohio Canal Company and federal assistance to continue a canal project in Ohio.[15] Five days earlier, on the May 19, he had signed the most controversial piece of legislation ever to reach his desk, the tariff of that year, quickly to become known more familiarly as the "Tariff of Abominations," the sobriquet given it by its enemies because it raised tariff rates to unprecedentedly high levels—in fact, to their highest levels before the Civil War.

As is well known, Adams and Clay had fallen victim to a pro-Jackson intrigue in the House of Representatives. Jackson's men there had arranged matters so that the tariff bill considered by the House probably would be too obnoxious to win passage but would still embarrass the administration. When the bill did pass, narrowly, Adams, who had pledged himself to the cause of protectionism, had no choice but to sign it into law, thereby providing Jackson with an ideal campaign issue and touching off the December crisis of the South Carolina Legislature's Resolves and John C. Calhoun's *South Carolina Exposition and Protest*. The loss by Adams in 1828 was in part a consequence of the tariff trap laid for him by Jackson's congressional allies, but Adams blamed not Jackson but, rather, Calhoun for what had happened. The country had been in good shape when he took office in 1825, as "the principle of internal improvement was swelling the tide of public prosperity." All began to change when "the Sable Genius of the South saw the signs of his own inevitable downfall in the unparalleled progress of the general welfare of the North, and fell to cursing the tariff, and internal improvement, and raised the standard of free trade, nullification, and state rights." Writing almost nine years after he had been defeated, and in the closing weeks of the second Jackson administration, Adams was understandably bitter: "I fell and with me fell, I fear never to rise again, certainly never to rise again in my day, the system of internal improvement by means of national energies. The great object of my life therefore, as applied to the administration of the government of the United States, has failed."[16]

On December 16, not quite six weeks after Jackson's victory in the presidential election, and only three days before South Carolina's Resolves, Congressman Thomas H. Hall, a North Carolina Democrat, fired a parting shot across the stern of the Adams administration and its congressional allies. Hall introduced a set of resolutions in the House to the effect that Congress did not have the constitutional power "to execute a system of internal improvements, within the states," by which he meant roads and canals.[17] The timing of Hall's resolutions may have been coincidental, though it is unlikely to have been so, and if Adams saw Calhoun's hand in such mischief, he probably was not mistaken. Adams perhaps could not be expected to have attributed to Calhoun any but the basest of motives for opposing the American System. But, of course, Calhoun was not alone in his opposition. Jackson, too, considered the American System to be unconstitutional, not because of its interstate ambitions but because of its use of general Treasury funds to undertake intrastate and local projects. In practical terms, as events would quickly demonstrate, Jackson's objections to the American System were more to its means than to its ends, a distinction out of which a good deal of confusion and trouble would come.

The campaign leading up to Jackson's popular and electoral victories over Adams had presented voters with a fairly clear choice between two candidates, two parties, and two philosophies of government. Questions about how the nation's affairs should be conducted, and to what ends, had emerged early in the campaign, although intelligent voters might be forgiven if they had difficulty in knowing just where Jackson the candidate stood on some of them, including the improvements issue.[18] That would change very soon after he became president.

JACKSON AND IMPROVEMENTS

Some of Jackson's contemporaries and many of those who came after him misunderstood or misrepresented his policy with respect to internal improvements. The root of their confusion lay in his famous veto, on May 27, 1830, of the Maysville Road Bill, which proposed to appropriate $150,000 to buy stock in the Maysville, Washington, Paris, and Lexington Turnpike Road Company. The company was to construct a turnpike entirely within Kentucky, and it was this fact and the provision for purchases of the company's stock by the federal government that Jackson found objectionable, just as Madison and Monroe had opposed such projects. Pointing to the proposed road's local, that is, intrastate, character, Jackson found federal financial support in any form to be unconstitutional and declared that friends of such projects, including canals, should seek an amendment to the Constitution that would permit the expenditure of federal funds on them.[19] This position was hardly original with him, although none of his predecessors—including even Madison, who had suggested as much—had been prepared to press the point quite so far as was Jackson.

He had given an oblique indication of his thinking about six months earlier, in December, in his first annual message to Congress, in which he had noted that there was a widely held opinion that the prevailing program of internal improvements, put in place by Adams and Clay, was unconstitutional. This program, which was an integral part of the American System, was funded from the general revenue in the Treasury. It had, as table 4 indicates, launched several canal and road projects, most but not all of which were extensions of the Cumberland Road or military roads. It was Jackson's contention that, inasmuch as most of these undertakings and others pending were within the confines of particular states, those states could fund them, if they wished to do so, with surplus federal revenue, which Jackson thought should be distributed among them in proportion to the sizes of their respective delegations in the House of Representatives.[20]

The Maysville Road veto was an application of these views. It was not, however, in any way a repudiation of the idea of federally funded internal improvements, and appropriations to remove hazards from rivers and harbors found ready approval from Jackson, provided the rivers and harbors were involved in interstate commerce. In fact, as the data in tables 4 and 5 make clear, Jackson signed more legislation for this purpose than any of his predecessors or successors before the Civil War. While John Quincy Adams had approved twelve rivers and harbors bills, for a total of about $1.2 million, during his four years in office, Jackson during his two terms approved fifteen such bills, which appropriated almost $5 million for river and harbor improvements. Moreover, Jackson signed thirty-two appropriations measures for roads, compared with Adams's eighteen.

The inverted contrast between the two presidents—the one identified with and condemned for a general system of improvements, the other widely viewed as that system's most determined and effective foe—is even more sharply drawn when one compares the amounts that each approved for roads: less than $2 million by Adams but almost $5 million by Jackson. As much was known at the time. And yet, many of Jackson's supporters then, and many who over the next thirty years would style themselves his political heirs, insisted that he stood opposed to the general, federally funded system of internal improvements. How did matters become so confused?[21]

As one might expect, some of the confusion owed its existence to Jackson's occasional practice of representing his views to different constituencies in different terms, the latter designed to address the preconceptions and interests of the former. In the opinion of one Calhoun Democrat, Congressman Isaac Holmes of South Carolina, such variability had been one of Jackson's more prominent traits as a leader. Indeed, Holmes "had never known a public man whose policy had varied so much and so often, in accordance with his exact position at the time, as Gen. Jackson's."[22] Some confusion also arose because of an honest misunderstanding of Jackson's beliefs by politicians who simply were not as subtle as their hero-president. But much if not most of the problem stemmed from the determination of Democrats to trim Jackson's views to fit their purposes, seizing upon one aspect or another as the need arose. As Jacksonian Democrats, they were determined to make Jackson into one, as well.

To see the logic of Jackson's position on the internal improvements issue, one needs to bear in mind his larger objectives as president. As leader of a rapidly emerging party, he was intensely interested in sustaining and enlarging upon its coalitions of regions, including land-hungry western states, and interest groups, such as antitariff merchants, manufacturers, and planters. To these ends, he was willing to work for reduced tariff rates and, to compensate for the consequent diminished revenues, to release unprecedentedly large amounts of public land for sale or as grants to the states. Of course, such positions also helped to pull the teeth from Clay's American System, reason enough to hold and advance them.[23]

As a Jeffersonian and a fiscal fundamentalist, Jackson was determined to dismantle the Hamiltonian legacy of a national bank, a sinking fund, and a mortgaged future by destroying the Second Bank of the United States and paying down the public debt. Add to these political and ideological considerations Jackson's rather idiosyncratic strict constructionist view of the Constitution, and one gets a sense of his thinking on internal improvements. Certainly the positions he took on the issue in the various forms in which it was presented to him seem consistent when viewed in that light. An ideological nationalist who built a political career by balancing and joining sectional and parochial economic interests, he had his hands full as mounting pressure for rivers and harbors projects in the West aroused strenuous opposition in the Old South.

By 1830, demands for tariff reductions were becoming insistent enough that Jackson's floor leaders in Congress began to press him for an indication of how he wished to proceed in the matter.[24] The source of much of this demand were the states of the seaboard South where suspicion of the linkage between protectionism and internal improvements was greatest. And yet, as Charles Sydnor has noted, while southern opposition to internal improvements was partly a function of southern opposition to a high tariff, it was also a reflection of the fact that the Old South had received little in the way of benefits from the program.[25] By the end of 1829, total expenditures for internal improvements in the states of the region came to only $37,000, of which a negligible $80 were spent in Virginia, easily the most politically important state in the South. Spending in the Northeast totaled $569,000, $586,000 in the states of the Northwest, and $152,000 in the relatively new southwestern states, such as Louisiana, Mississippi, and Alabama.[26]

In other and starker words, the Old South's $37,000 amounted to less than 3 percent of total national expenditures on internal improvements, not counting spending on the Cumberland Road and four large canal projects, while the states of the region—Delaware, Maryland, Virginia, North Carolina, South Carolina, and Georgia—held more than one-fourth of the seats in the House of Representatives.[27] This disparity, remarkable as it was, had little if anything to do with anti-South sentiment in Congress and a great deal to do with the geography and geology of the South.[28] Of no southern state was this more true than South Carolina.

South Carolina's opposition to federally funded internal improvements was in no small way a stand rooted in constitutional principle. That the state received only negligible funds for river and harbor improvements (apart from not inconsiderable amounts for light stations and buoys) stemmed partly from this integrity, but it was also a consequence of the absence of navigable rivers as arteries of interstate commerce. Although the Savannah River formed the border between Georgia and South Carolina and served as a waterway for the two states, it joined the sea at Savannah. Consequently, Georgia, and not South Carolina, derived the chief economic benefit from the river. And so it was in Georgia, then, and not in South Carolina, that the federal government spent more than $373,000 on improving the lower reaches, especially the mouth, of the Savannah River in the decades before the Civil War.[29]

With comparatively little to gain from federal appropriations for river improvements, South Carolina also had nothing to lose by opposing them for other states, including those watered by the Mississippi River and its major tributaries, the Ohio and the Missouri. Where internal improvements were concerned, it was perhaps easier for South Carolina than for other more richly blessed states to stand on principle. Accordingly, with the exception of a handful of its political men—most notably John C. Calhoun, who, until rather late in his career, embraced a larger, less parochial, and more national view of the internal improvements question than many of his fellow South Carolinians—the state's congressional delegation maintained an unwavering opposition to the "general system" of internal improvements.

Although South Carolina's congressmen were often strident in their opposition to the program, the intensity with which they expressed their views was by no means peculiar to them. So much was obvious in the sharp exchange in the House of Representatives on February 19, 1831, between two North Carolina Democrats, Samuel P. Carson and Daniel L. Barringer, over a river and harbor improvements bill. On this occasion, Barringer showed himself to be a model of restraint as Carson began the exchange that Saturday by asserting that "the liberties of my country are by this bill put up for sale. I for one will not be bribed to vote for it."[30] This was a thinly veiled accusation that, by including pork barrel items for particular districts, the bill's sponsors hoped to buy support for the overall measure, which, to Carson's mind, would lead to an arrogation of power by the national government at the expense of the states and the citizenry.

Barringer immediately rose in answer to Carson, mocking his hyperbole about "liberties of the country put up for sale!" "How?" asked Barringer. Carson had not only insulted the House then in session but also "every Congress from the foundation of the Government, and every Executive from the commencement to the present." That, of course, was a reference to Jackson. Moreover, Barringer pointed out, Carson had not even understood the nature of the bill of which he had so bitterly and intemperately complained. "This bill . . . presented no question of internal improvement, as that question is understood by southern gentlemen generally" (756). This last clause is of some significance, alluding as it does to the American System, because it indicates that, even at this early date, the improvements question had assumed a distinct sectional cast that, ironically enough, emerged clearly on this occasion in an exchange between two southerners.

If Carson appreciated the irony in the situation, he gave no such indication and instead pressed his attack on those who would vote for the bill. Comparing the United States to Rome, he warned that just as "[i]n ancient times the Roman leaders bought up the liberties of the people with the spoils of the conquered provinces . . . this policy of internal improvements and our high-handed tariff, are the means with which the liberties of this people are to be bought up" (756). He then implored the Speaker of the House, Virginia's Andrew Stevenson, to make common cause with the Old Dominion's other noble sons who had defended liberty and lead the state "as the nucleus around which the States are to rally to resist the usurpations of the General Government" by resort to interposition, as Jefferson and Madison had done for Kentucky and Virginia (758).

Late into his remarks, Carson sensed that he had perhaps gone too far in impugning the motives and integrity of his colleagues and ascribed any intemperance of speech to the fact that "he was sick"; he had taken to his bed and had only come to the House Chamber because "he had been sent for" (758). One wonders by whom. His indisposition may well have loosened his tongue beyond the usual bounds required by congressional comity. His sentiments, however, were those of John C. Calhoun, who was by then well along in his first transformation from economic nationalist into states' rights champion. This was the Calhoun of the *South Carolina Exposition and Protest* of 1828 and of the Jefferson Day Dinner toast directed at Andrew Jackson in 1830. And, however numerous and powerful were his enemies and detractors, he had a growing cadre of admirers, including many in Congress, such as Carson.

Yet, one could oppose the bill that was so obnoxious to Carson and still be in favor of a federal program of internal improvements. While Carson objected to the bill on the grounds that it violated the Constitution by going too far, John Blair, a Democrat from Tennessee and a member of the Committee on Internal Improvements, opposed the bill because it did not go far enough. His complaint was that the bill seemed intended to set up "a system of appropriations for bays and harbors, and the mouths of rivers and creeks; in short a system of improvements for the tide water, to the exclusion of the whole interior" (758). How could he, "as a representative of one of the districts in the interior, join in a crusade against that section of the country from whence I come?" (758). He could not, and he was one of fifty-three who voted against the bill, which passed anyway with 136 votes in its favor (759). Jackson signed it into law on March 2.

But this was only one bill, and the acrimony that attended its passage was a presentiment of worse to come. Up until then, the dispute over improvements bills had been more or less confined to the question of their constitutionality. Now, however, that consideration had been joined and, to some extent, eclipsed by the controversial connection between improvements and protectionism, accompanied by the first clear, but still somewhat muted, expressions of sectionalism and speculations on the mortality of the federal union. These very quickly grew louder when another rivers and harbors bill came before the House of Representatives on June 2, 1832, a Saturday. Two days later, debate began over a Saturday motion by Tennessee's James K. Polk that the enacting clause be struck, which would effectively gut the bill.[31]

As he had done during the internal improvements debate of the previous year, Samuel Carson spoke first and complained that the House was being asked by the sponsors of the bill (it had come out of the Committee on Ways and Means) to appropriate ever larger sums for river projects that never seemed to get completed. Where and when would this sort of thing end? And, as he had said the year before, he of course "meant to cast no imputation on the gentlemen who had introduced the bill" (3248-49). Whatever Carson's intentions, the form of his assault on the bill was something of a departure from the terms in which critics of improvements had heretofore couched their arguments. To qualms over the improvement program's constitutionality and objections to the program's reliance on high tariff rates for its funding were now added the accusations that the program was too expensive and unproductive. All of these themes would persist during the years to come and would be augmented by other charges. But, for now, these became the rack on which opponents of federally funded internal improvements hoped to break the program.

As debate continued that Monday, South Carolina's John M. Felder asked if anyone in the Chamber could tell him whether "any engineer in the employ of this Government had ever reported unfavorably in relation to any work that ever was proposed." Pausing for dramatic effect, and hearing no one rise to take his bait, he went on to ask if any member of the House had "ever been known to oppose any project, however wild, the appropriation for which was to be expended within his own district?" (3251). He had watched congressmen sell their principles for "forty or fifty thousand dollars" in appropriations, and, yet, "there was a portion of the House whose principles would not permit them to accept these bribes" (3251). And, besides, in the end, this sacrifice of congressional integrity on the altar of expediency did no good. He himself "had seen many of these famous works of internal improvement, and he would not give a chew of tobacco for any of them" (3251).

According to Felder, the improvements program was more pernicious than it would have been had it merely been a revenue-burning engine of waste. Its profligate expenditures on these projects had served only to divide the nation, one section of it from another, so that, "if the system went on, it went on at the hazard of the Union" (3251), Although this sort of strong sentiment, this casual conjuring of the specter of disruption of the Union by Felder, may appear extreme, if not eccentric at this early date, it was not so unusual, at least not for a South Carolina politician in 1832. Felder, John C. Calhoun's Yale classmate, law school classmate, and firm friend, was no less attached to Calhoun's states' rights politics.[32]

Felder's criticism of the improvements program and his warning, or threat, that the continuation of the program would put the Union in jeopardy was more than a little irritating to one of his House colleagues, former president John Quincy Adams. In the course of replying to Felder and, also, to Polk, Adams noted that the criticisms lodged against the general system of internal improvements were always the same: vague and plagued by errors of fact and logic. Felder's assault Adams dismissed as "a general philippic against the system" (3255). As for Felder's warning that the improvements program threatened the survival of the Union, "[t]here was hardly a question that now arose in the House, in reference to which they were not told the same thing." The real threat to the perpetuation of the Union, said Adams with considerable feeling, came not from the continuation of the improvements program but, rather, from its termination: if that happened, then "this Union will soon break in pieces; and I will add that it will not deserve to be preserved" (3256).

By the time that Felder and Adams held forth on the House floor that June, the dangerous contest over the protective tariff between Andrew Jackson and South Carolina, orchestrated behind the scenes by Calhoun, was already taking shape and would, with the passage in July of the still somewhat protectionist Tariff of 1832, burst into the open. When that confrontation was settled the following year, with South Carolina having drawn back in the face of Jackson's obvious willingness to use federal troops to enforce the tariff law, the sort of overheated rhetoric to which Felder had been given receded from congressional debate, at least for the moment. More enduring, at least in the short run, was the matter of the rivers and harbors improvements program's costliness, a criticism that had been broached by Carson and developed by Felder.

The question of cost was something of a new tack for opponents of river improvements, which did not, of course, mean that they had abandoned their older, well-tested grounds for opposing improvements bills. Coming when it did, before the surge in federal revenues from public land sales made alarums over rising costs look like hysterics, this new basis for opposing the improvements program held considerable appeal for congressmen as a gesture toward economy and accountability in government, especially in a general election year. Also, both Carson and Felder had raised the question of political integrity by asserting the corrupting influence of rivers and harbors bills on members of Congress when the prospect of federal spending in their districts was dangled before them. Felder, especially, had tried to equate opposition to improvements appropriations with civic and political virtue. Fortunately for friends of the program, a considerable majority in the House was willing to embrace sin, and the June bill passed handily by a vote of 102 to 73.[33] That victory was sealed on July 3 by Jackson's signing of the bill, which appropriated more than $663,000 for miscellaneous river improvements, slightly more than the appropriation of a year earlier.[34]

Perhaps by raising the matter of the sometimes irresistible pull of improvements spending on members of the House, Felder had hoped to shame his Democratic colleagues into maintaining their antiprotectionist, anti–general system of improvements principles and, more practically important, party unity. He had reason to be concerned. Members of Congress sometimes did deviate from party orthodoxy to vote for such spending measures, and, as will be seen with respect to a later Congress, this sort of behavior was occasionally widespread, notwithstanding complaints about it on the floor of the House.

For now, a useful purpose will be served by recalling that the indictment against the federal system of improvements was not limited to condemnations of its seductive pull upon otherwise honest men, its allegedly exorbitant expense, its wastefulness and inefficiency, and its potentially disruptive effect on national unity. The issue of its constitutionality, problematic almost from the program's inception, would continue to energize its opponents throughout the next three decades before the war, as would its manifest, and presumably unwholesome, relationship to the protective tariff. And yet the program survived, in the face of intense and sometimes effective opposition, a fact that calls for explanation. Part of that explanation almost certainly is to be found in the often fragmented ranks of that opposition.

At the same time that the rivers and harbors bill of June 1832 was making its way over the vociferous objections of its opponents, the Jackson administration's road-building program was proceeding through Congress. No doubt some of the president's more naive supporters who had not understood his Maysville Road veto were taken aback when, on July 3, he signed an ambitious bill that appropriated more than $425,000 to extend the Cumberland Road beyond Zanesville, Ohio.[35] What he had specifically objected to in the Maysville Road Bill was its provision that the federal government purchase stock in the Maysville Road Company to fund construction of a road within one state. The Maysville veto of May 27, 1830, had been followed four days later by the veto of a similar bill concerning the Washington Turnpike and Road Company in Maryland. About six months after that, he had made his celebrated veto of the bill authorizing the federal government to purchase stock in the Louisville and Portland Canal Company.[36] In each of these instances, Jackson had objected to what he perceived to be an unconstitutional involvement by the national government in intrastate and largely private capital enterprises. He had, as he would expansively demonstrate, nothing against the idea of the federal government constructing interstate roads. Quite the contrary, he was the greatest road builder of all the presidents before Lincoln, most of whose roads were, of necessity, military ones.

Jackson's energetic policy with respect to extending the Cumberland Road and constructing military roads was more than matched by his commitment to improve the great rivers of the interior. Indeed, it was during his two terms, even more than in Adams's one, that the river improvements program really became established. As noted earlier, by the time he left office in March 1837, he had approved appropriations for rivers and harbors totaling almost $5 million. More than $663,000 of that total was appropriated in a rivers and harbors bill that he signed on the same day that he approved the bill to fund the construction of the Cumberland Road beyond Zanesville.[37] Jackson's support of river and harbor measures was firm and unstinting, so long as they met his criteria for approval, that is, that they be interstate in scope and vital to the national interest, as he conceived it to be. In fact, he vetoed only three such measures, out of a total of eighteen sent to him. He cast the first of these vetoes on December 6, 1832, only five months after having signed an ambitious rivers and harbors bill, because the objectionable bill included "a class of appropriations . . . for the improvement of streams that are not navigable, that are not channels of commerce."[38] Knowing his views in such matters, Jackson's allies and his more prudent opponents in Congress generally sent him measures that he could and would sign.

For Jackson, the right sort of river and harbor improvements, as much as military and interstate road projects, would benefit the country and therefore ultimately its entire citizenry. This stance drew to his side supporters from some unlikely places. One such was Congressman Chilton Allan, a Henry Clay Democrat from Kentucky and therefore not a natural ally of Jackson.[39] During the tense debate in June 1832 over that season's rivers and harbors bill, Allan defended "with much pleasure" the administration "against the charge of useless expenditures for internal improvements."[40] The expenditures had not only been made for important work that had been executed efficiently; they were also expressions of civic virtue. Instead of some other, wasteful uses to which the national treasure had been put, such as "pompous intercourse with foreign courts, which has made us so giddy, [we should] betake ourselves to the plain republican task of clearing out our rivers, so that our farmers and mechanics can send their produce safely to market."[41]

IMPLOSION

It is perhaps easier to be virtuous when comfortably situated than when in desperate straits. The healthy condition of the national accounts, in great part due to the floodtide of revenue coming into the Treasury from sales of public lands, had given Jackson the means with which to project his improvements program, a program on a scale that matched his ambition. Abruptly, however, beginning in 1837, the fiscal underpinnings of that program collapsed. By then, of course, Jackson was out of office. Martin Van Buren, his friend, vice president, and successor, had the misfortune of presiding over a government that no longer had the wherewithal to realize its Jacksonian ambitions.

Van Buren's first year in office coincided with the financial panic and business contraction of 1837. The difficulties of the economic crisis were reflected by the inactivity of Congress, from which came not a single bill to fund the construction of even one lighthouse, marker buoy, road, or canal, or the improvement of any harbor or river.[42] The next year was a year of recovery, for the economy as well as the Congress, most of whose members were facing the voters in November. The year was therefore, as one might expect, one of large appropriations for improvements, including more than $536,000 for roads (most of this amount went for construction of segments of the Cumberland Road) and more than $1.4 million for river and harbor improvements. All the bills sent to the president met with his approval, as did the appropriations measures passed in 1839 and 1840. Unlike bills passed in these latter years for lighthouses and buoys, legislation for roads, harbors, and rivers reaching Van Buren's desk provided for appropriations that were far smaller—in some instances, almost negligible—than the appropriations of 1838.[43] The reason for the retrenchment was the severe contraction of the economy that had begun in 1839 and had deepened in 1840. The depression of the early 1840s was under way, and Congress was not in an expansive mood.

The first years of the depression were the economic backdrop to the general election of 1840, a bad year for the Democrats, a year when the Whig Party, having learned to wage a Jackson-like campaign for the White House, complete with its own military hero, came roaring back. The decade of the forties was going to be a Whiggish one. That, at least, was the expectation of the party's leaders who, having forged a union ticket between Ohio's William Henry Harrison—"Ol' Tippecanoe"—and former Democrat John Tyler of Virginia, believed that their candidates' victory in November meant that the electorate agreed with their indictment of Jacksonian banking, tariff, and public land policies as the causes of the economic crisis. They were probably wrong in this assumption, but that misjudgment was nothing in comparison to their more fundamental mistake in assuming that the ticket's vice-presidential nominee agreed with the party's platform and its standard bearer. John Tyler's presidency, as a result of the death of President Harrison a month after Inauguration Day, would make plain to them the extent of their error.

An early indication of trouble down the road came during the summer and culminated on September 11 with the resignation of almost the entire cabinet, a cabinet whose members were Harrison appointees and who, with two exceptions, were Henry Clay Whigs pledged to a revival of the American System. The resignations were prompted by Tyler's veto two days earlier of a bill to create the "Fiscal Corporation of the United States," essentially a bank, to regulate the Treasury Department's collection and distribution of revenue. In its initial incarnation, as proposed in a bill introduced by Henry Clay and vetoed by Tyler, the "Fiscal Corporation" was called the "Fiscal Bank." The "Fiscal Corporation" bill, which presumably had Tyler's conditional backing, was supported by the members of the cabinet, who, following his veto, charged breach of faith on the president's part and resigned—with the exception of Secretary of State Daniel Webster.[44] Dramatic as was this collision with Tyler over a goal that lay at the core of the Whig ideology, that is, the creation of what in effect would have been a third national bank, it was about the only indication of Tyler's thinking that the party's leadership had to go on in judging the new president, whom his enemies were already calling "His Accidency" because of how he got the office.

The cabinet implosion apart, the first several months of the Tyler administration were so uneventful that the front page of the *Congressional Globe & Appendix* for Tuesday, December 16, 1841, carried an announcement in large type from its editors who complained that "so little has yet been done in Congress, that we have not been able to print the first number of the *Congressional Globe* as soon as we expected." The *Appendix* to the *Globe,* which was supposed to "be made up principally of long speeches" would not be printed for seven to ten days because "no long speech has yet been made at this session of Congress."[45] Perhaps the doldrums in which Congress and the *Globe* found themselves were occasioned by the fact that Christmas was scarcely a week away. Whatever their cause, they were soon to end, and the editors of the *Congressional Globe* would have nothing to complain about in the way of a dearth of speeches and action on the floor of the House of Representatives.

On Monday, January 24, 1842, Congressman John Quincy Adams introduced "a number of abolition petitions." In keeping with the twenty-first ["gag"] rule, which was designed to prevent just such a thing, a number were "not received."[46] He then introduced a petition from "citizens of Haverhill, in the Commonwealth of Massachusetts," asking that Congress "immediately adopt measures peaceably to dissolve the Union of these States—." The ensuing debate sparked a demand, by means of a formal resolution introduced by the then-Whig (but soon to be Democrat) Thomas W. Gilmer of Virginia, a close ally of President Tyler, to censure Adams for "presenting to the consideration of this House a petition for the dissolution of the Union."[47] Gilmer's resolution generated a good deal of heated debate, during which Kentucky's Thomas Francis Marshall offered a resolution amending and stiffening Gilmer's own. Marshall's amendment demanded the "severest censure" of Adams for having insulted the House of Representatives and "the people of the United States" by, as Marshall subsequently explained, introducing "a proposition so monstrous . . . —a proposition and that, too, like this, in the midst of the difficulties, embarrassments, and confusion with which our public affairs were involved—amidst all the sources of public discontent we see around us" that "the Government of the United States should terminate its own existence."[48]

In the end, the resolution for censure was tabled and on Monday, February 7, in its stead the House voted to determine whether it would receive the petition introduced two weeks earlier by Adams that had set the entire train of events in motion. The lopsided vote—40 members in favor of receiving the petition and 166 opposed—was an indication of just how desperate most members were to lay to rest the entire matter of slavery and the dissolution of the Union. Marshall, who, along with some other members, had asked in vain to be excused from voting on the question of receiving the petition, recognized that the tabling of the censure resolution was a victory for Adams and his assault on slavery and would be seen as such in the country at large.[49] Adams, for his part, was determined to make the Democrats and their southern Whig allies pay, and continue to pay, for their sins, chief of which was the perpetuation of the slave system, but which also included the defeat of his and Clay's cherished American System.[50] As events were soon to make plain, Adams's revenge was to be limited to creating some momentary awkwardness for the defenders of slavery. And, as he already knew, his integrated system of economic development—including the protective tariff, a national bank, and ambitious and ubiquitous internal improvements—was dying, soon to be buried, hurried on its way, or so critics alleged, by, among others, "His Accidency." However merited were the charges of betrayal hurled at Tyler for his position concerning a national bank, the evidence is clear enough that he did not try to gut the improvements program. In fact, he supported almost every appropriations measure sent to him by Congress for improvements to rivers and harbors.

The second and third years of John Tyler's presidency were beset by the depression that had begun during Van Buren's third year in office. Van Buren had pursued Jackson's program of internal improvements, especially river and harbor projects, on a scale commensurate with the Treasury's reduced means. During his four years in office, he had signed bills for roads, canals, and rivers and harbors that appropriated only about 20 percent of what Jackson had approved during his eight years as president. Confronted by even smaller annual revenues than Van Buren, Tyler nevertheless approved almost $1 million in projects, almost all of it for rivers and harbors. Where he differed significantly from Van Buren was in the amount of money approved for road projects: Van Buren had signed seven pieces of legislation for such projects that appropriated just over $600,000 and had not vetoed one bill of that sort. Tyler also signed every one of the four such bills that reached his desk, but the four bills appropriated a total of only about $59,000 for roads. This differential—Tyler's total was just under 10 percent of Van Buren's—makes for a stark contrast, but then so does that between Van Buren's $600,000 for roads and Jackson's nearly $5 million, a differential of 12 percent (see tables 4 and 5).

Tyler cast his single veto of an improvements measure on June 11, 1844, when he rejected a rivers and harbors bill. The House of Representatives failed to override the veto in a vote that reflected party and sectional lines. But that same day, the president had approved a $655,000 appropriations measure for rivers and harbors, and, on June 15, he signed five additional but quite small rivers and harbors appropriations bills.[51] He signed almost as many bills for roads, canals, and rivers and harbors as Van Buren, and he only vetoed one, but that veto was tendered in such Jacksonian terms and tones as to confirm the opinion of Whigs, especially those loyal to Henry Clay, that the president was their great political enemy. The political damage to Tyler was not even partially offset by his observation in the veto message that he had, that very day, signed a big appropriations bill for the Mississippi River because that river "occupies a footing altogether different from the rivers and water courses of the different States."[52] There were any number of points in the veto message that were bound to offend Whigs. Tyler, for example, pointed out that, had some of the projects included in the objectionable bill been "separated from the rest," he would have approved them on an individual basis "however much I might deplore the reproduction of a system which for some time has been permitted to sleep with apparently the acquiescence of the country."[53] More offensive to Whig sensibilities was the president's concluding observation, which rehearsed an old canard against a general system of improvements: "Every system is liable to run into abuse, and none more so than that under consideration; and measures can not be too soon taken by Congress to guard against this evil."[54] The situation had long since been clear to the Whigs: the president was the enemy of their party and their programs. As matters turned out, however, his successor, James K. Polk, may well have made many Whigs almost nostalgic for John Tyler.

3
POLITICS
POLK AND POST-POLK

James K. Polk enjoyed his nickname of "Young Hickory" for its evocative and legitimizing power. Like "Old Hickory," President Polk stood firmly opposed to any bill that smacked of funding river and harbor improvements through a protective tariff. Also like Jackson, he found no implied power lurking in the Constitution that would justify federal funding of intrastate improvements. But, if imitation is flattery, it is not duplication, and there were significant differences between Polk's stand on internal improvements and that of his famous predecessor. Unlike Polk, Jackson had not been at all reluctant to support a general program of improvements, so long as its funding did not come from a protective tariff. As we have seen, no pre–Civil War president before or after Jackson committed as much political capital and spent as much of the public's money as he did on roads, rivers, and harbors. Polk's record on improvements, both as president and as a congressman, was more ambivalent and far more modest.

Notwithstanding his own efforts and those of his political friends to cast him in the role of someone who had an open mind on the subject of improvements bills, Polk had never encountered one such measure that had passed muster with him, even when he was a congressman. He himself recognized that his positions lent themselves to being misinterpreted. During debate over his motion to strike the enacting clause from a bill for river and harbor improvements in June 1832, he noted that "gentlemen seemed to throw out the idea that his motion amounted to an attack upon the whole system [of internal improvements]; and that, should it prevail, the system must be abandoned. But this was not by any means the case."[1] But, of course, that was precisely the case when he attempted to prevent an improvements bill from coming to a vote in the House, and John Quincy Adams, back in the House of Representatives, said as much.[2] There is no reason to doubt Polk's sincerity in opposing such measures on the grounds that they were unconstitutional. He proved, after all, to be an energetic advocate of state-funded improvements to Tennessee's rivers and roads when he became the state's governor in 1839. But proposals to use federal funds to construct those sorts of improvements within his home state, no matter how beneficial they might prove, were unacceptable to Polk, who viewed such proposals "with a dislike that was fanatical in its intensity."[3] That view, which was the prevailing view of his political party, may have been one of the reasons that he was a Democrat.[4]

On August 3, 1846, he vetoed the Rivers and Harbors bill, which had been passed on July 24. His veto of the bill and the inability of the measure's friends in Congress to override it were serious setbacks for the improvements program, though so much was not clear at the time. Even John C. Calhoun misjudged Polk's strength, predicting that the "veto of the harbor bill will ensure its complete success. The veto message is a poor document."[5] Perhaps the program's advocates should have seen the veto coming, something easier to do with hindsight. At the time, though, Polk's cultivated ambiguity at least held out the possibility that he might approve an improvements measure. As early as February 1846, when the rivers and harbors improvements bill was still before the House, John Wentworth, an Illinois Democrat from Chicago, a district with considerable immediate interests at stake in the bill, took the floor in an effort to encourage his colleagues to vote for it, despite their concern over the possibility of a veto. Even Illinois Whigs knew him to be a staunch advocate for the cause of river and Great Lakes improvements, and so he was listened to by members on both sides of the aisle.[6] He reminded them that the president "is known to be committed to the policy of General Jackson," that his nickname is "Young Hickory," and that there is "reason to believe that he will approve of such works as far as General Jackson did, and no further." But, Wentworth went on to point out, Polk had not made any explicit reference to river and harbor improvements, having only made an "allusion" to "the condition of the public works" and to "prominent objects of national interest" that had been included in the annual report of the secretary of war. What, then, were the members of Congress to make of that? There was no sure indication of presidential intentions to be had in the matter, and, in the end, Wentworth had to peer through a glass darkly, looking for some indication of what they might be. He urged his colleagues to do their duty and vote for the bill, knowing that in doing so they had done what conscience and good sense had demanded.[7]

Proponents of the improvements bill of 1846 did have some indication that a veto was in the offing, and predictions of one came as early as March 9, while the bill was still in debate. One prediction was made by Frederick P. Stanton, a Tennessee Democrat. Stanton's district lay along the Mississippi River and therefore stood to gain by passage of the bill. Even though Tennessee's state legislature had passed a resolution urging the bill's passage because of its inclusion of funding for improvements to the Tennessee River, Stanton said that he would probably vote against it because of the measure's intrinsic defects, notably its failure to require a systematic survey of the proposed improvement to the Tennessee or an estimate of its likely cost. The bill also included the disagreeable provision that the federal government provide financial support for the Louisville Canal Company by purchasing its stock. This proposal had often been a bone of contention in improvements bills, and Stanton was confident that it—as well as the flaws in the proposed provisions for Tennessee River improvements—would provoke a veto.[8]

A more exuberant and strident prediction of a veto came the next day from Alabama's William L. Yancey. South Carolina–born, with close familial ties to John C. Calhoun, Yancey had changed from being a Unionist who had condemned the South Carolina nullifiers to an advocate of states' rights and the South's sectional interests and an apologist for the nullifiers and slavery. In doing so, he followed a political and ideological trajectory that, in the years since Jackson's presidency, had become less exceptionable and more respectable in a South growing ever more defensive on the subject of slavery.[9] Of a mercurial, even violent temperament, he felt strongly about almost every issue that came before the House of Representatives, including a program of federally funded internal improvements, which he saw as an imposition on the South and a threat to the survival of the Union.

In the course of his set speech, Yancey praised the abandoned and discredited John Tyler, whom most Whigs had disowned and most Democrats could not bring themselves to trust, crediting him with having suppressed the incipient revival of a general program of improvements. But, he noted, now, in 1846, the program threatened to rise again, its friends apparently counting on getting Polk's approval of a bill sent to him for his signature. In thinking so, Yancey was certain they were mistaken: "I sincerely believe . . . that these gentlemen have counted without their host. The President has not improperly or unwisely been called 'Young Hickory.'" Polk had shown himself, to Yancey's mind, to be a reliable and fit bearer of the mantle of Jacksonianism, and Yancey reminded his colleagues of Polk's stalwart support of Jackson's veto of the Maysville Road Bill.[10] Polk's principles in 1846 were the same ones that he held in 1830. Had Yancey delivered his speech in mid-August, he could also have cited the president's support for the Independent Treasury Act that firmly established in law the Jacksonian determination to forestall the chartering of another national bank.[11] And, with respect to at least one issue, the tariff, he was even more Jacksonian than Jackson had been: "No President has ever done as much for free trade" as had Polk. The inference to be drawn from Polk's record on the tariff was clear enough, Yancey believed: the president would veto the improvements bill.[12] And, as matters turned out, Yancey had read Polk more astutely than had the bill's friends.

Other, more reasoned voices were also heard against the bill, and it was perhaps to these that its supporters listened most closely for an indication of how much strength they could muster to override a veto. The speech in the House by Alabama's George S. Houston on May 26 gave them a good idea of just what they were up against. Houston was no red-hot like Yancey and, unlike Stanton, made no predictions of what the president might do if and when a bill reached him. Instead, Houston devoted most of his remarks to an attack on the entire program of federally funded internal improvements and, in the course of doing so, rehearsed almost every one of the objections to the program that its enemies had long raised against it. If his speech did nothing else, it was noteworthy for having offered a cogent, comprehensive indictment of the program and its funding.

For Houston, the present bill was only the latest manifestation of an obnoxious system, and, notwithstanding the undeniable benefits to the people of his district from the bill's provision for removing obstructions to navigation from the Tennessee River, he opposed it because it imposed the needs of one state or section on the people of another: "What interest have my constituents in the improvement of the Hudson River; the canals and harbors of Illinois, Indiana, or Michigan; or the harbor at St. Louis? None. And it is the same with the eastern Atlantic States; they have no interest in the improvement of the Tennessee River." Houston also objected to the inducements to congressional greed held out by the bill. It was clear as day that the internal improvements "system is not only unjust in all of its bearings upon the different sections of the country, but it is kept up by log-rolling and bargaining—that it is the child of corruption, and ever has been." And, of course, no assault on the improvements program would have been complete without an effort to link it to the tariff. For Houston, what made the program particularly obnoxious was that it, "like a protective tariff, *feeds* upon its own *inequalities, and fattens* one portion upon the hard earnings of the other; as soon as you make it equal in its *exactions of* and *favors to* the *whole people,* it will tumble into ruins."[13]

Accusations of "log-rolling" and pandering to special interests had, of course, been heard before in connection with the tactics employed by proponents of improvements bills. Earlier during the House debate over the bill, J. A. Woodward, a Democrat from South Carolina, had deplored both practices and accused Whigs of using appropriations for "the Mississippi and its tributaries" to entice "a large body of western Democrats" into voting for the Whigs' "general system of internal improvements." Urging his "western friends to return to the true doctrine, rally around the standard Andrew Jackson planted, and suffer not the temptation of money to lead you astray," Woodward warned his colleagues that unless such pernicious practices were discarded, the House of Representatives would become "a rendezvous of demagogues, mere local job-agents, coming here with no higher purpose than to feed the greedy appetites of their constituents, whom they themselves have misled and corrupted."[14]

Again, such charges against an improvements bill had been made before and would be made again when subsequent bills came before Congress. And, Jackson's policies—or at least the interpretations placed upon those policies by others—had, for more than a decade, frequently figured in the improvements debate. If there was an element of novelty in any of what was being heard that session, it was the invocation of Jackson, who had died less than a year earlier, on June 8, 1845, as some sort of secular icon. In life a force to reckon with, he was just as formidable in death as an incarnation of civic virtue. That consideration, too, weighed in the balance, one that increasingly had been working against the improvements bill's chances once it left Congress. When the bill finally passed, the congressional debate had given Polk a variety of grounds on which to base his veto.

Woodward's reference to the narrow, sectional interests of western Democrats, who were being tempted away from the supposedly pure Jacksonian opposition of their party to a general system of internal improvements, touched on two related congressional phenomena that had been gestating for some time. One of these was a growing demand by western congressmen that their states receive a return from the federal treasury commensurate with the wealth that they contributed to it. The other, an increasingly vehement southern backlash to the internal improvements program, was potentially dangerous. The hardening of southern opposition to federally funded improvements was at least in part a reaction to the increasing scale of the improvements program, which was itself very much a result of relentless western pressure for it. But it also bespoke a growing irritation on the part of some southern Democrats with what they considered the West's importunate and occasionally extortionate posture when the subject of improvements came up. The day before Woodward spoke, William Yancey had vented his outrage and frustration at what he characterized as western threats "that the West would soon have the power, and let us be careful that there be no old scores to settle." For his part, Yancey said, such threats "can have no influence over a single vote I have to give." Besides, he went on, "I fear not the West. . . . Let empire take its way there."[15]

There was a rather direct connection between the rising chorus of demands for proportionality in improvements appropriations coming from the representatives of the western states and the growing populations and wealth of those states. As early as February 1831, Tennessee's John Blair, a Democrat, had objected to an improvements bill before the House because it slighted the states of the country's interior. Although not a western state by the geographical standards of a later generation, Tennessee was often considered one at the time Blair spoke, and he linked its interests with those of the other trans-Appalachian states.[16] A dozen years later, in 1843, the memorial by several citizens of Cincinnati, in justifying a call for improvements to the Ohio and other western rivers, asserted the economic, political, and geographic importance of their region: "The West is the centre of the Union, the citadel of its power, the great living fountain, whose boundless resources are destined to sustain and enrich the Nation."[17]

CALCULATIONS OF CALHOUN

This western claim of centrality received influential support from a somewhat surprising source in November 1845. John C. Calhoun spoke before hundreds of delegates gathered in the Methodist Episcopal Church in Memphis, Tennessee, for the first of the great commercial conventions held in the South.[18] As president of the convention, Calhoun gave the keynote address, a speech that was memorable chiefly because of what he said about the great rivers of the interior:

> [T]he invention of Fulton has, in reality, for all practical purposes, converted the Mississippi, with all its great tributaries, into an inland sea. Regarding it as such, I am prepared to place it on the same footing with the Gulf and Atlantic coasts, the Chesapeake and Delaware Bays, and the Lakes, in reference to the superintendence of the General Government over its navigation. It is manifest that it is far beyond the power of individual or separate States to supervise it, as there are eighteen States, including Texas and the Territories—more than half the Union—which lie within the valley of the Mississippi or border on its navigable tributaries.[19]

Calhoun's listeners could not have heard words more encouraging and reassuring, words made all the more important and portentous because of who had spoken them. Almost ten years earlier in a Senate speech, Whig Daniel Webster had used almost the same words when he called the Ohio River "one of those running seas which bear on their bosom the riches of Western commerce."[20] But such rhetoric was to be expected from a Massachusetts Whig, not from a South Carolina Democrat. John C. Calhoun, whose *South Carolina Exposition and Protest* remained the most cogent and explosive challenge to federal power and authority since Jefferson's and Madison's Kentucky and Virginia Resolutions, had proclaimed that the improvement of the Mississippi River and "all its great tributaries" was legitimately, that is, constitutionally, a federal responsibility. If he could make his view of the matter prevail in Congress by fashioning an alliance between the West and the South, then one of the most potentially dangerous constitutional issues and disputes over public policy would be laid to rest, and to the benefit of the West and its economic allies in every region.

Such an alliance between regions would, he thought, also blunt a growing abolitionist assault on the political and constitutional security of slavery. That hope was one of the reasons James Gadsden—Calhoun's close friend and advisor—had urged him to go to Memphis for the convention. Only Calhoun's participation in the convention, Gadsden insisted, could secure the "triumph of our principles & integrity" by removing the "delusions under which they [the People] are duped by a selfish clique." Gadsden reminded Calhoun that the West was crucial to their plans: "It is only in the West that we can have any hopes of a triumph. If they do not come to us we will be overwhelmed by the power that has conspired for our ruin."[21] Gadsden also thought that West-South unity would be a bulwark against any attempts in Congress to reinvigorate trade protectionism, and he assured Calhoun that "[t]he elements are in our favor and if we will not take advantage of the currents and the winds we cannot expect to work our Passage into the Port of Free Trade."[22]

Another benefit, purely personal, which is to say political, would also accrue to Calhoun: such an alliance would likely get him the presidency.[23] That consideration probably influenced his decision to make the speech at the Memphis Convention. But Calhoun had other, external concerns, as well. Delegates from nine southern states and two western states, Illinois and Indiana, were in attendance. Also present were delegations from the territory of Iowa and the Lone Star Republic of Texas.[24] Calhoun was especially interested in Texas, the annexation of which he believed would provide a bulwark for the South against the rapidly growing population, economic power, and political muscle of a free-labor North.[25]

As Tyler's secretary of state in 1844, Calhoun had successfully negotiated a treaty of annexation with Texas, a treaty that had failed to make it through the Senate, in large part because both Tyler and Calhoun had publicly noted the importance of ratification for the defense of slavery, something that had generated strong abolitionist opposition to the treaty. An attempt to accomplish annexation by means of a joint resolution had also failed. Although the treaty was dead, the issue of annexation was very much alive and figured prominently in the presidential election campaign. Calhoun at first thought that his identification with the Texas question might help secure the Democratic Party's nomination for him. Instead, the party chose Polk, whose modest popular vote victory over Henry Clay in November added some weight to Tyler's argument that the outcome of the election vindicated the annexationists. The idea of annexing Texas through a joint congressional resolution was revived, and the Senate and the House passed the resolution by narrow margins in votes sharply divided along partisan and sectional lines. On March 1, 1845, only days before he left office, Tyler signed the joint resolution, thereby formally annexing Texas to the United States and making it a state; it came into the Union as a slave state. As for Calhoun, the new president indicated that he wanted James Buchanan of Pennsylvania, and not Calhoun of South Carolina, as his secretary of state.[26]

Thus it was that Calhoun was available for a seat in the Senate (the seat became vacant with the planned resignation of South Carolina senator Daniel Huger), intent on preparing for a run at the presidency in 1848, and therefore also interested in championing the cause of federally funded improvements to the great rivers of the interior at the Memphis Convention in November.[27] His timing was superb.

Cotton planters and brokers, bankers, and merchants in the South, enjoying boom times after the end of the depression of the early forties, were receptive to proposals for a regional and, with the West, an interregional transportation network, with an emphasis on the rivers of the two regions, especially the Mississippi River. Calhoun arrived in Memphis with just such a proposal in hand, having traveled there from New Orleans, appropriately enough, by steamboat, following a triumphal voyage from Mobile to New Orleans, from New Orleans to Natchez, from Natchez to Vicksburg, finally arriving in Memphis on November 11 to the sound of celebratory cannon blasts.[28] Enthusiastically received by the delegates in Memphis, his grand design provoked suspicion and enmity in South Carolina, even from men who had been allies.[29] Sensing trouble, Calhoun abruptly changed course, steering closer to his usual public position on the issue and leaving observers of his behavior to wonder at his motives. He was, in the words of Bernard DeVoto, "the metaphysician.... All his stands were at the third remove of calculation."[30] In fact, Calhoun was always consistent in the end he pursued, if not in the means he adopted to achieve it. Intent on the maintenance of slavery and ever the pragmatist, he soon moved on to other issues, leaving the forging of an economic and political alliance between the South and the West in other, enthusiastic, but less capable, hands.

One of these champions was James D. B. De Bow, who in January 1846 began to publish a journal in New Orleans, the *Commercial Review of the South and West,* more familiarly known as *De Bow's Review.* De Bow had been a delegate to the Memphis Convention and left it convinced of the wisdom of a program of transportation development, including river improvements, for the South and West. He made his journal an organ for spreading this gospel of development, publishing numerous articles, notes, and announcements concerning the Mississippi River, steamboat navigation, and railroads. Although also read in the North, the *Review*'s influence was felt chiefly in the South and so—as had been the case before *De Bow's Review* had first appeared and before John C. Calhoun had spoken in Memphis—the fight for river improvements found its most fervent champions and greatest support in the West.[31]

Those westerners at the Memphis Convention who were cheered by Calhoun's words had heard what he knew they had wanted to hear. That he had not explicitly called for funding the removal of hazards to steamboat navigation from the United States Treasury was not noticed by his audience, though the omission turned out to be significant. In an extensive report in 1846 to the United States Senate on the resolutions of the Memphis Convention, he still maintained that the Mississippi had become "for all practical purposes" "a part of the gulf [sic], or an inland sea."[32] But, unlike his remarks at the convention, in which he had implied that he favored federal funding for hazard removal on the Mississippi and its major tributaries, his report to the Senate emphatically opposed any such arrangement, proposing instead that the cost of clearing the rivers be defrayed by permitting each of the states along them to levy tonnage duties on commerce within its territory.[33] Calhoun thought that he had squared the circle by at once upholding "the strictest State rights doctrines" and establishing "that the federal government has the right to improve the navigation of the [Mississippi] river." The combining of these two ostensible immiscibles, he hoped, would "remove the only barrier, that remains between the union of the South & West."[34]

THE WEST

By 1846, westerners in and out of Congress were determined that their region get its fair share of money appropriated and spent for the purpose of improving the nation's rivers and harbors. Their determination was partly an outgrowth of their conviction that the West for too long had been serving as the milch cow for the other sections of the country, a sentiment powerfully expressed by Illinois's John Wentworth in his speech to the House on February 10, when he claimed that "[T]hrough the tariff and the public land offices, the West is drained of its money by the General Government." And all that was returned to the region was "the little that is doled out to us for our harbors and rivers, after hard struggles on this floor."[35]

One reason for the increasing militancy of western congressmen was the torrent of memorials from their constituents that had been falling on the House beginning with the Twenty-fifth Congress, which had sat from 1837 to 1839 (see table D.14). In those two years, eleven memorials were received from the West and entered into the official record of the proceedings of Congress; an equal number came in from the western states during the Twenty-sixth Congress. During the same two Congresses, the rest of the nation sent in only eight and four memorials, respectively. Of 127 memorials sent to Congress from 1825 to 1861, 63, or almost half, came from the states along the Mississippi, Ohio, and Missouri rivers and Great Lakes. And these figures do not count the memorials from states on the Gulf of Mexico, such as Louisiana and Mississippi, which also had a direct interest in the condition of the rivers of the interior. Of the 70 memorials directed to Congress from the inception of the Twenty-fifth Congress in 1837 to the conclusion of the Thirtieth Congress in 1849, which were the twelve years when memorials arrived with the greatest frequency and in the largest numbers, 44 memorials, about 63 percent of the total, originated in those states (again, memorials from the Gulf states are not included in these calculations). The geographical, which is also to say political, pattern of the improvements memorials to Congress was one of their defining characteristics. Another was their relationship to congressional activity with respect to the improvements issue, especially funding.

During Jackson's presidency, his demonstrated willingness to approve ambitious and expensive appropriations for rivers and harbors projects encouraged their proponents in various states to send in a growing volume of memorials urging Congress to enact the necessary legislation. The correspondence between the level of appropriations provided in bills for river and harbor improvements and the number of memorials received by Congress is clearly illustrated in figure 3. For the most part, the two moved in accord with one another, though, beginning with the collapse of the Jacksonian-era prosperity in 1837, generally declining appropriations levels seem to have set the pace with which memorials arrived in Congress. This sort of pattern makes sense. Memorials to Congress were not really demands for action so much as pleas, not so much expressions of impatience as of hope. The fall-off of appropriations after 1837 likely dampened hopes and discouraged would-be memorialists. The brief but sharp rebound of appropriations in the Thirty-third Congress may have been the reason for the similarly sharp and short-lived surge in memorials, a rekindling of hope.

As one would expect, the tempo with which memorials came to Congress was at least in part determined by the level of interest and activity by representatives and senators with respect to improvements measures. It probably also responded to the perception outside Washington of the sitting president's sympathy or antipathy for them. In any case, almost all memorials introduced into the record of congressional proceedings were in support of such measures, and most expressed the interests, either parochial or general, of a region, state, district, county, or city in a particular project or in the improvements program as a whole.

Most of the memorials sent to Washington, from the Nineteenth Congress through the Twenty-third—a span that embraced the presidency of John Quincy Adams and almost all of that of Andrew Jackson—originated in the South, New England, and Middle Atlantic region. The states of the Old West, especially Indiana, Illinois, and Michigan, did not become active sources of memorials until the Twenty-fourth Congress and then not with any vigor until the Twenty-fifth and Twenty-sixth Congresses (1837–39 and 1839–41, respectively). The Nineteenth through Twenty-third Congresses were the ones that passed the first significant appropriations to improve the western rivers and Great Lakes, appropriations that always received Adams's approval and almost always got Jackson's. The successful campaign for each of these measures likely encouraged the drawing up and submission of memorials to Congress, most of which called for additional appropriations for already-funded projects or appropriations for new ones. Occasionally, a memorial would come in that commended or condemned Congress for funding any and all improvements.

As lobbying instruments, pro-improvements memorials sent to Congress attempted to establish the advantages that would accrue to the nation, as opposed to any one region, state, or locality, from the execution of a particular river, harbor, or road project. For example, one of the putative benefits to be realized from executing the river improvements requested in some of the memorials was an increase in the capacity of the nation to defend itself against foreign enemies. That point was made as early as 1835 in a memorial from Pennsylvania calling for the improvement of "the navigation of the Monongahela river" from Brownsville to Pittsburgh.[36] Observing that "[N]o nation can wisely calculate upon an exception from the evils of war," the Pennsylvania memorialists warned that the nation's "free institutions, our rapid improvements, our increasing strength, and the happy condition of our republican population, certainly gain us no good will among the powers of Europe. We may therefore well

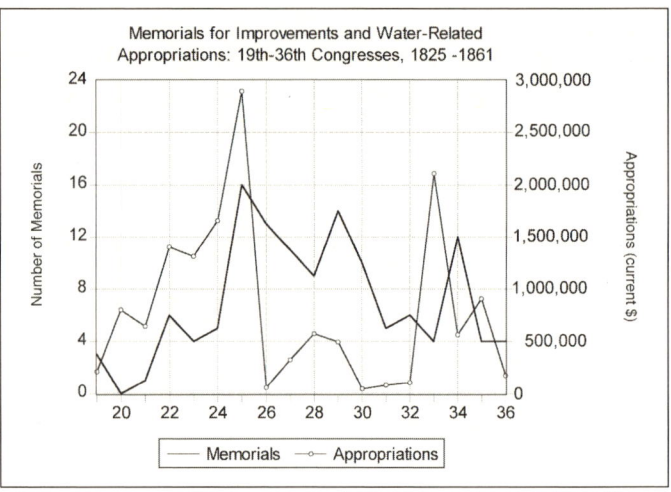

Figure 3. *Sources:* See notes to text.

expect that the time will come, and perhaps soon, when we may be involved in an arduous contest for everything dear to us as a nation."[37] When that time came, they continued, "the improvement of the Monongahela river . . . would be invaluable, for the speedy transportation of troops, ammunition and arms, from the great valley of the west to the seaboard, and from the seaboard to the interior."[38]

Such arguments acquired more immediacy as war with Mexico became a real possibility at the beginning of 1845. No doubt because of the relative proximity of Arkansas to Texas, the annexation of which was still pending in January 1845, the Arkansas state legislature sent a memorial to Congress expressing concern for the national security. Foreseeing eventual war with Mexico, whether or not the United States annexed Texas, and warning against "Indian depradation" [*sic*], the memorial urged the removal of the great raft on the Red River. Clearing this obstacle to navigation would facilitate the movement of troops and supplies to the frontier "to prevent aggression" or to defend the nation in the event of hostilities.[39]

A vigorous program of internal improvements would do more than augment the nation's ability to deter or resist foreign aggression. More important, such a program would strengthen the Union and counter any centrifugal forces that might tend to break it apart. A group of memorialists had advanced that argument in 1838, petitioning for "the improvement of the Alleghany river for steamboat navigation between Pittsburgh, Pennsylvania, and Olean, New York."[40] They buttressed their general point that the proposed improvement of the Alleghany was "so clearly of a national character" with several specific assertions, including that "in time of war" the improvement would provide the federal government with "a highly important avenue for the transmission of troops and munitions of war."[41] But the most important reason for funding the improvement of the Alleghany River was that better navigation on that river would not only promote the growth of commerce "between different and distant parts of the Union, and all the advantages of increased wealth and enterprise to every branch of industry . . . , but by it a community of interest, of feeling, and of friendship will be more directly cemented and perpetuated."[42]

A memorial from "a number of citizens of Bond County, Illinois," early in 1844 made a similar point, but in a very different way. The state of the Union was very much on the minds of these memorialists as they called for "the improvement of the navigation of the Western rivers and lakes, and for the completion of the Cumberland road."[43] Having pointed out the great benefits to the nation from an enlightened, fair, and generous improvements policy, the memorialists—a committee of nine "And five hundred others"—made a thinly veiled threat: "Your memorialists are not prepared to say that, unless they receive justice at the hands of the General Government, the Union will be in danger; for they believe no such consequence is to be apprehended."[44] Then why avert to a nonexistent peril to the Union? Not surprisingly, the memorials from the states of the interior expressed the same sorts of sentiments that members of their congressional delegations frequently voiced in the course of debate over bills to fund improvements of western rivers and lakes. A sense of grievance over the presumed regional bias against the West that operated when appropriations for such improvements were allocated was becoming a fundamental tenet of the western political perspective. Thus the Bond County memorialists insisted that, because "[t]he East and the West contribute alike by their labor and enterprise to fill and replenish the Treasury of the Union," the two sections should have "a just participation in its wise and judicious disbursement. It would have been a magnanimous policy on the part of our Atlantic brethren to have anticipated our demands, and yielded with a ready hand those appropriations which are justly our due," but that had not happened.[45]

It is difficult to say with anything approaching certainty that the tide of memorials to Congress on the general subject of improvements ebbed or flowed for this reason or that. As expressions of organized "popular" political sentiment, the tempo of their composition at home and transmittal to Washington presumably reflected the changing fortunes of the major national political parties, if only on the state level. Then, too, there were concrete economic interests which the memorials and their authors intended to advance. The fortunes of these economic interests, it is safe to say, waxed and waned with the fluctuations of the larger national economy, thereby influencing the frequency with which memorials reached Congress.

Thus it was that, in 1836, following the sharp reduction of the Jackson administration's budgetary engine of public land sales and the contraction of the general economy not long after, the number of memorials entered into the official records of Congress jumped from an average of six every two years during the Twenty-second, Twenty-third, and Twenty-fourth Congresses (1831–33, 1833–35, and 1835–37, respectively) to nineteen during the Twenty-fifth Congress. Beginning with that Congress, and continuing through the Thirtieth (1847–49), an average of roughly fourteen memorials was entered into each Congress's records. Not until the Thirty-first Congress (1849–51) did the number fall back to pre-1837 levels. By then, two of the developments that had helped to precipitate the sharp rise in the number of memorials on improvements had abated and even reverted to something resembling what they had been before the Panic of 1837. A party pledged to national economic development—in this instance, the Whigs—was back in power, holding both Congress and the White House, following the Mexican War. This political resurrection, however short-lived, was accompanied by the more enduring recovery of the economy and, with it, a swelling of the national treasury.

Although the frequency with which memorials entered the records of Congress in the years following 1848 was comparable to the pace characteristic of the years before 1837, the reasons for the low rate in each period were fundamentally different, and the difference was significant. During the years before 1837, a period of considerable organized political activity, there was little need for citizens, either individually or in groups, to memorialize Congress to request construction of roads, improvements to rivers and harbors, or even appropriations for canals (though partisans of the Louisville Canal tried, and failed, to convince Congress to override Andrew Jackson's opposition to continued federal funding for it). Both Congress and two presidents, John Quincy Adams and

Andrew Jackson, supported most such projects. Moreover, turnpikes, canals, and steamboats on rivers and lakes were the major arteries and means of transportation available to passengers and shippers, and the railroad was still in its early stage of development. By 1848, however, the transportation situation was very much changed.

Of only modest economic significance and even more limited geographical extent before 1837, the railroad had, in little more than ten years, become a large and growing industry. In those states in which railroad construction had been and continued to be significant, notably in the northern tier of states from the Atlantic coastline to the Missouri River, the railroad had become the great attractor both of investment capital and political attention. State legislatures and congressional delegations embraced railroads with an enthusiasm born of an infatuation with an industry and technology that were new, progressive, and potentially lucrative.[46] Although politicians in states with river interests still promoted improvements to advance those interests, some, if not much, of the energy that had once animated their efforts was now diverted to railroad-related projects, almost all of which involved funding on the local and state levels, and not by the federal government. For that reason, the number of memorials sent to Congress each year in support of internal improvements projects fell off dramatically.

Very likely, the successive presidential vetoes during the administrations of James K. Polk and his Democratic successors also discouraged individuals and interest groups from sending as many memorials as before. Still, memorials continued to come to Congress. Defeated in their attempt to get a rivers and harbors bill past Polk in the summer of 1846, as the war with Mexico got under way, advocates of the improvements program regrouped and waited for another chance. They waited until early 1847, when the war was still intensifying. The effort to pass the legislation was accompanied by a campaign to exert the pressure on Congress of a mobilized public opinion in support of the measure. Memorials and petitions calling for action flowed into Congress from state legislatures, cities, and groups of interested private individuals. This had, of course, happened before. Perhaps the most substantial of any of the memorials was the one forwarded in May 1848 by a committee of the Chicago Convention, which had met during the preceding July to call for "improvement of harbors and rivers by the general government."[47] The war was obviously on the minds of the Chicago memorialists, and they averted to it when they drove home the point that the improvements program was a bargain for the nation, the accumulated total of all federal appropriations for such improvements since 1789 coming to "less than eighteen millions of dollars—but little more than half of your annual revenue, and probably not equal to three months' expenditures in waging a foreign war."[48] They were substantially correct.

The Mexican War strained the federal government's fiscal resources, soaking up money that might otherwise have been available for other purposes, including river improvements. Spending on the army and navy accounted for about 80 percent of total expenditures in 1847 and 1848, up from about 50 percent before the war, while spending on river and harbor improvements in those years was minimal, approaching zero. No doubt the war necessitated a sharp reduction in spending on river improvements, but the war's unexpectedly large and rapidly increasing cost was not the only reason for the drop. Another reason was the way in which the federal government's finances were arranged.

During the early 1830s, the Jackson administration had been able to prosecute its military campaigns against various Native American tribes, including the expulsion of the Cherokees, and at the same time lift spending for internal improvements to unprecedented and, before the Civil War, unsurpassed levels. Jackson had been able to support improvements spending on this scale because of the infusion of revenue from the federal government's sale of public lands. The collapse of such sales during the last years of the 1830s once again made the tariff the primary and almost the only source of federal revenue. One consequence of this change was a revival of the linkage between the protective tariff and federally funded internal improvements as a sensitive political issue. Another consequence was that the federal government's finances were somewhat more straitened. Taken together, these circumstances provided cover for a succession of Democratic presidents who, averse to anything that smacked of the American System and determined to be faithful to the policies of Andrew Jackson, as they understood them, were all too ready to veto appropriations for internal improvements spending. All this was known to the delegates of the Chicago Convention, and they were understandably worried about this newest manifestation of an old bugbear. Their memorial concluded on a note more imploring than hopeful about the prospects for favorable congressional action: "The people are willing to trust their representatives: let not those representatives exhibit to the world the spectacle of refusing to trust themselves."[49]

Many, if not most, of the convention delegates came from parts of the country that had an immediate interest in hazard removal on the western rivers and in the creation or improvement of harbors on the Mississippi and the Great Lakes. Of the delegates and officers whose homes were not in congressional districts that stood to benefit directly from appropriations for those purposes, many were Whigs from New England, New York, and some of the other remaining strongholds of the American System, including Florida, Georgia, and Missouri. But in this group there were also some prominent Democrats whose immediate political interests vis-à-vis western river and Great Lakes improvements were not readily apparent. All in all, the conventioneers were an interesting lot.

The convention's president, Edward Bates, was a Missourian who had won election to the Twentieth Congress (1827–29) as a supporter of John Quincy Adams. In 1848, he was formally out of politics and was practicing law. He would reenter public life in 1856 as a prominent Whig and would serve as Abraham Lincoln's first attorney general. One of the seventeen vice presidents of the convention was also a lawyer, the Connecticut Whig John H. Brockway, who had served in the Twenty-sixth and Twenty-seventh Congresses (1839–43). Another vice president was T. Butler King, a Whig congressman from Georgia who had served two terms in Congress alongside Brockway. He lost his bid for a third term in 1842 but then regained his seat in 1844 and kept it until he resigned in October 1850 to become customs collector for the Port of San Francisco, a position to which Whig president Millard Fillmore appointed him. William Woodbridge of Michigan also served as a vice president, although by then he no longer took an active part in politics, having served a single term in the U.S. Senate, from 1841 to 1847. Woodbridge was probably one of the least partisan of those who attended the convention. He had won his Senate seat with support from both Whigs and Democrats, and, by the time the convention got under way, he had given up politics for horticulture. Two Democrats among the vice presidents were the Albany, New York, politico and iron maker Erastus Corning and Littleton Kirkpatrick, who had served a single term in the Twenty-eighth Congress, from 1843 to 1845.[50]

Although the Chicago memorialists had reason to fear for the future of river and harbor improvements, the political situation confronting the program in Congress in 1848 was not as bad as it might have been or as it would be in less than a decade. Party and sectional lines had not yet hardened beyond any reasonable hope of flexibility, and, in the year of the Chicago Convention and for a few years before and after it met, many Democratic congressmen deviated from their party's anti-improvements position and voted for river and harbor measures. This is not to say that partisan politics was not an important influence upon the outcome of internal improvements votes in the House of Representatives. So much had been obvious in the first significant clashes over the program in 1824.

Not surprisingly, as party identities crystallized during the 1830s and 1840s, the improvements issue, along with virtually every other matter, profound or trivial, became grist for party mills. At the same time, sectional interests began to assume increasing importance in and out of Congress, especially after the Mexican War, as the slavery issue explicitly and implicitly determined the terms in which almost all other national questions were couched. Between considerations of party and section—a whipsaw of loyalties of unprecedented and all but unimagined ferocity—one might think that there would have been little or no opportunity for politicians to find a third way. Resourceful and often supremely parochial in their outlook, a number of congressmen broke with party discipline and, before the mid-1850s, with sectional ties, as well. As Joel H. Silbey has shown, they voted their districts' local interests on internal improvements bills, especially those concerning rivers and harbors, as well as on such issues as the tariff, public land policy, foreign affairs, slavery, and war in the House of the Twenty-seventh through the Thirty-second Congresses, which span the years 1841–52.[51]

The reason for such maverick or anomalous voting behavior becomes clear when examined on the level of individual congressional districts in each of four Congresses (the Twenty-ninth through the Thirty-second): the geography of a district influenced the vote of the congressman who represented it. In the context of river and harbor appropriations, a congressional maverick was a Whig who voted against the appropriations bill or a Democrat who supported it. Generally, there was a fair degree of party loyalty and sectional unity on roll-call votes, but when river and harbor improvement bills came up, party and sectional ties often were less compelling considerations than whether a district lay along a stretch of river or one of the Great Lakes earmarked for federal funds. In such cases, many congressmen bucked their respective parties and voted to advance local interests. There were, in fact, some striking patterns to the maverick voting behavior in each of the four Congresses, the most obvious of which was the remarkable degree of party discipline on improvements bills among both slave-state Whigs and free-state Whigs, but especially among the latter (see table D.15). More often than not, free-state Whig mavericks were congressmen whose districts had no water interests targeted for improvement with federal funds. Just the opposite was true of the significant numbers of Democrats, from slave and free states alike, whose districts lay along rivers or one of the Great Lakes and who cast off party discipline to vote in favor of federally funded improvements bills that would benefit their districts.[52]

The connection between such voting behavior and local interests becomes even more apparent when examined with respect to two key issues, improvements and a protective tariff, which the Twenty-ninth Congress (1845–47) took up (see table D.16). The tariff bill before the Twenty-ninth Congress became the antiprotectionist Tariff of 1846, called the Walker Tariff after its chief Polk administration architect, Secretary of the Treasury Robert Walker. Whigs, especially northern Whigs, were pledged to the revival of some form of protectionism and fought hard to defeat the tariff bill before Congress. There was far greater Whig Party discipline with respect to the tariff issue, on which only one congressman went against the party's position, than on improvements. Of the seventy-seven Whigs in the House, only seven bucked their party, and six of those seven did so only on the improvements bill, voting against it. In almost every instance, these maverick southern Whigs, like their fellow Whigs from free states, hailed from districts without water interests that might receive federal improvements funding.

Proponents of federally funded river and harbor improvements could, through the early 1850s, count on almost all Whig votes and a fair measure of support from Democrats willing to defect from their party's orthodoxy on the improvements program. The victory of war hero and southern Whig Zachary Taylor of Louisiana in the presidential election in 1848 meant that, at least for a while, the tide of presidential opposition to the program had ebbed. Taylor's electoral triumph was not, however, matched by congressional Whigs, who actually lost some ground in the House and made only modest gains in the Senate.[53] The erosion continued from 1848 to 1850, when Whig strength in the House fell from 109 seats to 88. Two years later, in 1852, the party held only 71 seats, its disintegration and dissolution already well-advanced and apparent.[54] Fortunately for those in and out of Congress who backed the program, a coalition of Whigs and improvements-minded Democrats, primarily from the states of the mid- and upper-Mississippi Valley and those along the Great Lakes, prevailed during the debates and voting on the rivers and harbors improvements bill in the House and Senate in 1852. They were able to send the bill to Millard Fillmore, a New York Whig who had been Taylor's vice president and had become president upon Taylor's sudden death in July 1850. When the legislation reached Fillmore, he signed it. The omnibus measure appropriated more than $2 million for improvements to the Mississippi, Missouri, and Ohio rivers, the Great Lakes, and several other waterways, far more than had ever before been appropriated for such works by a single piece of legislation or would be appropriated until 1867. The remarkable victory scored by the champions of federally funded improvements to western rivers and the Great Lakes had been hardwon. The terms of debate were harsher than they had been on any occasion since the overheated exchanges of 1846, perhaps harsher than at any time since 1832. The tension that is apparent in the record of debate probably reflected the hardening of party and sectional lines as the toxic controversy over slavery and the territories began to poison every public question. And, yet, while the tone of congressional discourse was approaching new and potentially dangerous levels of incivility, there was nevertheless something all too familiar about the substance of the debate.

CLIMACTERIC

By the summer of 1852, almost every syllable that had or would be spoken on the subject of river improvements had already been uttered in Congress. Long-familiar arguments for and against the improvements program had been heard each time a new rivers and harbors bill was introduced or even suggested. That summer's debate was no different, and all the arguments, pro and con, were revived and mustered as the Thirty-second Congress considered the omnibus Rivers and Harbors Bill. With so much money under consideration, the jockeying for political advantage by the members of the House of Representatives, which considered the bill as a committee of the whole, became even more intense than usual.

During debate on the bill on July 29, Whig congressman William T. Ward of Kentucky offered an amendment to include $30,000 for removing snags on the Cumberland River. This was the amount that had been appropriated in 1832 and 1834 in the first two successful improvements bills to have included the Cumberland, and perhaps Ward counted on that fact to move his colleagues to support his amendment.[55] If so, he was wrong, and his amendment failed overwhelmingly by a vote of 72 to 51. Insisting plaintively over the vocal objections from his colleagues, "But, Sir, my district has rights—," he was interrupted by David Cartter, a Democrat from Ohio, who asked the chairman to rule Ward out of order. Ward persisted and went on to plead for "justice" and equality in the appropriation of internal improvements funds.[56]

Moments later, Ward's fellow Whig, Lewis D. Campbell of Ohio's Second District, located near the Ohio River, offered his own amendment to appropriate $25,000 for improved navigation above the falls on the Ohio at Louisville, Kentucky. He began by noting that he had always favored a program of federally funded improvements, but one designed "to embrace equality to all the various sections and interests of this great Confederacy."[57] But, he asserted, the bill before the House failed this test. As he spoke, his voice was quickly drowned out by loud cries of "Order!" "Order!" When the amendment died for want of a second amidst shouted demands by his colleagues that he be ruled out of order, Campbell's frustration overrode his judgment and exploded in an intemperate indictment of the states along the Great Lakes and the Atlantic coast. Raising his voice to be heard over competing shouts of "Order!" "Order!" and "Go on!" "Go on!" he insisted that the congressional delegations of those states had formed "a combination . . . against the great rivers of the interior," with the result that "the mighty valley of the Mississippi is again cast off with a mere pittance." Having finally made his point, he was then ruled out of order and instructed by the chairman to sit down. Before he did so, however, he excoriated those western Democrats who, "after all their Buncombe speeches in favor of western rivers," voted against more liberal appropriations for them. That last sally provoked William Hawkins Polk, the brother of the late president and a Democrat from Tennessee's Sixth District, to insist that Campbell be firmly ruled out of order. This was done, and debate continued, but along the same lines as before, as Alexander Evans, a Maryland Whig, echoed Campbell's sentiments.

Speaking over rising cries of "Order!" "Order!" Evans exclaimed: "Ah! you are endeavoring now to cry down and suppress the utterance of the truth; but I, like the gentleman from Ohio [Mr. Campbell], have been in just such crowds before, and mean to say my say." At that point, a frustrated chair said wistfully that were it in his power to order the sergeant-at-arms to enforce his rulings, he would do so; in any case, he ruled that Evans was out of order. Unimpressed, Evans responded that he "wanted to see these appropriations passed." This last remark prompted some wit to say, "Let the riot act be read," a comment that was greeted by laughter. The successive outbursts finally moved the chair to admonish his colleagues to remember that "[t]his is a branch of the American Congress, a body which is respected throughout the world, and the Chair hopes that the committee will throw itself upon its dignity for a few moments, and the bill will be reported and passed." He might have saved his breath. Evans responded to the chair's request for a return to decorum by complaining that he could not hear "what the Chair has been saying, there is so much noise in the Hall. I am in order. Let others be quiet." To this last outburst, the chair said simply: "The gentleman's time has expired."[58]

Such was the tenor of the debate in the House chamber that day. Although Campbell had invited the scorn of his colleagues when he injected the word *equality* into the debate over the rivers and harbors improvements bill, the question he raised, quite apart from its political implications, was an important one. What he probably meant would have been more closely approximated by the word *proportionality* than *equality*. He, like Ward, had railed against a distribution of funds for improvements that, he maintained, did not give adequate weight to the interests along the Ohio, Missouri, and Mississippi rivers. Implicit in his complaint was the question of how such funds should be apportioned to ensure that most of the money went where it was most needed. Campbell was hardly alone in his complaints and suspicions, although he perhaps entertained the latter more intensely than did most of his colleagues.

Clearly an outburst born of exasperation, Campbell's remarks were nevertheless somewhat odd, even by the standards of day-to-day debate in the House, especially because he was, by 1852, an experienced politician who was then a second-term congressman. So, what had shaped his views and sharpened them to such a point? While others, notably Ward, shared his conviction that the rivers and harbors bill then before Congress was marred by its failure to accord "equality" to all sections of the nation, Campbell's brief speech had more than a whiff of paranoia about it, what with its accusations of a "combination" of seaboard and Great Lakes states and its insinuations of a conspiracy by them against the river states of the interior. His characterization of the sponsors of the bill to which he so strenuously objected was an early expression of his larger, distorted conception of how things worked in Washington.

Within a few years of his failed attempt to amend the 1852 rivers and harbors bill, Campbell the Whig became a major figure in the nativist American National Party, more familiarly called the "Know-Nothing Party." In 1856, he ran for the party's presidential nomination, getting only one vote and thereby tying for last place in a field of eleven candidates that included Sam Houston and former president Millard Fillmore, the party's nominee in that year's election.[59] The year before, Campbell had failed in his bid to be elected Speaker of the House of Representatives as a Know-Nothing partisan and a firm opponent of the Kansas-Nebraska Act. Of 219 votes cast, he received only 4, coming in second to last out of five candidates.[60] A member and leader of an insurgent party that he hoped would carry him into political power, Campbell was, like his Know-Nothing party, out of step and out of touch with the main currents of American political, social, and economic life during the last decade before the Civil War. Perhaps he thought his district was a microcosm of the nation; at all events, his remarks during the improvements debate of 1852 revealed a narrow conception of issues and a maladroitness in dealing with them. But he was hardly alone in that.

The parochial character of much of the debate over the Rivers and Harbors bill is readily apparent from this congressional vignette and suggests the truth of the aphorism that all politics are local politics. In fact, a comparison of the 1852 omnibus bill's provisions with those of earlier and subsequent rivers and harbors bills suggests that the complaints by Ward and Campbell were of the most parochial sort, narrowly focused as they were on the interests of their respective constituencies. Contrary to their assertions, the Mississippi Valley states, especially those along the Ohio, Mississippi, and Missouri rivers, did fairly well in the omnibus bill, getting $580,000 of a total of $860,000 in appropriations for "miscellaneous" improvements to those waterways.[61] Where those states may be said to have fared relatively poorly was in the category of appropriations for specific projects, receiving only another $167,500 of a total of $1,186,290 in such appropriations. This unfortunate distribution mattered greatly.[62] What Ward and Campbell were really complaining about, then, was not that the Mississippi Valley states had failed to get adequate funding but, rather, that the bill had not made adequate provision for the specific, localized improvements projects that were a politician's bread, butter, and pork. Ironically, but not surprisingly, the bill met their demand for "equality" or proportionality almost precisely by assigning about 37 percent of the national total of appropriations for specific and miscellaneous improvements to the states along the Mississippi, Missouri, and Ohio rivers, states that together accounted for that proportion of the national population. This precise proportionality was, in fact, born of a meticulous regard by the members of the House of Representatives for the political arithmetic of appropriations. Time and again, they had practiced their art of crafting legislation so that the regional sums and quotients balanced. That they had managed to do so again was all the more remarkable and perhaps never more necessary in light of the rising sectional tension and deterioration of the Whig Party.

The intensity of the summer's exchanges over the omnibus Rivers and Harbors Bill was a cause of profound concern to many in Congress, who saw a growing danger in the overblown rhetoric and murky innuendo of the debate. The secession crisis of 1850 had been averted, thanks to Stephen A. Douglas's skillful salvaging of Henry Clay's failed attempt to construct a compromise between pro- and antislavery extension factions in Congress. Tempers and sensibilities in and out of Congress were still raw in the summer of 1852, and almost every issue of any consequence seemed to have the potential to touch off another crisis of national unity. Proponents of river and harbor improvements, having prevailed in the House of Representatives, were prepared to carry the day and perhaps even enlarge their gains in the Senate, to which the House bill went. It had, after all, been a long time—almost fifteen years—since appropriations for the program had even approached the level likely to be achieved in the omnibus Rivers and Harbors Bill. Western Democrats in the House of Representatives, making common cause with northern Whigs, had overcome the improvements program's opponents, primarily southern Democrats but, also, half of the free states' representatives. For astute political analysts, the conclusion was inescapable that, just as the Mexican War had aggravated sectional tensions within the Whig Party in the last years of the 1840s, the improvements controversy threatened to do the same thing to the Democratic Party. A desire to prevent such a thing from happening probably explains the actions of Stephen A. Douglas in the Senate late that summer.

On August 23, Douglas offered an amendment to the rivers and harbors bill that had come over from the House of Representatives. He proposed to permit states and municipalities to levy tonnage duties to fund the costs of river and harbor improvements within their individual jurisdictions. This arrangement would supplant the federally financed "general program" of improvements that had become increasingly controversial and that, over the past decade, had encountered opposition from a succession of presidents, all Democrats, with the exception of the ostensible Virginia Whig, John Tyler. Douglas perhaps imagined that his proposal would be greeted with an almost universal sigh of relief from a Congress beleaguered by the seemingly intractable dispute over a general program of river and harbor improvements. As a midwesterner, and one from Chicago at that, Douglas understood better than most of his colleagues the importance of river and harbor improvements to the commerce of the states of the Mississippi River valley and Great Lakes region. But, as a Democrat, he also understood how potentially disruptive the improvements issue could be for his party. Years earlier, when he had been a congressman, a fellow Democrat, Isaac Holmes of South Carolina, had ridiculed him as "the gentleman from Illinois (or *all noise*)" for urging appropriations for harbors on the Great Lakes, appropriations that Holmes insisted were unconstitutional.[63] Douglas wished to cut this knot of policy with an arrangement that he hoped would preserve the vitality of improvements efforts but would, at the same time, remove from the national political arena the question of how those efforts were to be financed. Any hopes he entertained that this attempt at compromise would be received with the same approbation that had greeted his success in fashioning a compromise to defuse the secession crisis in 1850 were about to be crushed.

There was little, if anything, new in Douglas's proposal, and, to be fair, he never claimed novelty to be among its virtues. Long before the first rivers and harbors appropriations were made in 1824, Congress had authorized states to collect tonnage duties, essentially user fees, from vessels that put into their customs port cities. The purpose underlying such authorization generally had to do with an unusual set of circumstances, as was the case when Congress had given Georgia, specifically Savannah, the power to impose a tonnage duty. Savannah had confronted a unique situation resulting from the fact that the mouth of the Savannah River, which formed the city's harbor, had been hotly contested during the Revolution. Both the British and Americans had deliberately sunk ships there in an effort to erect a barrier of wrecks to prevent hostile forces from entering. These wrecks had survived the circumstances that had sent them to the harbor's bottom and had long constituted peacetime hazards to navigation and obstructions to commerce. The tonnage fees collected in Savannah were to be dedicated to the removal of these obstructions. By 1816, many of these wrecks still blocked sections of the port, and tonnage duties were superseded by specific congressional appropriations to rid Savannah of its undesirable Revolutionary War legacy.

For many years after 1816, and especially after 1824, the question of tonnage duties seldom came up in serious discussions of how best to finance river and harbor improvements. During the presidency of John Quincy Adams, any suggestion of reviving the use of tonnage duties in place of congressional appropriations would have received short shrift. A general system of congressional appropriations, funded primarily from revenues generated by a protective tariff, was, after all, an intrinsic element of the American System. There was also little, if any, place or need for tonnage duties within the improvements program of Andrew Jackson, a program largely dependent on the proceeds from public land sales. The same may be said of the policy inherited by Jackson's hard-pressed successor, Martin Van Buren, a policy that he left substantially unchanged in direction but that he had to cut back severely in magnitude due to the Treasury's straitened circumstances. Indeed, not until 1846 did the idea of tonnage duties surface anew, and then it did so within the context of a sharp dispute over the direction of national commercial policy, especially the tariff. Its revival began that year during the course of debate in the House of Representatives over a proposed rivers and harbors bill.

On March 11, Congressman George Washington Jones, a Tennessee Democrat, took the floor to announce that, while his own constituents would benefit by a proposed appropriation to improve the Tennessee River, he would nevertheless vote against it and his own "interested selfishness."[64] His reasons for doing so were a by now familiar mixture of strict constructionism, antiprotectionism, and a muddled form of neo-Jacksonianism. Going on from there, Jones pointed out that the bill before the House also proposed appropriations to improve harbors, every one of which "is included in the prescribed limits of some one of the States." It followed, then, that "the money for that purpose should be raised by duties, levied upon the tonnage entered and cleared at the several harbors, by the Legislature having jurisdiction over the particular harbor proposed to be improved." That, he reminded his colleagues, was how such matters were arranged "until about the year 1816, when the protective tariff had its origin, when it became necessary to increase permanently, by all possible means, the expenditures of this Government; thereby affording the advocates of the protective policy the double argument of the necessity of levying high taxes to meet expenditures and give protection to domestic manufactures."[65]

That, Jones pointed out, was by no means all. Not only was the general system of internal improvements essentially a rationale for the imposition of a protective tariff, but the claim made by its champions that it was designed to address pressing national needs was without merit because the principal objects of the system, the great rivers of the United States, were "not national." "We are," insisted Jones, "not a nation, but a confederacy, composed of different and distinct sovereignties."[66] There could be no disputing this fact, he informed his listeners, by way of concluding his remarks. The Constitution must be considered the unyielding, unflinching arbiter in such matters: "The Constitution is correctly represented by the iron bedstead of Procrustes, and he who does not fit it should be stretched out or cut off, as the case requires."[67] As we have seen, President Polk pleased Jones and his allies and dismayed advocates of the general system by vetoing the rivers and harbors bill of 1846. And while his veto message indicated that the bill had failed the test of Procrustes, he might have added that he had other uses in mind for the money, anyway.

Absent from Jones's indictment of the general system of improvements and his call for a renewed use of tonnage duties was an assertion of the putative superiority of state-funded and state-directed improvements, as opposed to those undertaken by the federal government. Also conspicuous by its absence was the bromide of the greater efficiency of private capital over public spending for carrying out river and harbor improvements. Both points had been made in the past, and both would soon enjoy a new vogue. What immediately mattered to Jones was a decoupling of federally funded improvements from the protective tariff. Without the insatiable demand for money to support the former, what possible justification could there be for the latter? The use of tonnage duties by states and localities to fund improvements would put an end to the general system of improvements and cleanse the body politic of the corruption of an unconstitutional policy.

The revival of the idea of tonnage duties to finance river and harbor improvements was a major source of concern for advocates of an ambitious general program. In great part, it was the fact that tonnage duties were once again under serious consideration in Congress that had galvanized into action the delegates to the Chicago Convention of 1847. The convention's leaders and delegates were determined to make a compelling argument against tonnage duties in their memorial to Congress. The memorial, which was really a manifesto, was composed and signed by the convention's Executive Committee in May 1848 and offered two major substantive objections to the substitution of tonnage duties for congressional appropriations as the means of financing river and harbor improvements. First, the memorialists pointed out that tonnage duties were excluded from the "navigable waters in the vast valley of the Mississippi and in the great basin of the lakes, by the most solemn compacts," by which they meant the Northwest Ordinance and the subsequent acts of statehood that had created the several states out of the Old Northwest Territory.[68] Their second objection to tonnage duties was on the grounds of expediency: "The system is utterly inapplicable for the removal of obstructions in navigable waters which are common to several states."[69]

The convention's objections to the duties did not stop there. Calling the "principle itself of local duties . . . unsound and delusive," the critics at Chicago, in a sarcastic jab, let fly at the very idea of tonnage duties: "If the wit of man were taxed to devise a scheme utterly destructive of all trade, commerce and navigation upon these waters, a better one for the purpose than this, of artificially obstructing them by hosts of collectors of tonnage duties imposed by local legislation, could not be framed."[70] Their concern over the resuscitation of tonnage duties was well-placed. The adoption of a set of tonnage duties would, after all, be an expedient way for Congress to support the idea of river and harbor improvements but, at the same time, divest itself of the disagreeable matter of having to wrangle over how to pay for them. This was an arrangement that might very well appeal to politicians who for the last two years had been voting Mexican War appropriations so large that there had been little left in the Treasury for virtually any peacetime undertaking, especially something as controversial as river and harbor improvements.

Douglas was familiar with the history of tonnage duties, and he could hardly have been ignorant of the hostility with which many of his constituents regarded any proposal to impose them. And yet he introduced his proposal as an amendment to the rivers and harbors bill, anyway. Why? In the course of answering critics of his proposal during the debate on his amendment, he asserted that his plan was "preferable to appropriations by the General Government, for the reason that the money will be more economically expended if left to local authorities than by the authorities of this government."[71] As he later made plain in a letter written to his friend and political ally Illinois governor Joel A. Matteson, Douglas was certain that the federal government could never carry out public works projects as efficiently as could localities and entrepreneurs. The letter provided a clear and comprehensive statement of Douglas's proposal for tonnage duties and a critique of the existing federal system for funding river and harbor improvements.

Tonnage duties levied by states, with the permission of Congress, would accomplish the great and necessary work of internal improvement without the inefficiency, incompetence, incomplete execution, and corruption that, Douglas said, had become intrinsic to the existing system of improvements funded by the federal government.[72] Tonnage duties would permit a river city to improve its harbor by collecting tolls from shippers who used the port.[73] And, unlike the current system, which was continually beset by political wrangling and jeopardized by uncertain funding, the system of tonnage duties "would withdraw river and harbor improvements from the perils of the political arena, and commit them to the fostering care of the local authorities, with a steady and increasing source of revenue for their prosecution."[74] The federal system of improvements funding, Douglas insisted, was worse than "a miserable failure . . . because while it has failed to accomplish the desired objects, it has had the effect to prevent local and private enterprise from making the improvements under State authority, by holding out the expectation that the Federal Government was about to make them."[75] These sentiments reflected more than Douglas's Democratic ideology; although he revered and might invoke Andrew Jackson, Douglas was no Jacksonian Democrat, at least not of the economic fundamentalist stripe. A champion of the market and its larger corporate enterprises, particularly railroads, he believed that "[t]he operations of the Government have not been sufficiently rapid to keep pace with the spirit of the age."[76]

When Douglas introduced his amendment to authorize states to collect tonnage duties to finance the cost of river and harbor improvements, the proposal drew immediate fire from Senate Whigs. One, Truman Smith of Connecticut, was especially scathing in assailing the proposal and its author, asking rhetorically, "[w]ho is to pay these tonnage duties in case this ill-advised scheme is carried into effect?" The undeniable answer, according to Smith, was that the cost of this system "was to fall on the necks of the farmers of the West and Northwest. . . . The Senator has proposed to go into every log cabin in the Northwest, with his system of tonnage duties, and levy them on the hard-fisted and hardworking boys of that region of the country."[77] For Smith, Douglas's motive for introducing his amendment was plain enough. Surely Douglas knew that any duties levied on shippers for the use of a stretch of river or a harbor on a river or a Great Lake must be passed along in the form of higher prices. "This is inevitable," said Smith, and was a law of trade [that is, economics] "and here I would suggest that it might be well for the honorable Senator . . . to study more the laws of trade, and less those of presidential nominations" (1133).

Unrelenting, Smith went on to point out that the evils and defects of the federal system of improvements, cited by Douglas as justification for overturning that system and replacing it with tonnage duties, were entirely the doing of Douglas "and his political friends" (1134). For all Douglas's oft-stated concern for the interests of the "people of the Northwest," his record showed that he did not really care about advancing those interests. The fiscal means to accomplish great things were at hand, Smith insisted, available in the federal Treasury; what was required was the political will to apply those resources to the great task of river and harbor improvements. Indeed, money had long been available to address many of the most pressing needs in that line, but veto had followed veto of river and harbor bills, and whose fault was that? "Have not the Senator and his friends had the entire control of the two Houses of Congress for the last two years? Have they not constituted the committees, and had everything their own way?" asked Smith. And was not Douglas, having been "among the most active and efficient in elevating Mr. Polk to the Presidency," now attempting to secure the presidential nomination for Franklin Pierce, "a man of kindred spirit, who never yet voted for a river and harbor bill, and who is certain to apply sternly to every such measure THE VETO as did Mr. Polk?" (1135).[78]

It was clear enough to Smith that Douglas had introduced his amendment "for the purpose of blowing up the whole system of river and harbor improvements by the General Government." As such, the plan was mischievous, if "ingenious," but it "has not even the merit of originality." Smith pointed out that Calhoun had first suggested the systematic use of tonnage duties and that, in the course of giving an explanation in December 1847 of a pocket veto of a rivers and harbors bill, President Polk had offered tonnage duties as an alternative to federally funded improvements and as "the only safe and practicable method of making river and harbor improvements" (1136).[79] Moreover, Smith pointed out, Douglas had been anticipated in Congress by Robert Barnwell Rhett, who, while a congressman from South Carolina, had offered a bill to allow states to collect tonnage duties in lieu of a federal tariff. Sarcastically, he continued: "What a pity it is that Mr. Rhett has resigned his seat in the Senate! If he were here, we should have the conjoined powers of Illinois and South Carolina in full activity to lay on the farmers of the Northwest much the larger portion of these burdens" (1136). Such an unholy alliance would prey upon the various sections of the nation, but especially on the Northwest and Southwest, the great interior of the Mississippi Valley, to which the rest of the country owed so much. Smith denied "that there is any inequality in the operation of our Government. I deny it utterly. . . . But away with this miserable business of accounting and arithmetic! Can we not look over our own country—over this great Republic, and rejoice in its prosperity? And can we not believe that the prosperity of each part is the prosperity of the whole?" (1136). Smith wished to leave his colleagues in no doubt as to where he stood: "I am for this system; I will stand by it. I will suffer no insidious scheme (I do not say this scheme is insidious,) calculated to undermine and tumble it into ruins, to be brought forward, without an effort to reveal it to the country in all its enormity" (1137).

Douglas was highly skilled in the art of debate, as the electorate in Illinois knew and the rest of the country would in a few years have occasion to learn; he was also willing to go for an opponent's political jugular. He demonstrated both capacities in the course of replying to Truman Smith. Immediately upon getting the floor, Douglas attacked Smith's integrity, accusing him of having, "during the campaign of 1848 . . . held up the offices of this Government as bribes to the voters of Indiana, on condition that they would vote as he told them to vote" (1137–38). He also castigated Smith as "[t]he representative of Wall street [sic] operators," and as the mouthpiece for "certain mining companies upon Lake Superior; who changed that bill ["the Sault Ste. Marie Canal bill"] from a land to a money appropriation, not for the benefit of Michigan, and not at the request of her delegation." Rather, Douglas insinuated, the change had been made to expedite the construction of a canal to afford cheaper access to iron mines in which Smith had a direct personal interest. "The notoriety, if not the fame of the Senator, has extended to every corner of the Union" (1137).

Apart from the defects in Senator Smith's character and public life, defects that, Douglas said, rendered Smith unfit to lecture him on the matter of political integrity or to instruct his constituents in where their interests truly lay, Smith had not represented accurately the economics of improvements and tonnage duties. He had failed to point out that "[a]ll internal improvements, to the extent of their cost, are a tax upon the navigating or transportation interest; but at the same time they reduce the price of transportation over the lines of commerce, and hence prove an advantage rather than a burden to the people" (1137). As for Smith's warning that a web of tonnage duties—that is, taxes—

would add to the cost of commodities, that was another canard. After all, "when we are reminded that we shall be taxed by tonnage duties for these improvements, it must be borne in mind that we are also taxed to raise money under the tariff system to make the appropriations provided for in this and all other river and harbor bills" (1137). At all events, he, Douglas, was ready and willing to engage opponents of his amendment "and discuss it in every part of this Union." And, in a parting shot at Smith, Douglas warned "that our western people are not going to abandon their principles by having the screws of party put to them to deprive them of any improvements of their rivers and harbors until they put Whigs in power. You cannot seduce our proud and intelligent people to submission in that way" (1138).

Douglas was not alone in pushing his proposal for tonnage duties or in going at Truman Smith. His most vocal supporters were Democrats from the South. Andrew P. Butler of South Carolina warned that he would "make war on the [federal improvements] system in every form that I can, hereafter." That system could "operate in but one way. The Senator from Connecticut has given you a clue to the policy. It is, as long as he can, by the legislation of Congress, impose duties and collect revenues for the Federal Government, he is willing to distribute money derived from other people for his benefit." Butler went on to say that the explanation of Smith's opposition to tonnage duties and his call for the maintenance of the general system of improvements funded by tariff revenues, much of which came "from the commerce of the South and Southwest," was that he wanted to use that revenue to "cut a ditch and improve his mine" on Lake Superior (1140). Worse, Butler warned, if the general system is sustained, and the tonnage duty plan goes down to defeat, the result would be tantamount to "burning the cotton of the South." He would not rest in his opposition to the general system and intended to have his say, if necessary by filibustering and bringing the business of the Senate to a halt. "I will lose the appropriations bills, and I will go to the country and take the responsibility. . . . I have something to say on this subject, and I intend to say it" (1140).

Butler's threat of a filibuster was echoed by Walker Brooke, a Whig from Mississippi, who, saying that he had amendments of his own that he wanted to discuss, warned that "[i]f the intention is to sit us out and take the vote [on a rivers and harbors bill] to-night, I give Senators notice of what they have to expect" (1141). Brooke's concern was to get appropriations for his state's smaller rivers, rivers not covered in the rivers and harbors bill before the Senate. He liked

Douglas's tonnage duties proposal well enough so long as it could serve as a substitute for, and not as an amendment to, the rivers and harbors bill before the Senate. In its present form, that bill was defective on two counts: "the one, is that the amount appropriated is too small, and the other is that the amount is too large. They are two very distinct propositions. . . . If it is a local appropriation, the amount is too large; if it is a general appropriation to improve the rivers and harbors of the whole country, it is too small. I cannot vote for it in either event" (1143). In saying that he would vote for Douglas's tonnage duties proposal if it were made a substitute for the rivers and harbors bill under consideration, Brooke had acknowledged a flurry of parliamentary activity, the purpose of which was to accomplish just that.

Sensing that his amendment was going nowhere that day, Douglas had attempted to withdraw it. Missouri's David R. Atchison objected and, in the face of Douglas's reluctance to persist in the matter, assumed sponsorship of the amendment and offered it again, this time as amended by Robert Charlton, a Georgia Democrat, who proposed that the Douglas-Atchison proposition be substituted for the rivers and harbors bill. Urging the transfer of resources and power to make improvements from the federal government to the individual states and to private enterprise, Charlton asked his colleagues who among them "believes that the General Government can make internal improvements with as much efficacy as the States? Who believes that we should have so many lines of railroads extending through Georgia if they had been left to the fostering care of the General Government?" The evidence was plain enough: "it is to the energies of the States, as to the energy of the individual man, that you must look to work out their own prosperity" (1141). Atchison accepted Charlton's amendment, though others objected on the ground that forcing a vote on Douglas's original amendment to the original rivers and harbors bill would reveal to the American people where each senator stood on so important a matter (1142–43).

Truman Smith's opposition to Douglas's tonnage duty amendment had invited and received a slashingly personal rebuttal from Douglas and the Illinois Democrat's southern allies. Those attacks did not, however, discourage other critics of Douglas's proposal. John Bell, a Whig from Tennessee, also assailed the tonnage duty plan, having "rejected the proposition in my own mind long since upon the ground that it is utterly impracticable. It is bringing us back to the old state of the Confederation. One of the reasons why the Constitution of the present Union was adopted, was to remedy this . . . the great evils which existed in the irregularity and want of uniformity in the commercial regulations between the different States of the Confederacy." If adopted, Bell warned, Douglas's plan would pit the states against one another in a contest of beggar-thy-neighbor. He also pointed out that the Northwest Ordinance of 1787 prohibited the imposition of tonnage duties within the states formed out of the Northwest Territory (1139). Douglas's response to this telling objection foreshadowed the reasoning and flexibility that he would employ two years later when he and Atchison again combined forces, then to repeal the Missouri Compromise and pass the Kansas-Nebraska Act. Even if, he said, one were to concede the Ordinance "to be binding, which I doubt," he did not believe that it prohibited "such duties as were necessary to improve navigation, but . . . only prohibited such as would be a restriction upon navigation" (1139).

Just how much potential for mischief there was in Douglas's tonnage duties proposal became clear when two southern senatorial allies, Robert Charlton and South Carolina's William De Saussure, got into an argument over how the plan would work in the case of the Savannah River, the border between their two states. De Saussure wanted to know whether, because the river's course lay within Georgia, that fact would mean that Georgia alone would be empowered to levy tonnage duties upon shippers using the Savannah? Charlton pointed out that, because Savannah River commerce entered and cleared through the port of Savannah, improvements to the river were clearly in the interests of Georgia and Savannah, but hardly in those of South Carolina and the city of Charleston. So much had been plain enough to the merchants, brokers, and planters of South Carolina when they constructed a railroad from Charleston to Hamburg on the Savannah River in an attempt to draw upcountry cotton away from the river trade to Savannah and bring it to Charleston. It would, he said, "be for the benefit of South Carolina in general to put up obstructions in every foot of that river" (1141). He immediately assured De Saussure that he did not mean that South Carolina "would do so if she had the power; I have great respect for the generosity and character of the State of South Carolina; but why put us in the condition of asking as a favor what we are entitled to as a right?" (1141–42).

The dispute between Charlton and De Saussure over a thus far hypothetical situation was quickly seized upon by Michigan Democrat Lewis Cass to drive home his opposition to the tonnage duty proposal. Cass warned that the argument between the senators from Georgia and South Carolina was a presentiment of worse to come. Could there "be a stronger illustration of the practical difficulties of adjusting this important matter than is now before us? . . . It is not the commencement of the end; it is hardly the commencement of the beginning. It should caution us. It speaks in loud tones, that we should not now adopt the plan of the Senator from Illinois" (1142). Cass's warning left Louisiana's Pierre Soulé unmoved. A Democrat of the Calhoun stripe, that is, a states' rights, southern nationalist Democrat, Soulé supported Douglas's amendment and condemned the general system of federally funded improvements. Calling it "monstrous," he asserted that "it organizes plunder, and makes corruption the main lever of this Government." If Congress did not discard the system, it would threaten the solvency of the federal Treasury and inevitably "sink under its own weight." In the end, Congress would have to take action to end the system because "[t]he country will force us to surrender it; it cannot bear scrutiny; it accuses, and in the end must destroy itself" (1142).[80]

Debate over Douglas's proposed amendment to the rivers and harbors bill, now taken in hand by Atchison and perfected by Charlton's amendment to make it a substitute for the bill, continued the rest of the day and into the next. On August 24, California Democrat John B. Weller, formerly a congressman from Ohio, warmly endorsed Douglas's idea of replacing the general system of improvements with state-levied tonnage duties. Weller was not so much concerned with the proposal's mechanics or, for that matter, with the specific defects of the general system as he was mindful of the danger to the Union if the improvements controversy were not settled once and for all by shifting the source of improvements funding away from the federal government to the individual states (1169). He informed his colleagues that, had he been in the Senate chamber the day before, when Douglas's amendment had failed in a vote, he would have voted in favor of it "for I hold that man a public benefactor who takes out of the arena of Federal politics any one of those questions which are calculated in their nature to divide and distract the people. He who can diminish the number of questions upon which the States are divided, is entitled to the thanks of every well-wisher of the Republic" (1169). The existing general system of improvements had increasingly tended to finance what, Weller insisted, were in truth local projects, projects that were properly the responsibility of individual states. This sort of arrangement inevitably precipitated "a struggle between the States as to the amount of money which they shall receive." For this reason, the improvements "question above all others" was "calculated to disturb the peace and harmony of the States" (1169). It would, he warned, bring the states "directly into conflict with each other," engender "bitterness of feeling, which, in the end, must . . . alienate them from each other. . . . How long do you suppose the Government will stand after the States have lost all affection for each other? Destroy that bond of love, of sympathy, which now binds them together, and the Union will soon cease to exist" (1170).

Weller, who had served in the army in the Mexican War, was a self-described "extensionist"—that is, a proponent of "the doctrine of extending the limits of this Union"—and he believed that the general system of federally funded improvements made the task of preserving domestic political tranquility difficult enough in a nation of thirty-one states. "There are already sufficient elements of discord to endanger our safety," he said, and "the larger the number of States the more difficult it would be to reconcile conflicting interests and maintain the Union." For Weller, federal appropriations for improvements were inimical to the realization of the nation's manifest destiny because, inevitably, a system of such appropriations would destroy the Union (1170).[81] That idea was, of course, not original with Weller; Jefferson Davis had made essentially the same point in 1846 during debate in the House over a rivers and harbors bill when he said that "[t]he extent of our Union has never been to me the cause of apprehension; its cohesion can only be disturbed by violation of the compact which cements it."[82]

It is clear from the record of the debate in which Weller participated, and from the bitter debate of a half-dozen years earlier over the rivers and harbors bill of 1846, that too many men spoke too often and too casually about the dangers of disunion or the prospects for it. Some of those who spoke in such terms were among the more excitable and intemperate members of Congress. Representative William L. Yancey, for one, had threatened secession by the South in March 1846 if westerners did not temper their demands for congressional appropriations for river improvements. Southerners, he said, would regret departing "from the hallowed graves of their [westerners'] sires, and not that we shall sever a connexion with such degenerate sons, as such a course of conduct, on their part, would clearly indicate them to be."[83] Others, such as Jefferson Davis, chose their words with greater care and restraint when they suggested, as Davis did about a week after Yancey's outburst, that the perpetuation of the general system of river and harbor improvements put the Union at risk of "dissolution."

The maintenance of that system negated the three elements required to "bind the people of our Confederacy perpetually together": "[s]tate sovereignty unshorn of its attributes, and private interests freed from undue interference; [and] mutual advantage."[84] Under the general system, "error and misdirection in appropriations may be expected constantly to occur, whilst corruption and dissension will attend the division of the spoil, wrung from taxed and toiling millions for works unconnected, it may be antagonist [sic], to their individual interest. The means thus proposed to preserve our Union," Davis warned, "will more probably generate disaffection and discord, like the teeth of the dragon, have an offspring for family strife and destruction."[85] There were, as Barbara Tuchman said of another, later antebellum period, "intimations ... of Neroism in the air."[86]

Despite the often bitter debate and dire warnings of the mortality of the federal union, the Senate passed the omnibus Rivers and Harbors Bill of 1852, which became law when President Fillmore signed it on August 30. It was of unprecedented scope and ambition, and there must have been friends of the bill in Congress who suspected that they would not again see its like anytime soon. As the summer drew to a close, the political prospects of the Whig Party became increasingly unfavorable. The stress and tearing brought about by the Mexican War, the fissioning of the party into its two great sectional wings over the issue of the extension of slavery into the territories, and the rise of "Know-Nothingism" contributed to the decline of the national Whigs and the growing strength of the Democratic Party. In the presidential election that November, the Democrats overwhelmed the Whigs, both in the popular vote and the electoral vote, especially the latter. The Democratic candidate, Franklin Pierce, carried twenty-seven states, losing only the Whig strongholds of Kentucky, Massachusetts, Tennessee, and Vermont to his opponent, Winfield Scott. Pierce had emerged on the forty-ninth ballot as the Democrats' nominee at their convention in Baltimore. A Yankee from New Hampshire, he had nevertheless been the choice of the southern wing of the party. He was known to be a states' rights Democrat and to favor the Compromise of 1850 and vigorous enforcement of the Fugitive Slave Act.[87] In 1852, Pierce was a still a Jacksonian, as were many other Democrats from New Hampshire, an island of Jacksonianism in New England. So closely identified was he with Jackson that, eight years later, following the unfortunate Democratic convention of 1860 in Charleston, South Carolina, where a Pierce boomlet had failed to develop into anything more rewarding,

Vanity Fair magazine had occasion to twit him for it, saying: "It might not be generally known that there was one delegate at Charleston who, through thick and thin, voted for Franklin Pierce." While the editors said that they did not know the delegate's identity, they had "a theory that it is a delegate from one of the interior counties of Pennsylvania, where, as we are told, they still continue to vote at every election for General Jackson."[88]

In going after Pierce, the editors of *Vanity Fair* had picked a large target. By the time he ran for president in 1852, he already had a substantial record as a public figure. He had been a congressman and a senator and had served as a colonel and then as a brigadier general in the Mexican War. As a congressman, "[h]e had even out-Jacksoned Jackson in his relentless opposition to internal improvements."[89] In the Senate, his opposition to all manner of internal improvements measures was just as solid and included votes "against appropriations to continue the Cumberland road, against a bill for the benefit of the Alabama, Florida and Georgia Railroad and against the rivers and harbors bill—all of which passed."[90] That sort of record perhaps suggested to some voters that Franklin Pierce—"General Pierce," as he was called—was Andrew Jackson redux. Anyone who entertained such thoughts soon enough had reason to discard them.

Pierce did not distinguish himself as president. Not only was he not another "Old Hickory" or another "Young Hickory," he was not even another John Tyler. On Pierce's watch, as is well known, Congress passed the Kansas-Nebraska Act of 1854, which, crafted by Stephen A. Douglas with the help of David R. Atchison, overthrew the Missouri Compromise, opened a Pandora's box of political and moral equivocations over slavery, precipitated irregular warfare in Kansas, and generally heightened sectional discord. As he confronted this rapidly evolving situation, Pierce resolutely hewed to his long-standing states' rights convictions, encouraging his southern allies, frustrating growing numbers of northern Democrats, and swelling the ranks of the Republican Party. But as late as the end of 1853, this troubling, unfortunate situation was not yet on the horizon for even the most astute of politicians. In a letter written in mid-November of that year to the editor of the *Illinois State Register*, Stephen A. Douglas sized up the Pierce administration and the challenges before it: "There is a surplus Revenue which must be disposed off [sic] & the Tariff reduced to a legitimate Revenue standard.... The River & Harbor question must be met & decided. Now, in my opinion is the time to put those great interests on a more substantial & secure basis by a will [sic] devised system of Tonnage duties. I do not know what the administration will do on this question, but I hope they will have the courage to do what we all feel to be right."[91]

In addressing the matter of western river and harbor improvements, the president displayed the same resolution and imagination with which he handled the issue of slavery in the territories, and he achieved similar political results. During his four years in office, Pierce vetoed nine pieces of legislation, six of which were rivers and harbors measures. Andrew Jackson, by way of contrast, vetoed twelve bills in eight years, and only three of his vetoes were of rivers and harbors bills.[92] Unlike Jackson's vetoes, all of which were sustained in the Congress, five of Pierces's six vetoes of internal improvements measures were overridden. In his veto messages, Pierce advanced no new objections to the appropriations measures sent to him by Congress; his reservations about the bills' provisions were the same ones expressed by Jackson and Tyler when they had vetoed similar legislation. As did some of his predecessors, he invoked Andrew Jackson, whom he called "one of the greatest men of the Republic," by way of justifying his approach to a general system of rivers and harbors improvements, as in his veto of a the bill sent to him in 1854.[93] In at least one instance, the improvements legislation sent to Pierce came with the grounds for a veto prominently displayed within it. Opponents of the rivers and harbors bill of 1854, the most ambitious of the six improvements bills turned back by Pierce, managed to include in its body a number of local projects, that is poison pills, which guaranteed a veto by Pierce. On August 1, the Senate passed the measure, but only because some of those voting for it wanted it subsequently killed by the president. Receiving the legislation that day, he had only a few days in which to notify Congress of his decision in the matter before the current session ended.[94] On August 4, he sent Congress a brief notice of his veto in which he said that he would "present to Congress at its next session, a matured view of the whole subject."[95] Pierce's message reached Congress the next day and set off an explosion of western bitterness and anger. Congressman Harry Hibbard, his friend from New Hampshire, attempted to defend the president by explaining the constitutional scruples that had moved him to veto the rivers and harbors bill. Seeking to defuse the tense situation, Hibbard went on to minimize the potential impact of the announced intention of westerners in Congress henceforth to ignore distinctions of party and to forge a regional political alliance to ram through legislation needed by the West. The reaction to his words was not what he hoped to elicit: "Within a few hours many of his hearers were on their way home to contribute to the organization of the Republican party."[96]

There were some river improvements measures that received Franklin Pierce's approval. His political problem with respect to the overall issue of such improvements was, however, plain enough. By the middle of August 1856, almost exactly two years after his collision with Congress over the rivers and harbors bill of 1854, he had signed five river bills, which together appropriated $931,000. Only one of these, a measure "[f]or continuing the improvement of the Des Moines Rapids in the Mississippi River," addressed a need of the states of the Northwest. The other four appropriated money for improvements projects in slave states: North Carolina's Cape Fear River (July 22, 1854), the Savannah River (March 3, 1855), the mouth of the Mississippi River, and the Patapsco River in Maryland.[97] Not without reason had many westerners concluded that President Pierce was no friend of their region.

On Monday, June 2, 1856, the Democratic Party's convention got under way in Cincinnati, Ohio. As balloting began that Thursday, a growing number of delegates became convinced that Pierce, if nominated for a second term, could not win reelection. The next day, Harry Hibbard withdrew his friend's name from consideration, Stephen A. Douglas released the votes pledged to him, and the delegates nominated James Buchanan of Pennsylvania by acclamation.[98] Word of Pierce's political unseating was a source of satisfaction in the West. For the *Detroit Free Press*, the decision at the Democratic Party's convention was cause for rejoicing: "We thank God that President Pierce's term of office is drawing to a close."[99]

The election and administration of Pierce's successor, James Buchanan, would give westerners little reason for further exultation. During his term in office, demands for appropriations for rivers and other improvements got nowhere. Only two river bills passed Congress, each early in 1860, and both were turned back by Buchanan's vetoes, which were sustained. As was often the case when Congress sent river improvements bills to a president, these arrived on his desk on the last day of the session, in this case the last day of the Thirty-fifth Congress's second session. If the last-minute arrival of the legislation was a tactic intended to pressure Buchanan to approve the bills, the tactic failed of its objective. In his message to the Senate giving his reasons for not approving one of the bills, an appropriation for work on Michigan's St. Clair River, the president explained that putting him in a position where he had to approve legislation presented to him without adequate time to consider its merits would be to "convert him into a mere register of the decrees of the Congress." He therefore deemed it necessary to retain, that is, to pocket veto, the legislation because "it was not presented to me until the last day of the session."[100] Buchanan went on to provide a detailed list of objections to the bill, and these were essentially the same grounds on which Tyler, Polk, and Pierce had objected to rivers and harbors bills sent to them: that the work had already been largely completed; that the appropriation was unconstitutional—here he invoked James K. Polk's veto message of December 15, 1847; that the proposed project represented an intrusion by the federal government into a matter that was properly the concern of an individual state; that the use of federal funds to make improvements might corrupt the political system; and that the appropriation measure was inequitable and injurious to the peace of the Union because it gave to one state that which had not been given to others.[101] He went on to urge that, instead of relying on federal appropriations to improve the St. Clair River, Michigan be authorized to levy tonnage duties to raise the money necessary to undertake the work. Nowhere in his message did the president allude to the straitened circumstances of the federal government.

Economic and political conditions during Buchanan's term in office severely limited what he and Congress could do to address a variety of pressing issues, including river and harbor improvements. The business contraction associated with the fall "Panic of 1857" sharply reduced federal revenues. Consequently, the health of the Treasury, explored in detail in the next chapter, was poor, and the budget remained deeply and persistently in the red for the remainder of Buchanan's presidency. His refusal to lend a sympathetic ear to industrialists' complaints that inadequate tariff protection from foreign competition had helped to precipitate the economic crisis further eroded his political base in the North. And the deepening sectional crisis over slavery frustrated efforts at coalition building in Congress, efforts that in past years had produced improvements legislation.

The occasional memorials still arrived in Congress from the West, asking for appropriations to fund improvements, but their numbers were sharply reduced—down from twelve during the Thirty-fourth Congress (1855–57) to four during the Thirty-fifth Congress (1857–59) to just one during the Thirty-sixth Congress (1859–61). Three of the four memorials sent to the Thirty-fifth Congress to ask for improvements appropriations came from Michigan. That state sent no memorials concerning improvements to the Thirty-sixth Congress but instead forwarded a memorial on a very different subject. On February 2, 1861, Michigan's state legislature conveyed four joint resolutions to the second session of the Thirty-sixth Congress on "the state of the Union." Noting that "certain citizens of the United States are at this time in open rebellion against the government," Michigan's legislature assured Congress "[t]hat Michigan adheres to the government as ordained by the Constitution, and, for sustaining it intact, hereby pledges and tenders to the general government all its military power and material resources."[102] The secession crisis of the winter and spring of 1861 would quickly put the bitter improvements controversy in perspective.

One reason for the intensity of the struggle over the improvements issue, especially over river and harbor improvements, was that it involved conflicting interpretations of the Constitution, and therefore irreconcilably different conceptions of the federal union. Another reason was that the improvements controversy engaged two mutually antagonistic approaches to political economy. One of these was a Jeffersonian brand of laissez-faire private enterprise, according to which any public assistance rendered to business was to be provided on the level of local and state governments. In conflict with this approach was that of a neo-Hamiltonian capitalism in which business enterprise and agencies of the national government, including a national bank, worked in concert with one another to forge a national market. Still another reason that the improvements debate was often so bitter was that ultimately the improvements issue became inextricably bound up with the festering dispute over slavery and was hostage to it. Consequently, even when the resources were available to finance extensive improvements projects, there was often a failure of political will to undertake them. At other times, the resolution to do the work was abundant, but the fiscal means were lacking. Always, critics of the general system of improvements assailed the program for its costliness, arguing that the work could be done for less and with greater efficiency by the states or by private contractors. That was hardly the case, as an examination of the record clearly shows.

PART II

◢ 4 ◣
WAYS AND MEANS

Advocates of a comprehensive general program of internal improvements believed that the federal government's authority to undertake the program was implied in the Constitution and that its financial resources were adequate to accomplish the task of clearing the western rivers of their hazards to navigation. Their opponents were just as certain that the federal government, whatever its capacities in the matter, had no constitutional basis for undertaking such a task. Further, they did not want a government that could exercise that kind of power, and their fears of it were as large as the ambitions of those who championed the program. These conflicting understandings of the proper role to be played by the federal government beg the questions of just how large was that government and how extensive were its resources. Fortunately, these are not difficult questions to answer.

Some of the basic pertinent information is laid out in table 6. As is plain from the table, the size of the federal government's civilian workforce more than doubled every twenty years between 1821 and 1861.[1] Most of that growth was concentrated in the Post Office, which grew almost sevenfold. The increase reflected the proliferation of post offices and the awarding of patronage, both of which occurred as the nation's population grew and towns and villages were established. The postal service was undoubtedly the most immediately and continuously felt extension of the existence, if not the power, of the federal government. Moreover, it was one of the few manifestations of that power that can be said to have expanded during the last four decades before the Civil War. Whatever the national state was, it was no leviathan or juggernaut, certainly not by the standards of any year after 1860–61.[2] Yet, it was also not a puny thing when considered in light of its finances and its capacity to undertake projects of considerable magnitude, especially war.

OVERVIEW OF FEDERAL REVENUE AND SPENDING

From 1821 through 1860, the total receipts and expenditures of the federal government generally increased. As figure 4 illustrates, until about the mid-1830s

Table 6
Some Measures of the Scale of the Federal Government, 1821–1861

Year	Congress	President	No. of States	No. of Reps.	Size of Cabinet	No. of P.O.s	Total Civilian Employment	Post Office Employees	No. of Bills Introduced	No. and % Acts Passed
1821	16th	Monroe	24	213	6	4,650	6,914	4,766	480	200 (42%)
1831	21st	Jackson	24	240	6	8,686	11,491	8,764	842	360 (43%)
1841	27th	Van Buren	26	223	6	13,778	18,038	14,290	1,146	495 (43%)
1851	32nd	Taylor	31	234	7	19,796	26,274	21,391	1,011	269 (27%)
1861	36th	Buchanan	34	241	7	28,586	36,672	30,269	1,595	323 (20%)

Sources: Total civilian employment and number of Post Office employees are abstracted from: U.S. Bureau of the Census, *Historical Statistics of the United States, Colonial Times to 1970,* electronic ed., ed. Susan Carter et al. [machine-readable data file] (Cambridge: Cambridge University Press, 1997), Government Series Y 308–317, "Paid Civilian Employment of the Federal Government: 1816 to 1970," pt. 2, 1102–3. Number of states and representatives are from ibid., Series Y 215–219, "Apportionment of Representatives among the States: 1790[–1860] to 1970," 1084; the number of bills introduced and the number of acts passed are from ibid., Series Y189–198, 1081–82; the number of Post Offices (P.O.s) is from ibid., Series R 163–171, "Postal Services—Post Offices, Revenues and Expenditures, Postage Stamps, Stamped Envelopes and Postal Cards Issued, and Pieces of Mail Handled: 1789–1970," 805; the size of each president's cabinet is derived from *Biographical Directory of the American Congress, 1774–1949,* "Executive Officers, 1789–1949" (Washington, D.C.: U.S. Government Printing Office, 1950), 13–19.

both receipts and expenditures grew only gradually. Following a sharp decline from the levels reached in the latter half of that decade, however, revenue and spending grew dramatically from their respective lows of about $10 million in 1843 to more than $70 million of revenue in 1856 and an equal level of spending in 1858. Until 1837, revenues exceeded expenditures by healthy margins, averaging about $17 million each year prior to 1836. For much of the next quarter century, however, the federal budget often ran a deficit, notably during the troubled economic times following the Panic of 1837, the Mexican War and its immediate aftermath, and the years following the Panic of 1857. The pool of red ink was at its deepest during the war, when it reached $25 million in 1847, and was only slightly shallower at $23 million in 1858, the first full year of the depression that began the previous year.[3]

SOURCES OF REVENUE

Throughout most of the four decades prior to 1861, the tariff was the single-largest source of revenue for the federal government, a reliance that had begun at the outset of government under the Constitution in 1789. That year, and during the next two years, customs receipts accounted for well over half of the United States Treasury's inflow of funds (see table D.17). Over the course of George Washington's two terms and John Adams's one term, the tariff provided an annual average of 55 and 75 percent, respectively, of all federal revenue; the share of revenue generated by the tariff rose to an annual average of 92 percent under Thomas Jefferson. While protective tariffs after 1816 became exceedingly unpopular, revenue tariffs—that is, tariffs with rates set with the aim not of discouraging consumption of foreign goods but of raising the funds necessary to finance the operations of the federal government—generally enjoyed wide support. Such tariffs were always more popular than excise taxes as a means of raising revenue. Even under Jackson, during whose administration annual receipts from public land sales averaged just under one-fourth of the Treasury's revenue stream, almost three-fourths of that revenue each year came from the tariff, and a protective tariff at that.

During Jackson's first term in office, receipts from the sale of public land provided an annual average of just under 10 percent of total federal revenues. As figure 5 indicates, that situation changed dramatically after Jackson won re-election. In 1833, public land sales generated more than 14 percent of all federal

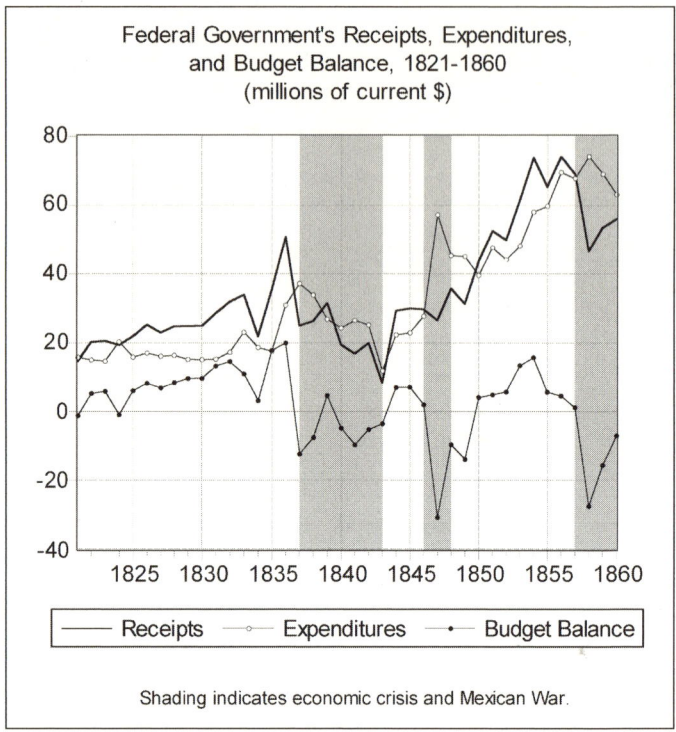

Figure 4. *Sources:* See notes to text.

revenues; the next year, their share was almost 30 percent; in 1835, that figure had climbed to over 45 percent and, in 1836, the peak year for public land sales, receipts from such sales accounted for just under one-half of all federal revenues. Of course, after 1836 the bottom fell out of the land market, and, apart from infrequent surges of land sales in subsequent years, revenue from that source seldom represented even 10 percent and, more often than not, much less of total federal receipts. In 1860, public land sales by the federal government amounted to well under 5 percent of the Treasury's intake for the year. That year marked the last full one of James Buchanan's presidency, during which the Treasury's reliance on the tariff, almost always a revenue tariff, had soared to its highest level ever, generating 93 percent of total federal receipts. That development capped a trend that had been underway ever since the end of Martin

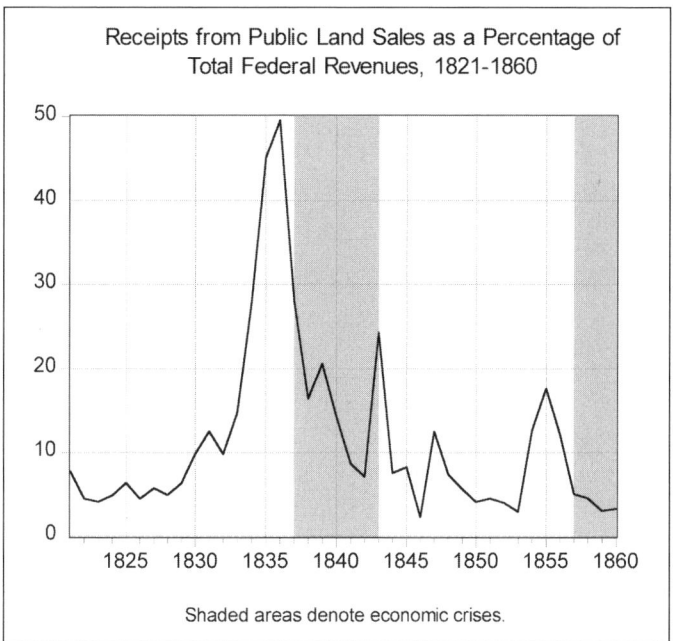

Figure 5. *Sources:* See notes to text.

Van Buren's unhappy tenure in office, during which time just over half of all revenue came from the tariff.

At the same time that the tariff once again began to serve as the primary source of revenue, the contribution to the Treasury from the sale of public lands fell sharply (the single sustained exception to this trend in the years following Van Buren's presidency occurred during Franklin Pierce's administration). The central problem for the federal government's fiscal operations therefore became one of finding a source of revenue to compensate for the decline in public land sales that occurred at the same time that tariff rates fell from their once protectionist levels to those characteristic of revenue tariffs. Beginning with Van Buren's administration, one president after another solved this problem by resorting to borrowing and to issuing Treasury notes (see table D. 17).

This dependence on borrowing and note issues was in stark contrast to the utter absence of any net new public debt under Jackson. Such borrowing was out of the question for Jackson, whose aim it had been to eliminate the public debt, and who almost succeeded in doing so. Under Van Buren, new debt issues accounted for 20 percent of all federal revenues each year. His successor, John Tyler, raised that annual proportion to 37 percent during his term in office, much of which was blighted by the depression of the early 1840s that had begun in 1839.

James K. Polk was compelled, both by his Jacksonian ideology, which brooked no protectionist tariff, and by the exigencies of his war, which turned out to be far more expensive than anticipated, to resort to borrowing to generate an average of 28 percent of each year's federal receipts. As a Democrat who proudly bore the moniker of "Young Hickory," the very un-Jackson-like turn to borrowing must have been supremely distasteful to Polk. He could at least console himself by looking back at the even more extreme resort to borrowing by James Madison, another Democratic war president. In any case, by the end of Whig Millard Fillmore's term in office, new debt had fallen back to an average of 20 percent of total federal receipts, still a fairly high proportion.

With the election of Franklin Pierce, who seemed determined to be more Jacksonian than Polk or, for that matter, even the venerated Jackson, the average annual share of federal receipts generated by borrowing fell to under 2 percent, while, at the same time, the contribution of receipts from sales of public land averaged just over 11 percent and in one year, 1855, hit 18 percent, a level not seen since 1843. Pierce's successor, James Buchanan, maintained Pierce's antiborrowing policy, even in the face of once again depressed public land sales, this time in part because of the Panic of 1857, through the expedient of extreme reliance on the revenue tariff.

THE COURSE OF RIVER AND HARBOR IMPROVEMENTS APPROPRIATIONS AND EXPENDITURES

Under ordinary circumstances, Congress and presidents practiced a form of fiscal economy to ensure that the nation lived within in its means. To that end, as figure 4 indicates, there were only fourteen years out of the forty within the period 1821–60 when the federal budget was in the red, and in two of those fourteen lean years, the deficits were negligible. But a deficit in a particular year was not necessarily a reflection of the spending decisions made in that year. Spending in any given year might reflect an appropriations measure passed that year, but also one passed the year before, two years before, or even three years before.

For this reason, the provisions of an appropriations measure tended to reflect the then-current condition of the Treasury and, especially, the flow of receipts into it. Also for that reason, expenditures, while given to sharp fluctuations, tended to be more evenly distributed over time than were appropriations.

By 1861, Congress had appropriated $18,544,222 for improvements to the nation's rivers, lakes, and harbors, improvements that fell into two groups. The first of these concerned construction projects and other work designated for individual states and were designed to improve shipping conditions along specific stretches of rivers, lakeshores, or harbor coastlines by removing snags, rapids, sandbars, and other obstructions. From 1791 to 1861, Congress appropriated $13,282,468 for such projects compared, for example, with $18,087,978 for canals and about $549 million for the United States Army. The second category of internal improvements appropriations consisted of projects, designated "Miscellaneous" by Congress, which were directed at a river system, an entire river, or a substantial part of a river. From 1824, when the first such measure was passed, through the end of 1860, Congress appropriated $5,261,754 for miscellaneous projects along the Mississippi, Missouri, Ohio, and other rivers.[4]

The course of these two types of appropriations for river and harbor improvements is charted in figure 6, which illustrates three noteworthy points. First, 1839 marked the rather abrupt end to sustained funding of river and harbor improvements; thereafter, significant funding occurred only on a sporadic basis. Second, in the general election year of 1852, after almost eight years of having made almost no significant appropriations, Congress responded to pent-up demands for appropriations from the states of the Ohio, Missouri, and Mississippi river valleys by passing the largest appropriation measure for river and harbor improvements prior to the Civil War. Finally, as one would expect, state-specific appropriations generally exceeded those made for miscellaneous projects, the former being more politically sensitive and rewarding than the latter; the major exceptions to this pattern occurred during the first half of the 1840s and the latter half of the 1850s. In both instances, the larger miscellaneous appropriations were for projects to clear the western rivers of their obstructions and to maintain the passes at the mouth of the Mississippi River.[5]

Presidents could spend on the improvements program only what Congress had appropriated for it, a fact of no little significance. That important distinction between the two parts of federal funding determines how each is used here. Of the two, appropriations constitute the better measure of the changing

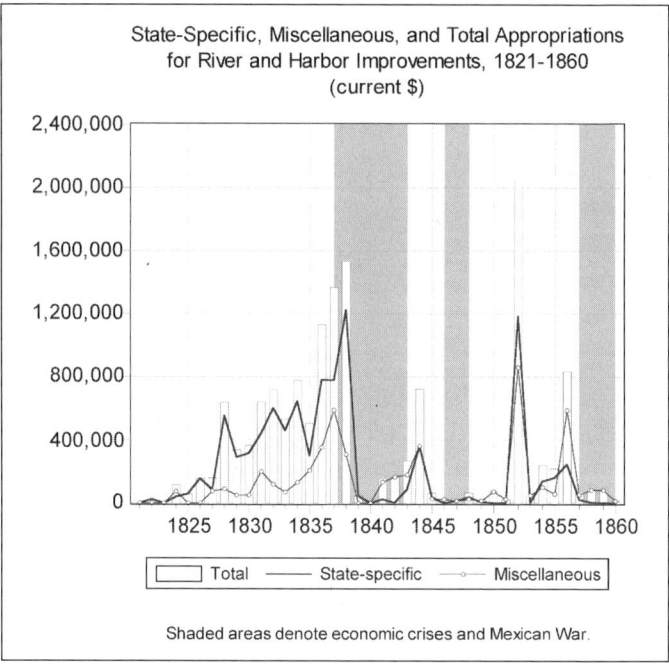

Figure 6. *Sources:* See notes to text.

fiscal commitment of Congress to the program because each appropriations bill, no matter how narrow or comprehensive in scope, was passed in a particular year, often in response to a demand for action either from within Congress or from its constituents. Expenditures, on the other hand, because they often were spread out over two, three, or more years, were a measure in dollars of the work actually done to remove or reduce hazards on the nation's waters each year. As indicated in figure 7—a chart of internal improvements appropriations and expenditures over time—the program's intensity fluctuated considerably. The fluctuations were, however, not random.

In their study *Western River Transportation*, Haites, Mak, and Walton observed three periods between 1838 and 1861 when federal appropriations and spending for improvements to western rivers fell to negligible levels. They attributed these relatively barren years—1840–41, 1846–51, and 1856–60—to a combination of strict constructionist-minded presidents (Van Buren, Polk, Pierce, and Buchanan) and shrinking federal revenues due to recessions.[6] Although,

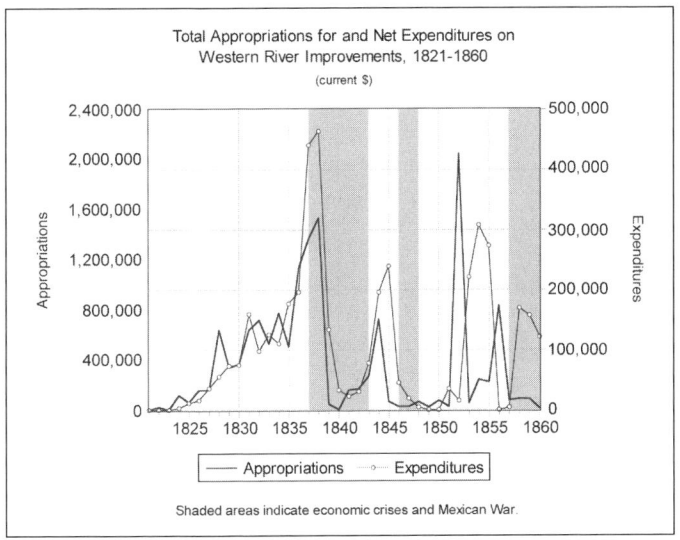

Figure 7. *Sources:* See notes to text.

as figure 7 suggests, Haites, Mak, and Walton somewhat overstated the chronological extent of this barren period, these were, in fact, rather strange years when compared to the 1820s and 1830s. The early 1840s were years of depression, as were the last years of the 1850s. From 1846 through 1848, the Mexican War diverted the national government's attention, energies, and wealth from their prewar paths and precipitated the eruption of the intense sectional antagonisms of the next dozen years over the entangled questions of territorial expansion and slavery. But, bad as was the fiscal situation, the main reason for the sharp reductions suffered by the program, according to Haites, Mak, and Walton, was that it lay outside the narrow interpretation of the Constitution imposed by most of the era's presidents. This proposition is open to at least a rough empirical test.[7]

Haites, Mak, and Walton demonstrated that annual congressional appropriations for western river improvements during the period from 1824, when the program may be said to have begun, to 1840, by which time appropriations made with any regularity had ended, were strongly correlated with the magnitude of the federal government's receipts during the previous year.[8] Although their results are valid, their model obscures an element of the decision-making process that figured in the crafting of appropriations measures. While receipts and anticipated revenues were of undeniable importance in congressional deliberations, Congress almost always made funding for the military, especially the army, a budgetary priority.

A close examination of the record of federal spending during the antebellum decades reveals that, for the 1824–60 period, annual appropriations for improvements to rivers, lakes, and harbors were closely related—even more closely related than to just receipts from the previous year—to the federal deficit and to appropriations for rivers, lakes, and harbors of the immediately preceding year.[9] It should be borne in mind that the deficit of a preceding year reflected not only that year's receipts but also its expenditures, chief of which were generally those on the army and navy. These circumstances bear directly on our understanding of the tightened circumstances of the western rivers improvements program: that is, that the harsh fiscal reality that confronted Andrew Jackson's successors and their congressional counterparts may have been as important a consideration in their approach to the program as was their ideological hostility toward it. The Committee on Commerce of the House of Representatives acknowledged as much in its report in 1858 during the depression that followed the Panic of 1857. The committee's members recognized the need for substantial and sustained aid to complete several important improvements projects that had already been begun but noted that the sums required had forced, "by their vast aggregate," the committee "to relinquish the hope of such extensive aid at the present period of financial distress."[10] In fact, the thirteen barren years of appropriations for western rivers noted by Haites, Mak, and Walton were almost invariably years of pronounced budget deficits. Moreover, with the exception of the two years, 1840–41, and the roughly four years, 1857–60, the budget deficits of the barren years were largely due to sudden increases in military spending, first for the Mexican War (1846–48) and later for the troubles in Kansas and Utah (1854–57).

And, yet, as we have seen, the constitutionally rooted opposition to river and harbor improvements was profound and extensive, perhaps sufficiently so to explain why, beginning in 1840, river and harbor improvements received so little political support from Congress and the White House. The deficits of the barren years must have emboldened many opponents of a general system of improvements, their argument having been, in effect, that we would not fund the program even if we could, and we cannot. The avowed concern over deficits by opponents to river and harbor improvements might account for the many disappointments experienced by advocates of the improvements program during the last two decades before the secession and war. But were the deficits, in fact, so large that the fiscal means for funding the rivers and harbors program were not at hand?

During the four decades between the inception of the program to improve the western rivers and the outbreak of the Civil War, successive Congresses appropriated increasing amounts of funds to construct light stations, marker buoys, and beacons at points along the Atlantic coast, the Great Lakes shorelines, mouths of rivers, and harbor entrances. The great public end served by these undertakings was the protection of shipping, especially shipping engaged in foreign commerce. These fixtures were used to indicate safe shipping channels, harbor entrances, and semistationary hazards such as bars, and, from 1791 through 1860, Congress appropriated $11,048,736 to provide them.[11] And, because such projects were location-specific, unlike the miscellaneous category of improvements to a river or river system, they were, like state-specific river improvements, inherently and intensely political in character. A report on the nation's "light-house system" to the House of Representatives in 1838 by the Committee on Commerce observed that there was evidence to support the suspicion that the reason appropriations for lighthouses passed the House so readily was because they "served to whet the appetite of men desirous of this denomination of public patronage."[12] Light stations were so nakedly politicized that they later became the object of the acid attentions of Ambrose Bierce, who defined one as "A tall building on the seashore in which the government maintains a lamp and the friend of a politician."[13] Not surprisingly, then, as figure 8 illustrates, appropriations for light stations and other markers, which, before 1840, had been overshadowed by river and harbor improvements, became the major objects of congressional largesse after 1840.

But there was another reason for the almost continuous growth in appropriations for the light stations, buoys, and beacons, one that is indicated by figures 9 and 10, which trace those annual appropriations and, respectively, the shipping tonnage engaged in the coasting trade and the tonnage clearing American ports in the foreign trade of the United States. As is evident from the figures, appropriations for light stations, markers, and buoys moved almost in unison with coastal and foreign shipping tonnages, suggesting that the ris-

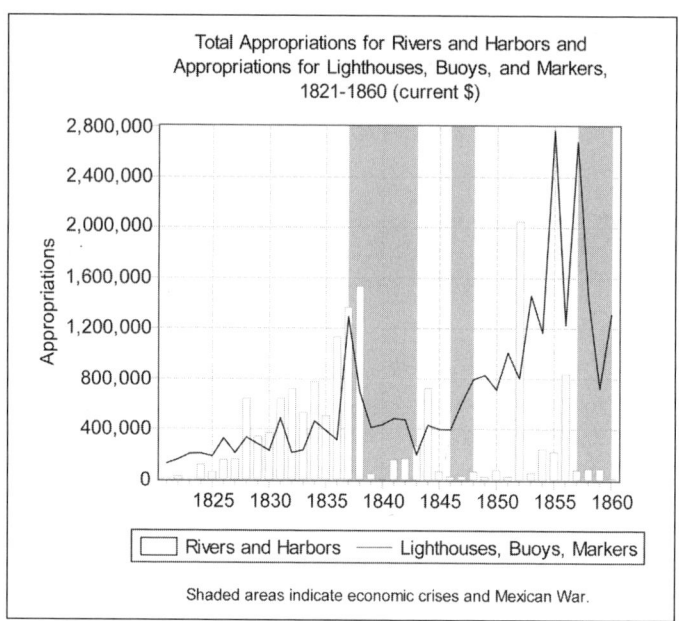

Figure 8. *Sources:* See notes to text.

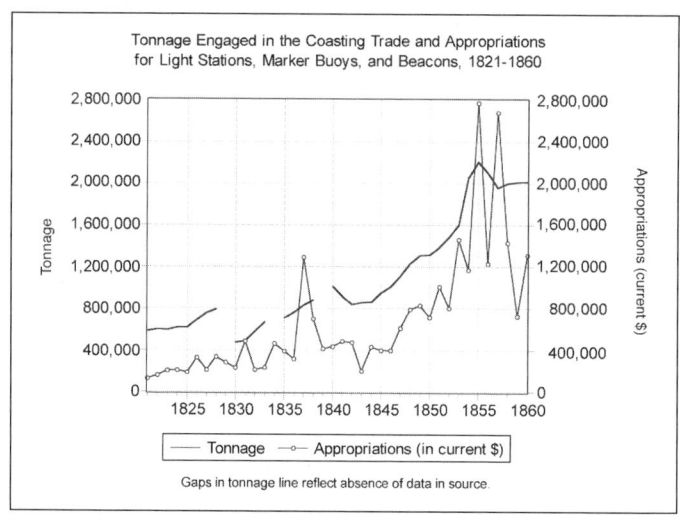

Figure 9. *Sources:* See notes to text.

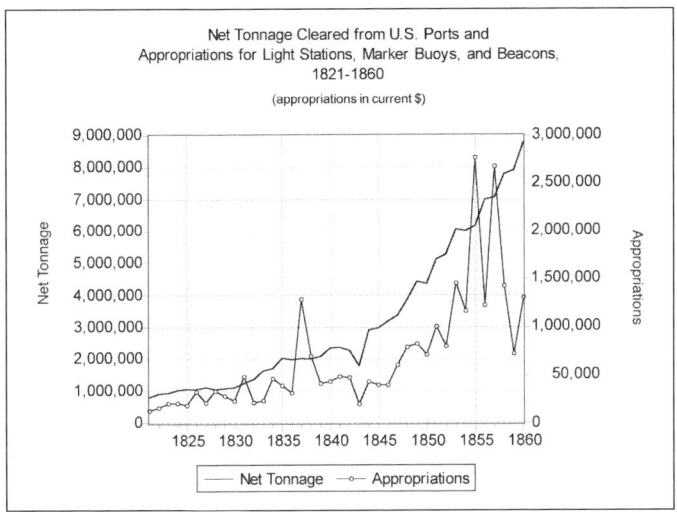

Figure 10. *Sources:* See notes to text.

ing appropriations for light stations and other markers were a response to the growing need for such facilities along the Atlantic, Gulf, and, after 1848, Pacific coasts.[14] Faced with the need to protect foreign and domestic shipping along the nation's coastlines and in its ocean harbors, and inclined to favor that form of commerce that generated tariff revenues over inland navigation, much of which did not do so, Congress understandably appropriated the funds necessary to construct the light stations, buoys, and beacons. In any given year, however, a consequence of that decision was that considerably less money was available for river improvements. During periods of wartime stringency and economic crisis, there was even less, sometimes nothing at all.

Ideological and political considerations of course carried considerable weight in congressional and presidential decisions concerning funding for river and harbor improvements. But one of the implications of the preceding analysis is that no Congress and no president, no matter how Whiggish, could have supported large-scale funding for such improvements in the face of frequent deficits, irresistible pressure to increase military spending, and the necessity of safeguarding an ever-rising volume of coasting and oceangoing commerce. In any event, for reasons that will be gone into later, too much can and has been made of the presumed deleterious effects of the near-absence of appropriations for the western rivers during the thirteen barren years. During the twenty-one-year period (1840–1860) within which they fell, about 65 percent of all naturally caused wrecks occurred during the barren years. But that is only to be expected. The thirteen barren years represented about 64 percent of the time covered during the entire period; that is, the losses during those years were almost precisely in proportion to the barren years' share of the period's time span.

MILITARY SPENDING AND SPENDING ON RIVER AND HARBOR IMPROVEMENTS

During almost every presidential administration, from Washington's through Buchanan's, at least one-fourth, and usually between one-third and one-half, of the federal government's gross expenditures went to servicing the national debt (see table D.18). Generally, debt reduction on a significant scale was a luxury to be indulged in during times of peace and prosperity. What was left after making payments on the debt were the budgeted net expenditures, that is, those made to conduct the current operations of the government, including the support of the army and navy. Although the actual amount spent each year on the military fluctuated, depending on economic and political circumstances, military spending typically accounted for between 50 and 60 percent of net expenditures. The two great exceptions to this pattern were the years of the War of 1812 and those of the Mexican War. In each of these periods, spending on the army and navy soared, as did the services' shares of gross and net total federal expenditures. But no episode before the Civil War compared with the Mexican War for its impact on the federal budget's spending levels and its balance.

The first significant increases in military spending during the Polk presidency occurred before hostilities with Mexico required them. In 1845, Polk's first year as president, expenditures for the army and navy were about $12,050,000, or roughly 52 percent of the federal government's net expenditures, which was only slightly less than the military's share of federal spending during Andrew Jackson's eight years as president. The spending level of 1845 was, in fact, only a little higher than that of the year before and, like it, reflected the budgetary decisions made during the administration of John Tyler. Expenditures on the army and navy had dropped sharply in the wake of the collapse of federal revenues following 1836 and had not bottomed out until 1843. Thereafter, they began to grow, but, even after the increases of 1844 and 1845, military spending in the latter year was still only 59 percent of its 1837 level. That situation changed very quickly.

The fiscal picture in 1846 was quite different from that of the year before and was a reflection of an improving economy and Polk's influence. Of course, the beginning of the war ensured still higher spending levels for 1847, when expenditures for the army and navy exceeded $46 million, or far more than double what they had been just a year earlier. This record high level of military expenditures accounted for more than eight of every ten dollars of total federal spending. War-related expenditures were only slightly lower in 1848, the last year of the war, but returned to prewar levels the following year. The end of the war also permitted the return of the officers of the army's Topographical Corps, most of whom had served in Mexico, where the majority "were maimed with wounds, or sick from the fatigues and exposures which their duties required."[15] By the early 1850s, spending on the army and navy was lower, having fallen to just under 40 percent of total spending. Beginning in 1855, however, military spending, particularly on the army, began to increase sharply in response first to the troubles in Kansas and then, in 1857, to the "Mormon War." Quickly and cynically dubbed the "Contractors' War," after the men who made money supplying the army detachments sent to Utah to cow Brigham Young's "Inland Empire," the affair was essentially an expensive episode of sound and fury but no fighting—"a war without battles" but with many casualties.[16] In terms of treasure and lives spent, however, it paled in comparison with the war against Mexico.

Increases in military expenditures immediately before and especially during the Mexican War had a profoundly adverse impact on spending for river and harbor improvements. Although military spending had had a somewhat similar effect upon spending for improvements while Tyler had been president, its impact had been nowhere near so pronounced as it became during Polk's presidency and his war. In John Quincy Adams's administration, both military spending and internal improvements spending had increased sharply, buoyed by revenues from the protective tariff. Spending for both continued to grow under Jackson, though military spending grew at a slower pace. When, during Van Buren's presidency, spending cuts became necessary, neither the army nor the navy was spared. But even as spending on the army and navy fell by two-thirds, from $20,330,000 in 1837 to $6,685,000 by 1843, the last year of the depression, the military's share of total expenditures remained almost constant and even increased slightly from about 55 percent in the former year to 56 percent in the latter one (figure 11).

Appropriations for the army and navy seldom provoked controversy in the House of Representatives and Senate, which considered both services, but especially the army, necessary instruments of national policy and guarantors of national sovereignty. Arsenals, forts, coastal lighthouses, and roads primarily identified as military routes had similar general support in Congress, and, for them, the expenses involved in their execution, almost irrespective of their magnitude, were of little or no consequence. Proposals to improve the western rivers enjoyed no such approbation. Controversial for constitutional and political reasons having nothing to do with their intrinsic value, river improvements projects were also assailed for their costs, despite the fact that the amounts necessary to carry them out were usually far smaller than the costs of military-related public works. In fact, as figure 12 indicates, the money spent during any given year for river improvements projects was a small fraction of the aggregate expenditures of the federal government, almost never amounting to even 1 percent of the total. Notwithstanding the improvements program's minor share of the budget, its political vulnerability made it a prime target for the budgetary axe men in Congress. Thus, net spending on rivers and harbors improvements, which reached its pre–Civil War high of $463,736 in 1838, fell by 95 percent to only $22,500 in 1841. Although it quickly began to recover, reaching $33,237 in 1842 and $77,742 in 1843, even the latter year's spending level was only 17 percent of what it had been just five years earlier.

The cuts in federal spending that occurred after 1837 reflected a falling-off of revenue that had begun the year before when receipts from sales of public lands imploded upon the issuing of the Specie Circular. Those sales, which in 1836 had generated just under one-half of all federal revenues, accounted for only 7 percent of total receipts in 1842; by 1846, proceeds from land sales constituted only slightly over 2 percent of the Treasury's intake. The collapse of land sales receipts meant a sharply increased reliance by the Treasury on its other major source of revenue, the tariff. Thanks to the recovery that had begun in late 1843 from the depression and the protectionist rates of the tariff of 1842, duties collected on imports soared. By 1844, for only the second time since 1836 (1839 had been an isolated exception), the federal government's budget was in the black, if only by just under $7 million. It remained so during the next two

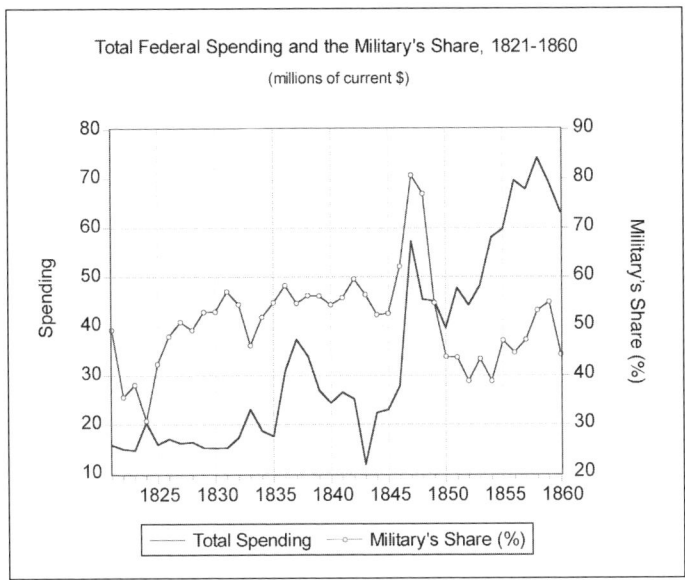

Figure 11. *Sources:* See notes to text.

years, though the surplus fell sharply in 1846 to a little over $1.9 million, reflecting the effect of the significantly reduced rates of the Walker Tariff of that year. A robust foreign trade nevertheless sustained the surge in tariff revenues, thereby partially mitigating the Walker Tariff's effects. But in 1847, the budget was again deeply out of balance, this time by close to $31 million, the worst annual deficit ever prior to the Civil War. That red ink was, of course, part of the cost of the Mexican War.

The budget deficits brought about by war spending rapidly swelled the national debt, which climbed from $15,550,000 in 1846 to $38,827,000 in 1847 to more than $47 million in 1848. The debt continued to grow until it peaked in 1851 at $68,305,000. Such were the wages of war. If Polk had not actually sought the war—and that he had done so was generally suspected in Whig Party circles—he quickly embraced it as the likeliest instrument with which to realize his territorial ambitions.[17] And, as we have seen, he also cited the war and its drain on the Treasury to justify his vetoes of river improvements bills in August 1846 and December 1847. The war did impose an unprecedented demand on the nation's fiscal resources, and Polk would not be the last president to en-

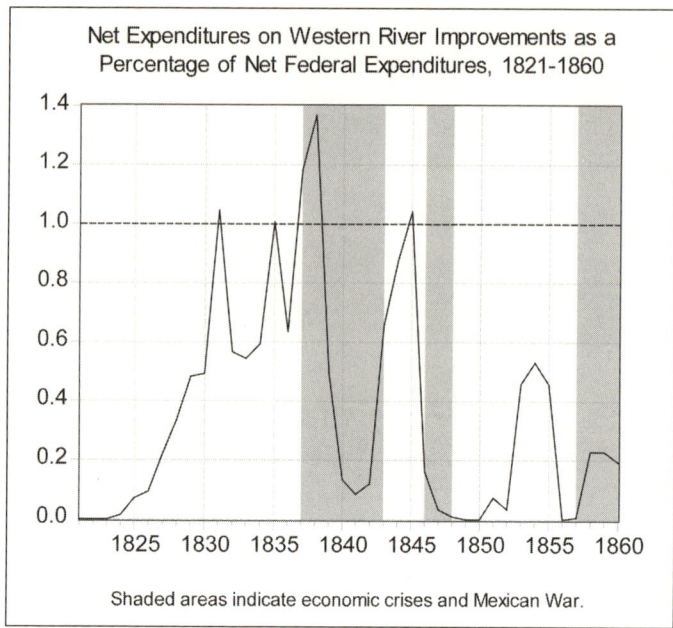

Figure 12. *Sources:* See notes to text.

join the nation to sacrifice butter for guns. And, yet, there was something both cynical and disingenuous about his invoking the war in his veto messages.

Increases in military spending required by the war with Mexico soaked up most of the federal government's revenue, which meant that far smaller amounts of the federal government's finite resources were available for other purposes. That this was the case with respect to spending on improvements after 1845 is evident from figure 13. It was the war's draining away of money that could otherwise have been spent on river and harbor improvements that had provoked Garrett Davis, a Whig congressman from Kentucky, to lambaste Polk in a February 1847 speech on the floor of the House of Representatives.[18] And Davis was hardly alone in holding that view. Another Whig, John A. Rockwell of Connecticut, was scathingly sarcastic in his criticism of the war and its domestic consequences. The war with Mexico "was, doubtless, a very splendid affair; and some gentlemen thought that all the money the country could raise ought to be expended in 'thrashing the Mexicans;' in 'whipping Mexico,' and in causing our gallant arm to 'revel in the Halls of the Montezumas.'" So voracious was the war's appetite for money and so willing were too many men in Congress to feed that appetite, he warned, that no other purpose, no matter how worthy, could win support: "if a bill should be brought here giving money for the deaf, the dumb, the blind—ay, for the insane—it would be refused, because all our money must go for this magnificent war." War and war finance, said Rockwell, had become the only concerns that animated Congress: "All our money must go to kill Mexican women and children, till Mexico paid her debts: so we could not pay our own, but must tax our creditors for money to carry on the war with."[19]

Even the president had admitted that the war's demand for money precluded most nonmilitary spending and had even hinted that he regretted the strictures upon improvements spending imposed by the exigencies of war. His critics, of course, did not take his expressions of regret at face value, according them little if any value at all. In any event, whether one deplored or applauded the impact of heightened military spending upon funding for the improvements program, the undeniable fact was that the relationship between the two was essentially an inverse one, a fact that had profound consequences for the operation and maintenance of the program.

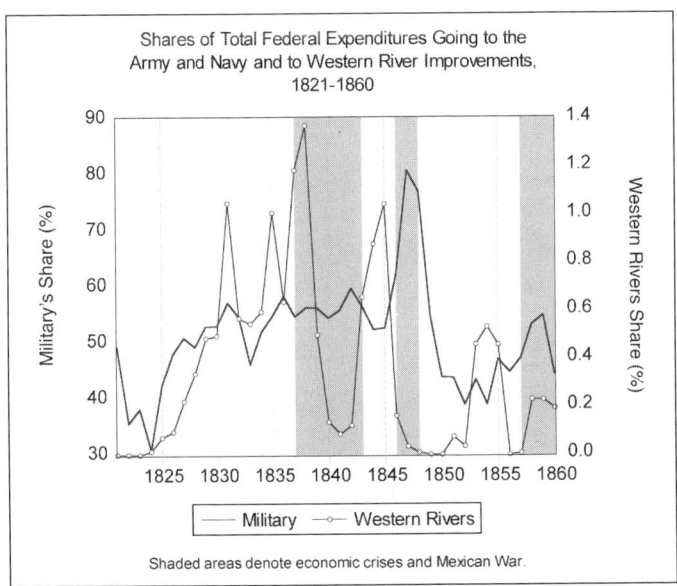

Figure 13. *Sources:* See notes to text.

THE PRIVATE SECTOR, PUBLIC LANDS, AND RIVER AND HARBOR IMPROVEMENTS FUNDING

The sums appropriated for internal improvements depended immediately as well as ultimately on decisions made in Congress and by successive presidents. As we have seen, these decisions were shaped by the fiscal condition of the federal government, specifically the amount of revenue flowing into the Treasury and spending levels dedicated to the military and to debt reduction. There were, however, other factors that influenced these decisions, if only indirectly and remotely. One of these was the almost total lack of incorporation of steamboat enterprises on the western rivers (see table D.19), which meant that the political influence of steamboat owners upon Congress likely was considerably less than that of an industry in which the corporate structure was common, such as the railroad. The controversy surrounding the bridge over the Ohio River at Wheeling, Virginia, provides a good illustration of this point. Despite a decision on May 27, 1852, by the U.S. Supreme Court favorable to the steamboat interests of Pittsburgh, in which the Court found the bridge to be an unwarranted obstruction of river traffic, a Congress favorably disposed to the railroad industry eviscerated the Court's decision on August 31 by passing legislation that declared the bridge and its roadway, built to accommodate railroad tracks, to be part of a post road and therefore a necessary element in the movement of the United States mail.[20] No doubt the technological luster of the railroad helped to sway the minds of many in Congress; railroad fever was already well-established in the United States. But the railroad corporations' lobbying efforts, the railroads' presence in several congressional districts, and the fact that ownership of the lines was widely held through bonds and shares of stock were probably the decisive considerations in moving Congress to pass a piece of legislation that was an obvious post-hoc contrivance.

The near absence of incorporated steamboat companies is hardly surprising when we consider the comparatively modest levels of capitalization, typically well under $30,000 and even $20,000, involved in the construction of a vessel. Turnpikes, canals, and railroads, by contrast, were objects of considerable expense for those undertaking their construction, and incorporation was a practical and necessary means of spreading around the costs and attendant risks. Capitalization levels of turnpike companies commonly approached

$100,000, and costs could run as high as $10,000 per mile of construction.[21] Canals and railroads were even more expensive, with construction costs per mile of right-of-way as high as $25,000 during the 1830s and 1840s.[22] Unlike steamboats, moreover, turnpikes, canals, and railroads were spatial systems, projecting capital and equipment over distance, and therefore, by their very nature, they required the benevolent attention of government to grant them what were, in effect, spatial monopolies. This fact, in turn, also necessitated incorporation, under the legislation of a particular state. As capital undertakings, steamboats required no such protection and, more to the point, would have frustrated them, in any event. The most well-known effort to organize and sustain a steamboat monopoly involved an attempt to control a particular route of service in New York State and ended in failure with the 1824 landmark U.S. Supreme Court decision in *Gibbons v. Ogden.*

Although the rarity of incorporated steamboat companies did not help the industry's campaign for congressional funding of river improvements, that campaign still enjoyed considerable support in and out of Congress, especially from the states of the Old West. The reasons for this staunch support, even from states in which turnpike and railroad construction was growing rapidly, had a lot to do with the success of another federal policy, the sale of public lands. As we have seen, public land sales provided the revenue from which hazard removal on the western rivers was funded in many years. The influence of those sales on the course of spending for improvements was, however, not limited to that.

One of the consequences of road building in the western states and hazard removal on the western rivers was greater access to public land in the interior, land that was then offered for sale directly to individuals by the federal government or given as federal grants to the states for subsequent sale to the public. There was, in fact, something of a mutually reinforcing relationship between river improvements and public land sales. The sale of public land led to settlement, population growth, farming, the formation of towns, and economic development that, in turn, led to growing demands for further road construction and continuation of efforts to improve navigation on the rivers.[23]

While the general nature of the relationship between improvements and public land sales has long been understood, it has thus far eluded attempts to define it with any degree of quantitative precision. What has been needed is a measure of economic development that can be tied to sales of public lands and to spending by the federal government on western river improvements. Annual data on town formation in states along the western rivers provide just such a measure. The number of towns formed is superior to the number of acres of public land sold as an indicator of actual economic development on what, before the sale of the government land, had been the unorganized, uncultivated public domain. Towns should not be confused with townships, such as those created under the Northwest Ordinance. Such townships were not necessarily settled places but were, instead, survey lines on a map, and the number of towns formed far exceeded the number of townships created. Thus, while the laying out of townships did not necessarily mean that the land was being taken up and populated, the rate of town formation constitutes a rather good indicator of economic development.

Town formation involved surveys, titles, streets (however rude), and the building of houses, one or more shops, one or more churches, and perhaps a school building. Such structures and the town itself represented a capital investment as much as a unit of social organization. Moreover, there is reason to believe that town formation in the river states was fairly sensitive to interest-rate fluctuations in national and regional capital markets.[24] The sale of the land, the taking up of the land, and putting it under the plow to grow food and other crops were possible only because migrants to the vast American interior were able to get there and were able to send their crops to nearby and distant markets.[25] The program of internal improvements, including road construction, but especially the advent of the steamboat and the campaign to remove hazards to navigation from the western rivers, not only made this migration practical, but also made it profitable.[26]

The essential relationship between sales of public lands and town formation in four of the states along the western rivers—Ohio and Indiana on the Ohio River, Missouri on the Missouri River, and Mississippi on the Mississippi—is charted in figure 14. Immediately apparent is the rather pronounced parallelism between the movements of the two measures. As land sales in these states climbed during the late 1820s and especially during the first half of the 1830s, the total number of towns formed there rose, as well. This remarkable increase in the rate of settlement and social and political organization was a direct result of Andrew Jackson's policies, and it is perhaps not to overstate the matter to say that Jackson had realized the Jeffersonian dream of a vast inland

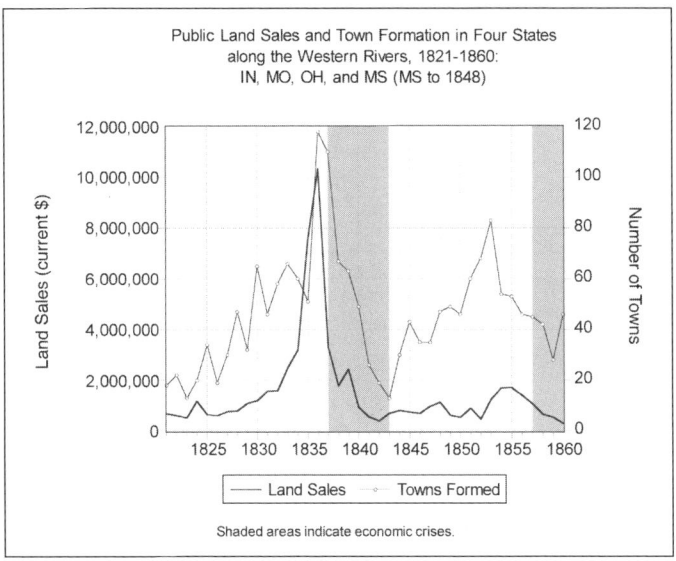

Figure 14. *Sources:* See notes to text.

republic through the application of federal power. Alexander Hamilton would also have been pleased.

When the boom turned to bust, following the collapse of land sales that had been triggered by the Specie Circular in 1836, the rate of town formation collapsed almost as catastrophically. Interestingly enough, this pattern did not persist after the recovery in 1843 from the economic crisis of the preceding four to five years, as the sharp rise and equally sharp fall in the number of towns formed during the 1840s and 1850s ceased to follow the lead of what became almost flat land sales. Beginning in the late 1840s, as figures 15 and 16 indicate, another influence was driving the rate of town formation: the railroad was coming into its ascendancy. In Ohio and, more so, Indiana, as well as in other states of the interior, the projection of railroad lines was accompanied by the establishment of towns at junctions with canals, turnpikes, and other lines and at depots, a process that continued well into the last decades of the century and into the next.[27] The eclipse of public land sales by the railroad as an influence on the rate of town formation was an instance of the displacement of the public sector by private capital as an engine of economic growth and development. But, as figure 17 suggests, this displacement was not total, and that other significant application of public policy, federal spending on western river improvements, persisted even as public land sales declined and almost certainly continued to play a significant role in stimulating town formation.[28]

MAKING IMPROVEMENTS

The point of all river improvements appropriations and expenditures was the funding of operations to remove natural hazards to and enhance navigation on the western rivers. That having been the case, the obvious question arises: What sorts of undertakings did all that money buy? While the improvements program included construction of breakwaters, wing dams, and harbors on the rivers and Great Lakes and other western lakes, the two chief activities involved in hazard removal on the western rivers and their major tributaries were dredging and snag pulling.

Dredging operations were conducted to deepen river channels and river, lake, and ocean harbors and involved the use of specially designed boats that were outfitted for the work, which might entail deepening Atlantic coast harbors, the passes of the Mississippi River, or the low-water channels of the western rivers. The typical steam-powered dredge boat and attendant scows, onto which the dredged material was deposited, cost about $14,000 and could

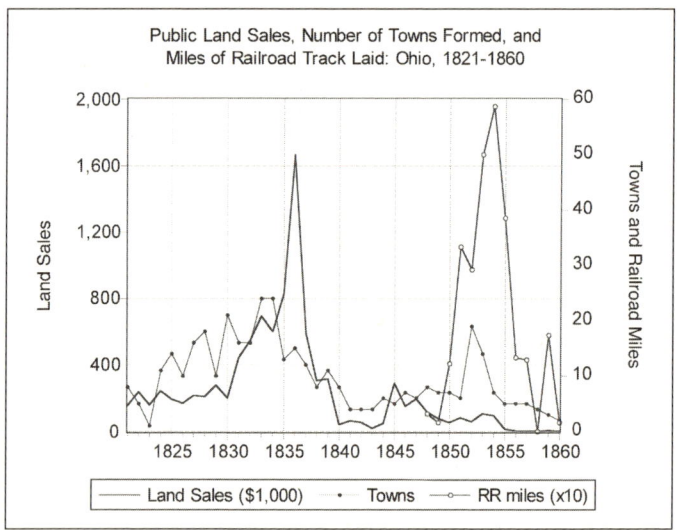

Figure 15. *Sources:* See notes to text.

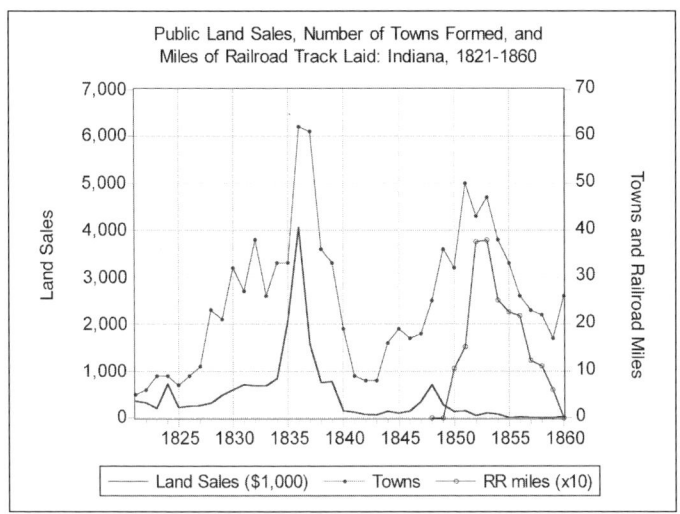

Figure 16. *Sources:* See notes to text.

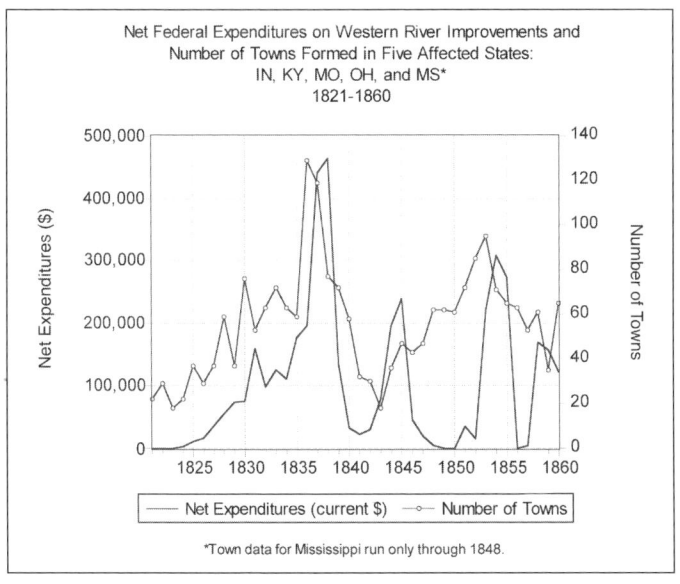

Figure 17. *Sources:* See notes to text.

remove dirt from a river or bay in ten feet of water at a rate of about one hundred cubic yards per hour.[29]

Because the most dangerous natural hazards on the rivers were snags, their removal was the most important objective of the river improvements efforts. Stretches of river that were infested by snags were generally the targets of one or more snag boats, which were either built nearby or brought up- or downriver to the problem area. The most celebrated snag boat was a twin-hulled vessel, built after the design of the first successful one, for which Captain Henry M. Shreve had received a patent in 1838. The use of single-hulled, or "light-draught" snag boats, which were considerably smaller than the twin-hulled type, was generally confined to "shoal and narrow channels," where they enjoyed a decided advantage over their larger counterparts.[30] Most snag pulling, however, was done using the twin-hulled snag boat, which steamed over a snag so that the hazard was between the boat's hulls. The snag was then roped and winched up out of the riverbed and then loaded onto an accompanying raft or barge for subsequent towing to shore, where it was either cut up for firewood or otherwise rendered harmless to steamboats. Usually accompanying a twin-hulled snag boat was a quarter boat, so called because it served as the "accommodation of working parties employed in cutting trees, roots, &c. on dry bars, and in felling impending trees on the shores and banks."[31] Single-hulled snag boats often worked in conjunction with a machine boat, the winches and windlasses of which, contrary to what its name implies, were not steam-powered, but which instead used the human muscle power of its crew, which had quarters onboard.[32]

Typically, a snag boat began its work at one location along the river and then worked up or down the waterway, removing snags as they presented themselves. The orders from Charles Fuller, supervisor of efforts to improve the Ohio River, to the captain of the *Terror,* a light-draft snag boat designated No. 5, in 1853 give a good idea of the scope of such undertakings: "My instructions . . . were mainly that he should proceed as rapidly as possible to the head of the Ohio, and thence work down, removing all obstructions in the shape of logs, snags, &c., in and near the channel, and to return to Louisville in season to pass over the falls, with the usual June freshet, and to operate in the lower Ohio. Unfortunately no 'June freshet' of sufficient magnitude occurred, and the boat was compelled to remain above."[33] During its more than nine weeks of operations along the Ohio, the *Terror* removed 140 snags and 7 logs, blasted 28 root masses, raised 23 sunken flatboats, and cut down 10 "impending trees." The latter were

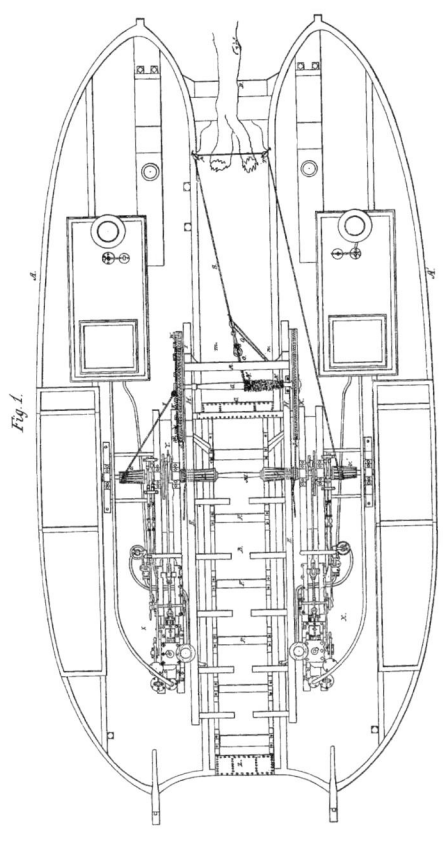

Drawing in Henry M. Shreve's snag-boat patent application, No. 913.
Source: U.S. Commissioner of Patents, 1838.

trees that, had they not been cut, might well have become snags if swept into the channel when the river undercut the banks on which they grew.[34]

At about the same time that the *Terror* was operating on the Ohio River, a mission of similar character, but on a larger scale, was under way along the lower reaches of the Missouri River, from its junction with the Mississippi to some 160 miles upstream. There, from August 10 to about September 20, two snag boats removed snags and other obstructions from the river. Between them, the two snag boats pulled "upwards of 500 dangerous snags from the low water channel, besides the cutting of trees and logs on the bars, and shoals, felling impending trees, &c." Two circumstances determined that the mission would conclude where and when it did. First, at about 160 miles upriver from the Missouri's mouth, the two snag boats came to a formation in the river called Smith's Bar, "beyond which they could not ascend for want of a sufficient depth of water in the deepest channel." Second, by the beginning of the third week in September, "the river had subsided so much as to render a retreat to the Mississippi quite precarious."[35]

Conditions similar to those encountered by the snag boats along the lower Missouri could have more serious consequences than bringing snag removal to an abrupt end. In early August 1853, snag boat No. 4 was working its way some sixty miles upstream from the mouth of the Arkansas River, pulling snags "from the low water channel." The vessel had remained on the Arkansas too long and, "[i]n attempting to retreat from the river, . . . grounded on a bar a little below the White river cut-off, and was detained in consequence during an entire month, or 31 days." No. 4's captain had either not known or had chosen to disregard that the "properest season for prosecuting the snag business" on the Arkansas "terminates about the last of July."[36]

Five snag boats worked along the Mississippi, Missouri, Ohio, and Arkansas rivers during the spring, summer, and early fall of 1853 and together they accomplished a remarkable amount of work, having removed 2,600 snags, of which 1,768 had menaced the channels of the Mississippi River alone. The crews of the boats also cut down 993 impending trees along the Mississippi, and another 418, almost all of which had grown along the Missouri and Arkansas. The greater danger along the Ohio, apart from the 182 snags pulled from its bed, were the twenty-seven steamboat wrecks that the No. 5 snag boat, *Terror*, had raised and cut up.[37]

Snag removal was hard and dangerous work, and service on a snag boat was decidedly unpleasant. Apart from the arduous nature of the work, both aboard and ashore, the job carried with it the extra dangers of venomous snakes, poisonous vegetation, clouds of insects, and disease, suffocating heat, and humidity. Disease often interfered with work on river improvements in a decidedly unpleasant manner, as the understated official report of one such instance makes clear: "The unhealthiness of the climate during the summer months, together with the appearance of cholera among the men, has somewhat retarded" the removal of the great raft on the Red River in Louisiana in 1855. Of the original crew of some three dozen men assigned to the work, thirty remained, following the deaths of five from cholera and one by drowning.[38]

The men who comprised the vessels' crews were hired because they were "in all cases robust, healthy, strong, active, and industrious." Ideally, they were also supposed to be (but one has difficulty imagining very many who were) well-mannered and well-behaved, with "quiet and peaceable tempers, kind dispositions, and . . . [were] temperate and orderly in all respects."[39] In view of the problem of drunkenness in American society during the last three or four decades before the Civil War, there would seem to be little likelihood that snag boats were manned by crews of abstainers, notwithstanding the fact that, as federal vessels, they were supposed to be run according to "the rules and regulations prescribed for the government of the western river service."[40]

Living conditions aboard the vessels probably militated against any such close observance of the rules. An inspection in April 1855 of snag boat No. 5, *Terror*, which had seen hard service on the Arkansas River that spring, revealed that the hull was out of alignment and that it and the main decks needed to be recaulked. As a result, in a heavy rain "the officers' rooms are so leaky that their beds and furniture become thoroughly saturated." Bad as the officers' quarters were, those of the crew, under the boat's forecastle, were worse; not only did these leak, but they were darker and danker.[41] *Terror* had cost probably about $30,000 to build; the lieutenant colonel who inspected the vessel estimated that the repairs necessary to bring it back into reliable service would cost "about $1,500," which if spent would put the boat "in as good a condition as when new."[42]

The repairs to *Terror* were never made; in June the vessel and the other four snag boats that had rendered such valuable service on the western rivers were put up for sale by the War Department, which was attempting to economize in the wake of low congressional appropriations that year for western river improvements.[43] The next year, as figure 6 indicates, such pessimism would seem misplaced. "But," as a War Department agent stationed at Keokuk, Iowa, observed in his August 1855 report, "of all things, the appropriations by Congress are the most uncertain."[44]

5
FACTORS OF DESTRUCTION

The fickleness of congressional and presidential support for internal improvements after 1840 precluded sustained funding, and therefore sustained operations, to remove hazards on the western rivers. Inevitably, this on again, off again basis for undertaking river improvements meant that some vessels that might otherwise have remained in service were wrecked by snags, bars, or other hazards to navigation, hazards that a steady program of improvements would, in the normal course of things, have removed. But there were other things, besides inconstant federal support, that affected the number and rate of steamboat wrecks.

Although, at the time, the full range of these other influences would have been beyond anyone's ken, historical hindsight permits their identification here. They included market forces, such as railroad competition and interest rates on borrowed capital, and, paradoxically, the evolution of steamboat design and engine technology. Politicians were fascinated by one of them, the railroad, and only dimly understood the other, steamboat technology. Of the two, the more important, because it exercised the most direct and immediate influence on the loss rate on the western rivers, was the technological development of the steamboat.

THE PERVERSITY OF TECHNOLOGY

The influence of technological change upon the rate of loss of steamboats was by no means simple. We are accustomed to having problems solved or at least ameliorated by technical improvements. That, however, is not what happened as a result of the technological development of the western river steamboat. Perversely enough, even as the technology of snag removal and dredging advanced, advances in hull design and engine power contributed to increasing, not decreasing, the toll taken by snags, bars, and other natural hazards, as well as by collisions. Although this is not the place for a comprehensive account of the western river steamboat's construction, operation, and performance, there are some aspects of this interesting subject that merit attention here.

About twenty years after the introduction of the steamboat to the western rivers, the vessel's design began to undergo a gradual evolution. In its initial form, it resembled the steamboat used on the rivers of the East, which itself derived from the oceangoing sailing ship, with its narrow beam, deep hull, and keel.[1] In the West, the latter feature almost immediately showed itself to be superfluous and worse. The comparatively shallow rivers of the West, with their often gradually sloping banks, made a keel an obstacle to efficient navigation by limiting how far upstream a steamboat could go and how closely it could approach a riverbank. Over the years, the keel, vestigial appendage that it was, disappeared. For similar reasons, boat builders progressively reduced the draft of the western river steamboats by decreasing the depth of the hull.

As boat builders gradually reduced the hull's depth, they broadened the beam and lengthened the hull, thereby increasing the size of the "water plane," that is, the area of the hull in contact with the water. These changes increased the steamboat's inherent stability and its above-deck carrying capacity, an important consideration as shrinking hull depths reduced already modest below-decks cargo space. The magnitude and timing of each of these aspects of the steamboat's evolution are evident in table 7.[2]

By the eve of the Civil War, the hull depth of the western river steamboat had decreased from an average of just over 7 feet in the last half of the 1830s to 5.5 feet, a reduction of about 27 percent, while the craft's average breadth and length increased during the same period by 14 percent and almost 20 percent, respectively. Changes in the hull dimensions of steamboats were also reflected in vessel tonnages, which increased by almost 68 percent as the average size of the water plane increased by just under 36 percent, despite a simultaneous reduction in hull depth. These increases in the size of the water plane occurred irrespective of differences in vessel displacement, differences that could be substantial, as, for example, between smaller steamboats of 100 to 150 tons and the larger ones of 400 to 500 tons.[3]

The increasing size of the water planes and the concurrently decreasing drafts indicate that a standard vessel configuration emerged gradually but rather early in the course of the western river steamboat's development.[4] Other data on steamboat hull design and construction lend additional support to the notion that the western river steamboat evolved over the decades in such a

Table 7
Hull Dimensions and Tonnage of Steam Packets, 1831–1860 (five-year averages)

	1831 1835	1836 1840	1841 1845	1846 1850	1851 1856	1856 1860
Depth (ft.)	6.70	7.17	5.76	5.90	5.69	5.57
No. packets	1	4	76	147	214	218
Breadth (ft.)	22.0	26.5	25.7	27.7	30.9	32.0
No. packets	1	5	77	147	217	224
Length (ft.)	146	151	170	176	180	181
No. packets	1	5	77	147	217	224
Area (sq. ft.)	3,212	3,977	4,493	5,017	5,733	5,982
No. packets	1	5	77	147	217	224
Tons	178	230	250	264	306	312
No. packets	1	5	76	146	216	220

Source: Steamboat Packet Database, derived from Way, *Way's Packet Directory, 1848–1994*.

Note: The number of packets for which data on hull dimensions and tonnage are available varies slightly in the source.

fashion as to converge on a standard design. Not only did steamboat drafts decrease while beams and lengths increased between 1820 and 1860, but ratio measures of depth of hull to breadth of beam and also to length, as well as of breadth of beam to length, approached uniform values for each tonnage class. An even stronger indication of this sort of standardization of hull design is the convergence of these ratios, irrespective of tonnage class, illustrated by figures 18 and 19, which chart the ratios until 1880.[5] Perhaps during the latter half of the 1850s, but certainly before 1860—precision on this point is somewhat elusive—steamboat hull designs arrived at the configuration and proportions that would define them through the Civil War and for decades after it (see figures 20 and 21).[6] The value of the ratio of hull depth to breadth, for example, fell from more than 0.3 at the beginning of the period to between 0.16 and 0.17 by its end.

There were some other regularities worth noting between the hull dimensions and displacement of the western river steamboat. Along with those already discussed, they essentially defined the parameters within which the vessel's design evolved toward a standardized model. One of the most striking

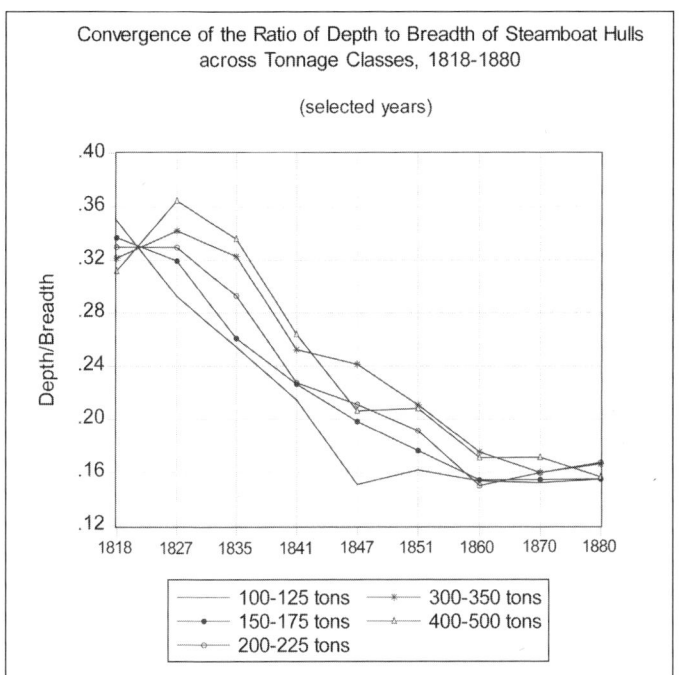

Figure 18. *Source:* See notes to text.

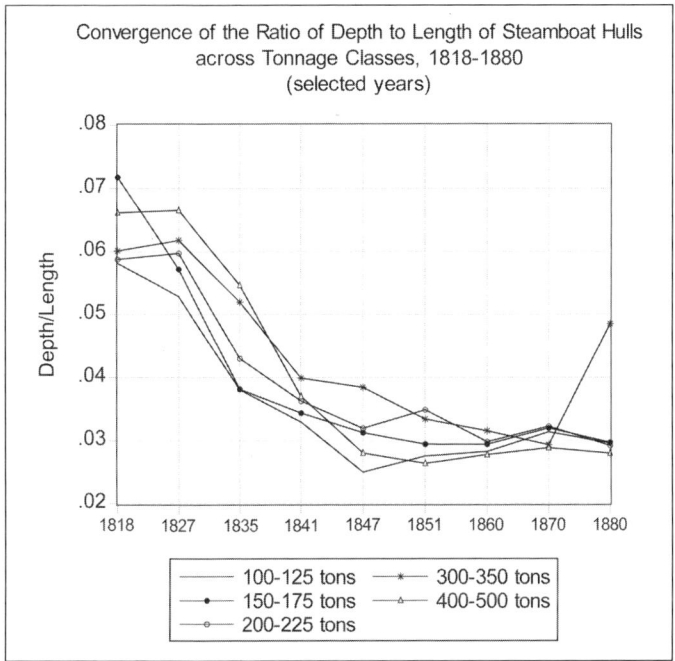

Figure 19. *Source:* See notes to text.

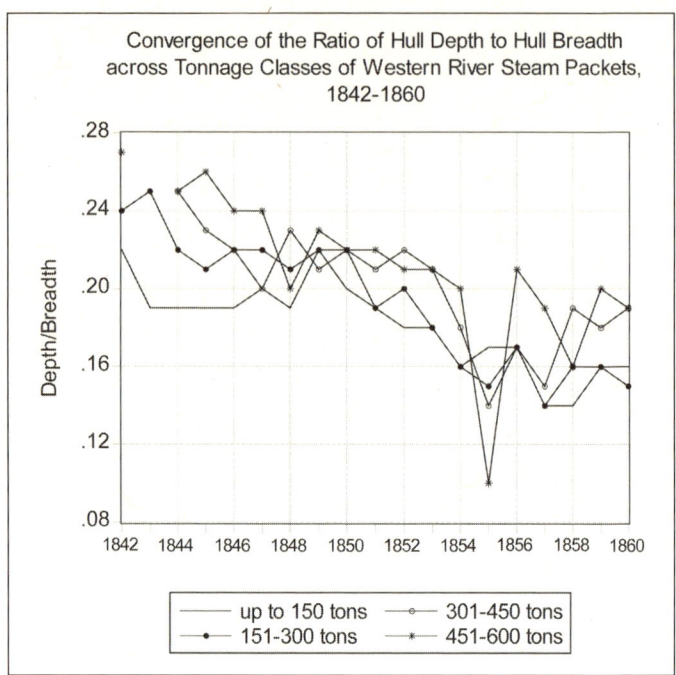

Figure 20. *Source:* See notes to text.

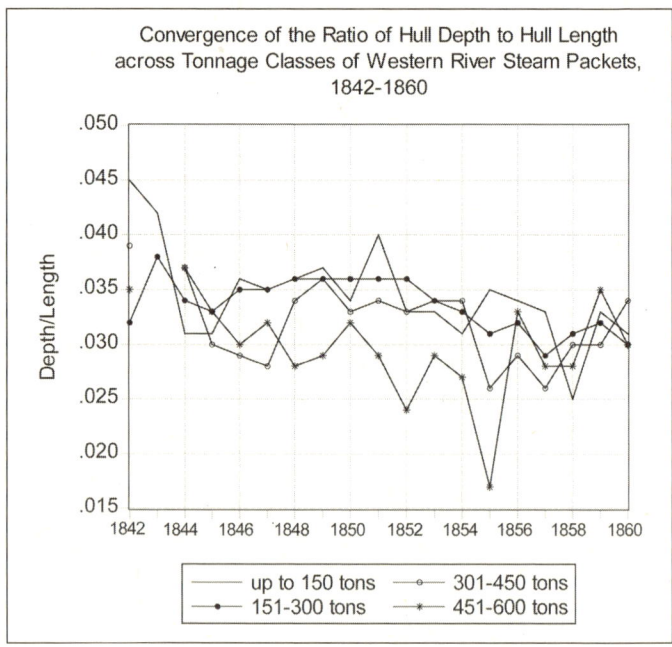

Figure 21. *Source:* See notes to text.

regularities was that between a steamboat's displacement, measured in tons, and the length of its hull, as shown in figure 22. An even stronger relationship was that between displacement and the area of the water plane, shown in figure 23.

At the same time that steamboat designs were converging on a standard model, American ship and boat builders concluded that the speed of steamships and steamboats was, in part, dependent on the length of a vessel's hull; that is, in a race between two steamboats equipped with engines of equal power and in all respects the same except that one vessel was longer than the other, the steamboat with the longer hull would win. Thus, in 1843, steamboat interests in Cincinnati asserted: "After many years experience, in the navigation of our rivers by steamboats, it has been ascertained that boats of a great length are those of the greatest speed, and best suited to the navigation of our rivers and the character of our trade."[7] The more general principle, established first by trial and error and then given formal mathematical expression, is that, all other things being equal, "speed is a function of boat size."[8]

The conviction, born of experience, that longer boats were faster boats, and the convergence of hull dimensions and configurations on a narrow set of standards point to the development of a new steamboat engineering aesthetic. This new aesthetic was, of course, conditioned by practical construction and operating considerations. Reductions in vessel drafts, especially in relation to their water planes, meant that steamboats could navigate the shallower reaches of larger rivers and ply tributaries that had been too shallow for the deeper-draft boats, as well as approach riverbanks more closely to take on and leave off cargo and passengers. And, at the same time that hull designs became increasingly uniform, the rate of loss suffered by steamboats due to natural hazards, especially snags, on the western rivers began to fall (see figure 24). The available evidence suggests that the decrease would have been greater had hull depths not been reduced, particularly relative to hull lengths and breadths.

When steamboats navigated shallower stretches of rivers or approached riverbanks, the larger vessels ran a much greater risk of falling victim to snags than would have been the case had they stayed in deeper water. There was, in fact, a fairly strong inverse relationship between the ratio of the number of steamboats tons snagged to the number of tons in operation and the ratio of hull depth to hull length (see figure 24). That is, as hull depths, relative to overall vessel size, decreased, the incidence of losses due to snagging increased. Shal-

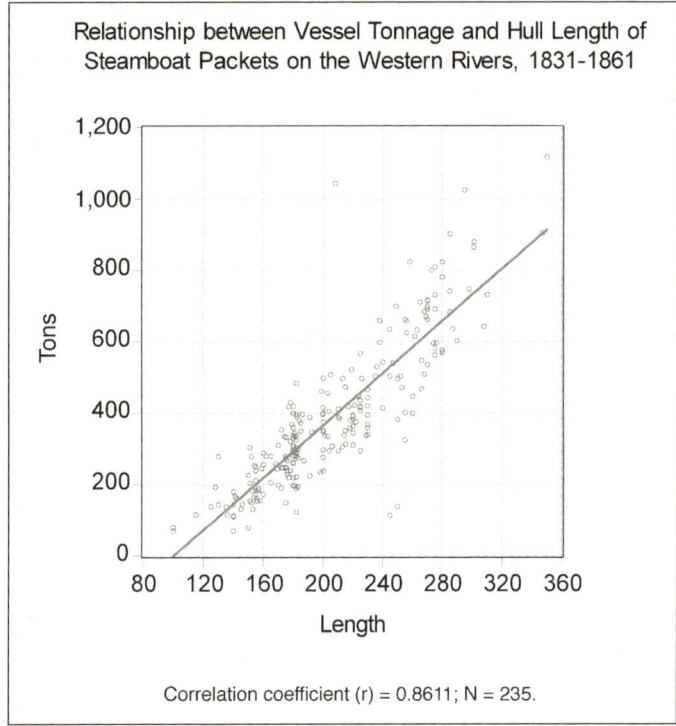

Figure 22. *Source:* See notes to text.

lower drafts enabled pilots and captains to risk more dangerous waters, and inevitably some lost their gambles. This situation was a consequence of the very advances in steamboat hull design that, well before 1850, had converged on the standard configuration of lower draft and greater beam and length, and that made possible the more efficient use of the steamboat.

Changes in hull design were so pervasive throughout the industry that the ratios of hull depth to beam and hull depth to length were more or less uniform not only across tonnage classes but also irrespective of the rivers on which the steamboats plied their trade. In part, this uniformity was the result of engineering and financing constraints and imperatives. It was also a consequence of the economic geography of river commerce and steamboat construction. This latter point is worth considering in some detail. It is not often recalled that steamboat builders, of whom there were not a great many, generally constructed vessels for particular rivers and even for specific stretches of a river. Such was

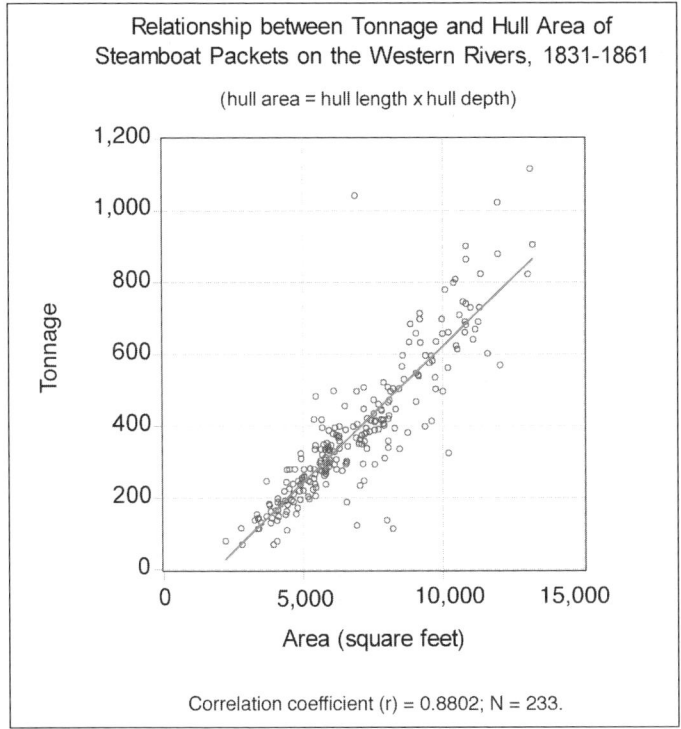

Figure 23. *Source:* See notes to text.

the case with respect to the Ohio River at Louisville, where steamboats used the Louisville Canal to circumvent the falls there. That circumstance had a direct influence on the size of steamboats built for the upper Ohio River trade.

The size of the locks of the Louisville Canal limited the length of steamboats on the upper Ohio River to less than 183 feet.[9] Less completely understood, however, was the significant impact of the Louisville Canal on the geography and scale of steamboat construction. Boatyards in Pennsylvania, which accounted for about 34 percent of all steamboat packets built during the last two decades before the Civil War, turned out only about 19 percent of those longer than 182 feet. Coincidentally and more impressive, 81 percent of the packets constructed in the state's yards were built to fit in the canal's locks. A similar but less extreme concentration on building canal-capable packets was characteristic of Ohio's and Kentucky's boatyards.[10]

So important to the inland commerce of the United States was the Ohio

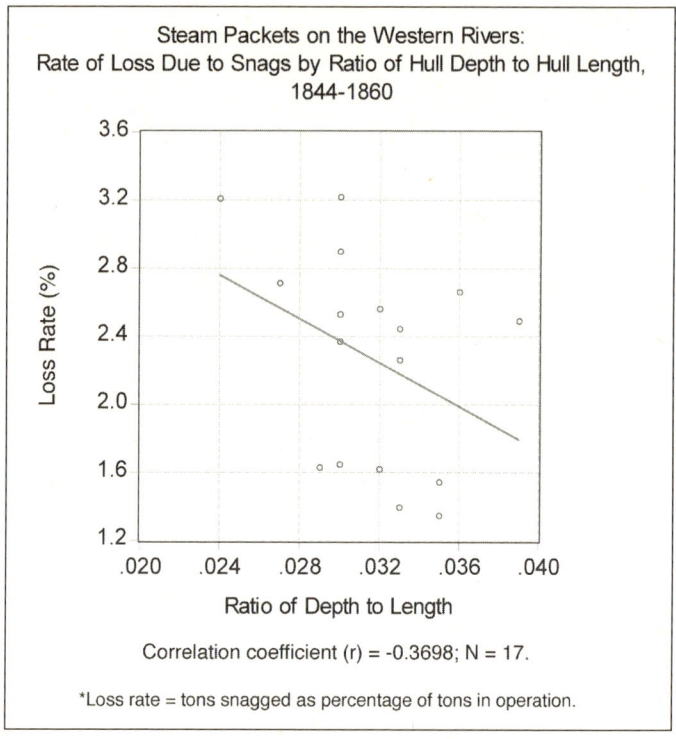

Figure 24. *Source:* See notes to text.

River trade, including its upriver component, which was the segment that had to contend with the falls, that most of the steamboat packets on the western rivers built in the twenty to twenty-five years before the war were 183 feet in length or shorter. This was true of vessels built in almost every state and also of those that plied every river except those on the lower and middle reaches of the Mississippi (that is, below St. Louis), the lower Ohio (below the falls and the canal), and the lower reaches of the Missouri, Tennessee, and Red rivers.

Alterations in hull design and dimensions, by far the most significant changes made to the western river steamboat, were not the only aspects of the vessel that underwent a profound transformation. The boilers and engines of the vessels also changed, but, as Louis C. Hunter noted, little of this change came about through patented invention. For example, while the application of steam power was a field of considerable interest for American inventors, boiler patents accounted for only 1 percent or less of all patents prior to 1860.[11] Inventions, patented and especially unpatented, resulted in increases in horsepower from better boilers and improved engines and contributed to dramatic increases in average upstream and downstream speeds to about ten miles per hour.[12]

Although data bearing on the development of steamboat power plants during the period are fragmentary, enough exist to suggest that the development of this technology occurred rapidly. One manifestation of this technological change was the dramatic increase in the relative power of the steamboat engine, inversely measured here as the ratio of vessel displacement (in gross tons) to engine horsepower.[13] This increase in relative power probably occurred fairly early in the evolution of the steamboat. Thus, the value of the ratio was as high as 6.5 for one steamboat in 1819 and 2 for another in 1829. By 1849, however, the ratio was just above 1.5, and, after a slight increase during the next two decades, it began to decrease until it stabilized at a bit above 1.6 during the last three decades of the century. This pattern of technological maturation was not confined to steamboat engine development and may have been characteristic of the development paths of any number of types of machine innovations. This was certainly the case with respect to two other motive power technologies, those of the steam locomotive of the nineteenth century and the twentieth-century farm tractor fueled by gasoline and powered by an internal combustion engine.

As figure 25 indicates, the trajectories of the ratio values of engine weight to horsepower for steamboats and steamships, steam locomotives, and farm tractors are quite similar to one another.[14] In each case, the greatest gains in relative power were realized during the first two or three decades of the innovation's development. This initial rapid technological progress was followed by a period of more gradual improvement, until the relative power of an engine approached its technologically imposed limit. This sort of pattern makes sense when we recognize that the initial gains in engine horsepower in each of these three cases occurred at a rate that was far greater than any change in the size of the respective conveyances. The reason that successive increases in power fell off after the first decade or so following the introduction of each technology is that initial enhancements made by engine builders had exploited most of the opportunities for significant gains in engine efficiency, leaving a much narrower range for potential improvements to subsequent designs.

The increases made in the relative power of steamboat engines had much to do with the design of the engines themselves (see table D.4 for specific mea-

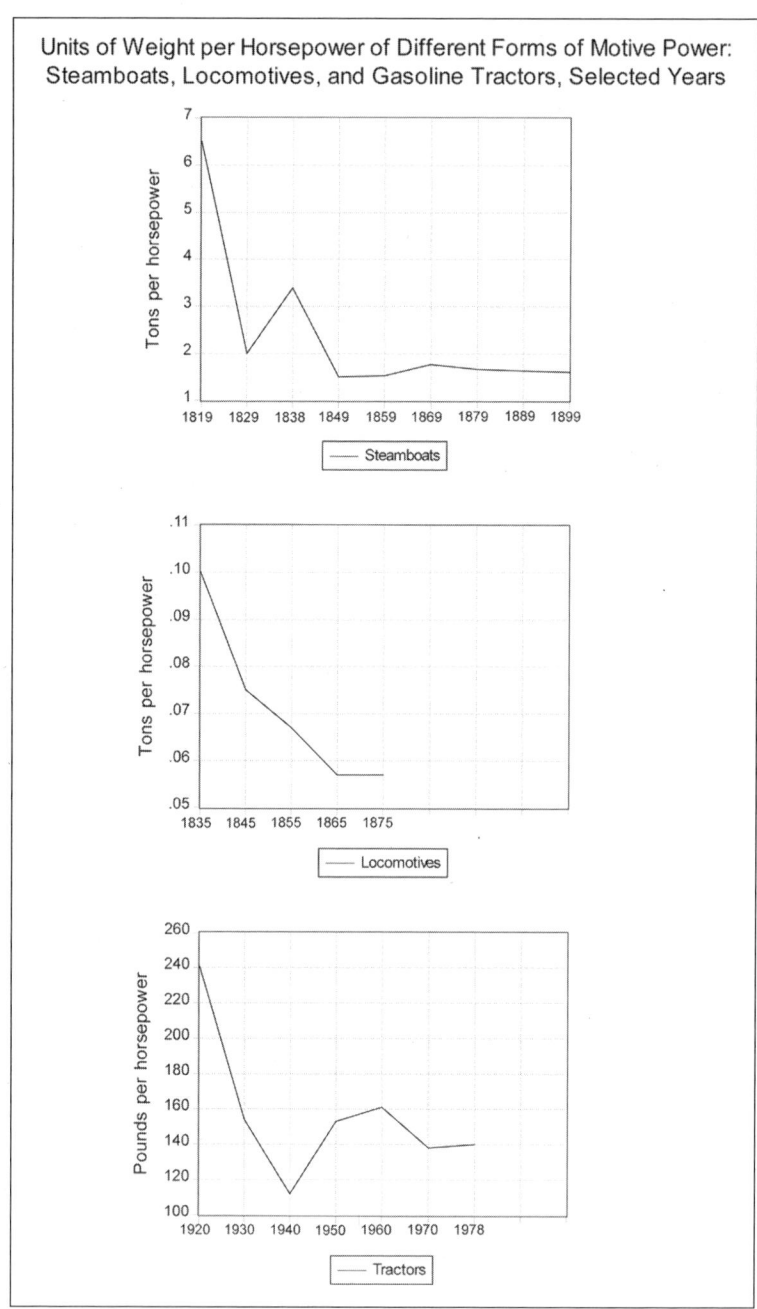

Figure 25. *Sources:* See notes to text.

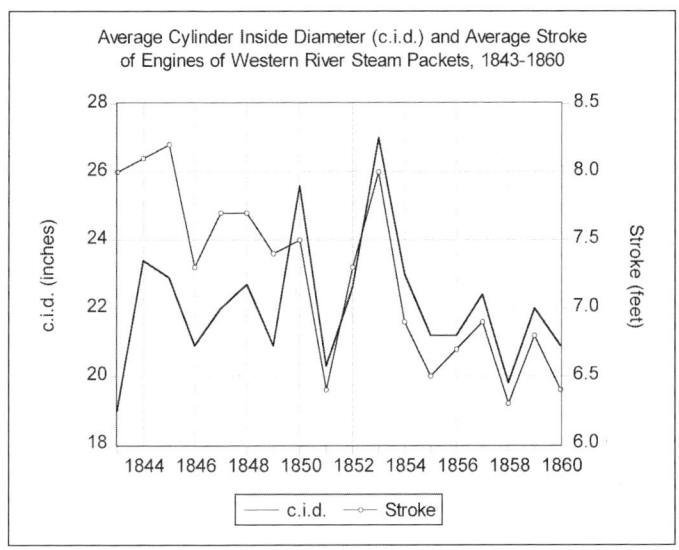

Figure 26. *Source:* See notes to text.

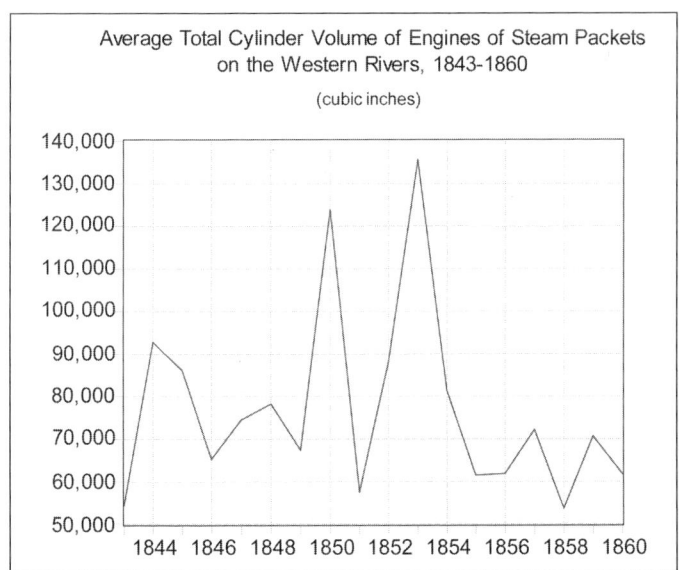

Figure 27. *Source:* See notes to text.

sures of engine size and relative power). During the decades before the Civil War, that design converged on a set of standardized sizes and power relationships, a development that is illustrated in figures 26 and 27. Both the length of the stroke of the engine's cylinders and the diameter of their bore decreased over time, so that the overall size of steamboat engines fell and then stabilized at between 60,000 and somewhat less than 75,000 cubic inches. But the most pronounced change was the reduction in the length of the stroke of the piston.[15] It was this reduction that yielded most of the significant saving of space realized by boat builders when they substituted the temperamental high-pressure steam engine for the long-familiar and relatively safe low-pressure engine.

The high-pressure engine had other advantages over the low-pressure model besides being less bulky. Smaller engines were also lighter engines, a significant consideration for western boat builders who had to design vessels that might have to navigate a river channel that, in severe low-water conditions, could be little more than the proverbial "heavy dew." High-pressure engines were also considerably cheaper to build than low-pressure engines, perhaps only 40 percent as expensive, and cost much less to operate.[16] It was also less prone to malfunction, and, in the event that it did, a steamboat's engineer might very well be able to repair it, a considerable advantage for steamboat captains and owners whose vessels operated in areas where foundries and machinists were not readily at hand.[17] Moreover, the high-pressure engine was a more versatile power plant that was capable of operating over a range of pressures and that could, if called upon, supply a sudden burst of power to permit a pilot or captain to negotiate a tricky maneuver to avoid a river hazard.

There were, however, some decided disadvantages to the use of high-pressure engines. First, they consumed far more fuel per hour of operation than did low-pressure engines, a fact that mattered much more on the rivers of the East, where wood fuel was costly, than on the western rivers, where wood was cheap. A more important deficiency of the high-pressure engines was that they were simply far more dangerous than were low-pressure engines.[18] Champions of the high-pressure model might insist that it was no more prone to explosion than the low-pressure engine, but the record of steamboat explosions on the western rivers made their claims ring hollow. For steamboat builders and operators, however, the many advantages of the high-pressure engine outweighed the risks inherent in its use, and it quickly became the standard power plant for the western river steamboat.

The packet *Jacob Strader*, "A Model Western Boat" of the 1850s, proclaimed its safety advantage of "Low Pressure" on its side.

Sources: Lithograph from Lloyd, *Lloyd's Steamboat and Railroad Directory, and Disasters on the Western Waters,* 145, and photograph from E. B. and N. Philip Norman Collection, Mss. 1084; Louisiana and Lower Mississippi Valley Collections, Louisiana State University Libraries, Baton Rouge. Courtesy of Louisiana State University Libraries' Special Collections.

At the same time that the shift to high-pressure engines reduced the average size of steamboat engines, the average tonnage displacement of western river steam packets increased. The more or less simultaneous occurrence of these two changes—the one in engine size and the other in vessel tonnage—resulted in a steady and pronounced decrease in the ratio of engine size (measured here in cubic inches of cylinder volume) to vessel tonnage. The dramatic drop of 50 percent in the value of this ratio, from 302 cubic inches per ton in 1844 to 151 cubic inches per ton in 1860, is charted in figure 28 (see also table D.4). The decrease might well have been even more pronounced had there not been an understandable desire by boat builders to equip vessels with the larger variety of high-pressure engines in pursuit of greater motive power and thus greater speed, though the extent to which the builders were able to gratify their desire has undoubtedly been overstated.[19]

One result of the concurrent but inverse changes in tonnage and engine size was an increase in the amount of hull and superstructure space for passengers and cargo. In this way, as well as in others, the technological development of the steam packet contributed to increasing the productivity of the vessel and of western river shipping, generally. In this instance, however, it probably also had the effect of increasing the steamboat's susceptibility to severe damage from snags and collisions. Increased engine power meant that when a steamboat hit a snag or a rock or another steamboat, the encounter probably was at a higher speed and more forceful and therefore more destructive than it would have been in the days of the low-pressure engine. So much is suggested by figure 29, which plots the rate of steamboat loss due to snags on the western rivers against engine volume per ton and indicates that lower values of the latter were associated with higher loss rates.[20] The evidence suggests therefore that steamboats powered by high-pressure engines not only incurred a higher risk of explosion but also ran a greater danger of destruction by natural hazards or collision.

CAPITAL MARKETS AND RAILROAD COMPETITION

The advantages of the high-pressure engine over the low-pressure type urged its adoption by boat builders, who, during the last two decades before the Civil War, tended to install ever more powerful versions of the design in their steamboats. But, as figure 28 indicates, the progress in that direction, while pro-

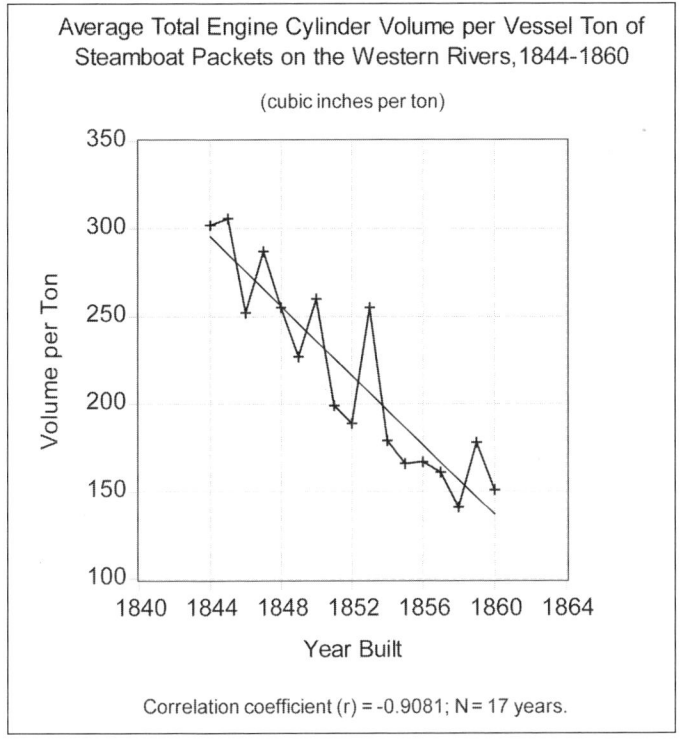

Figure 28. *Source:* See notes to text.

nounced, did not proceed with any consistency or even without frequent reversals. Of the three major components of a completed steamboat—the hull, the cabin or superstructure, and the machinery, which included the engine itself—the machinery was often the most expensive, accounting for as much as one-third to one-half of total construction costs.[21] The fluctuation in the pace at which increasingly powerful high-pressure engines were installed was probably not random but was more likely a response to conditions in the capital markets, specifically to movements of the rate of interest on borrowed money, which are charted in figure 30.[22] The impact of the Panic of 1837 and the severity of the depression that began in 1839 are indicated by the plunge in 1843, the depression's last year, of interest rates to 3.4 percent, far below the normal level, which was between 4.5 and 5 percent (no data were reported for the preceding year). The low rates that prevailed from 1839 through 1843 reflected the much lower-

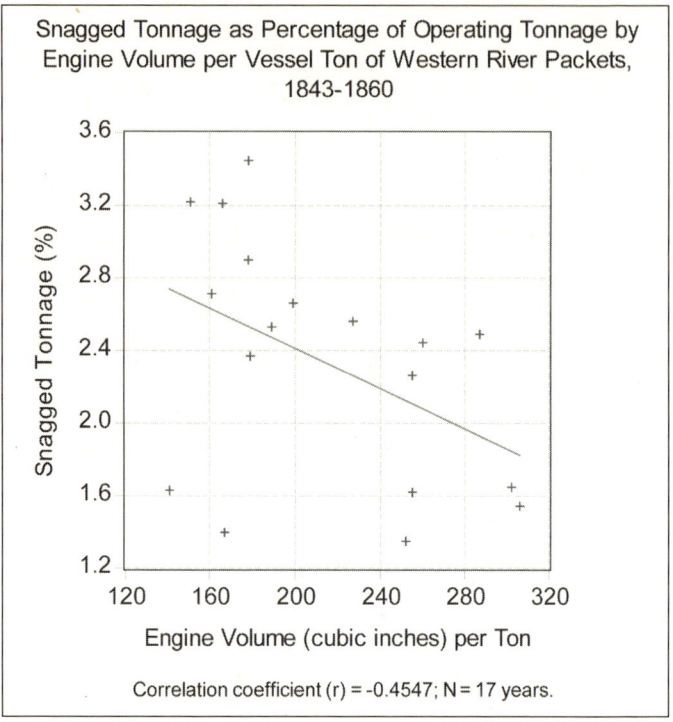

Figure 29. *Source:* See notes to text.

than-normal demand for capital during the depression. Similarly, the sharp rise in interest rates from 1852 through 1856 was probably an indication of an expanding economy's heightened demand for capital, a demand that was accentuated by that period's railroad boom, a boom that ended with the onset of the Panic of 1857.

The sensitivity of relative engine size and power to movements of interest rates is captured in two different ways in figures 31 and 32. When one examines figure 31, the reason for the sawtooth pattern of average relative engine size, displayed in figure 28, becomes clear: a rise in interest rates in one year was followed by a drop during the next year in the relative size of engines installed in western river packets because larger engines cost more than smaller ones.[23] The strength of this inverse relationship is shown in figure 32, which plots relative engine size against the rate of interest. This relationship assumes even more

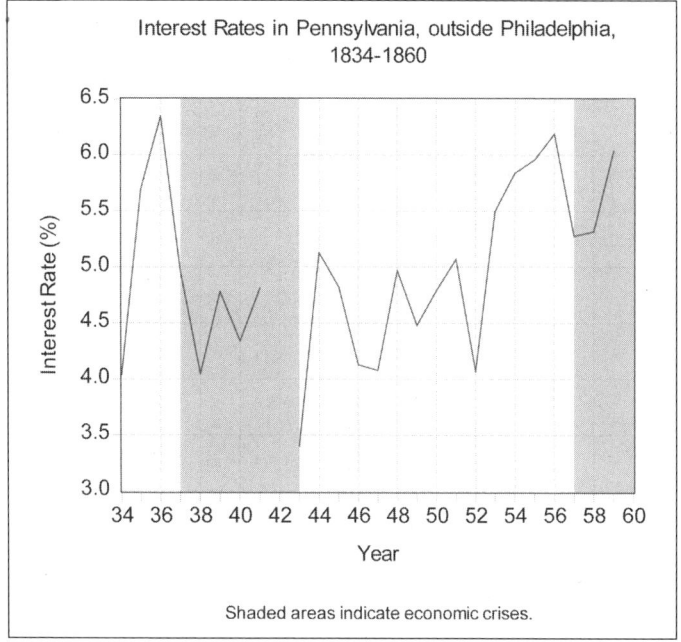

Figure 30. *Source:* See notes to text.
Note: Break in line indicates absence of data for 1842.

significance when we recall the point made earlier with respect to the connection between the loss rate due to snags and relative engine size, charted in figure 29.

It is now clear that the cost of borrowing money, given by the interest rate, was a driving force in determining the move away from larger, more expensive low-pressure engines to the more powerful and cheaper high-pressure design, and also the move toward high-pressure engines of ever-smaller relative size. This shift to the smaller high-pressure engines made the packets and other steamboats in which they were installed more likely to suffer serious damage from an encounter with a snag, bar, rock, or other steamboat. Market forces were, in other words, operating at cross purposes to the hazard-removal efforts of the river improvements program. But the price of money was not the only market force whose operation frustrated the efforts of the politicians who championed the program and the engineers who executed its attack on the hazard-plagued western rivers.

The rapid economic development in the Mississippi, Ohio, and Missouri river valleys and the ever-expanding volume of waterborne commerce were made possible by an increase in the number of steamboats plying the western rivers and their tributaries. The larger number of steamboats in service increased traffic density on the rivers. With increased traffic density, the risk of collisions, groundings, and snaggings became commensurately greater. Although the reason why greater traffic density should have led to more collisions is clear enough, the connection between more traffic and higher incidences of groundings and snaggings may be less so. The explanation has to do with the character of the channels of the western rivers, most of which meandered to a remarkable degree. Many of these rivers also had long stretches in which the deepest and safest parts of the channel were narrowed by island-bars and shelflike bars projecting outward from one bank or the other, and, in the face of increased traffic, many steamboats were forced to the margins of these comparatively safe passages where greater dangers lurked.

We can get some idea of the nature of these problems by considering map 2, which shows a navigational chart from 1849 of the lower end of the "Graveyard," an infamous section of the Mississippi River between Cairo and St. Louis.[24] Five

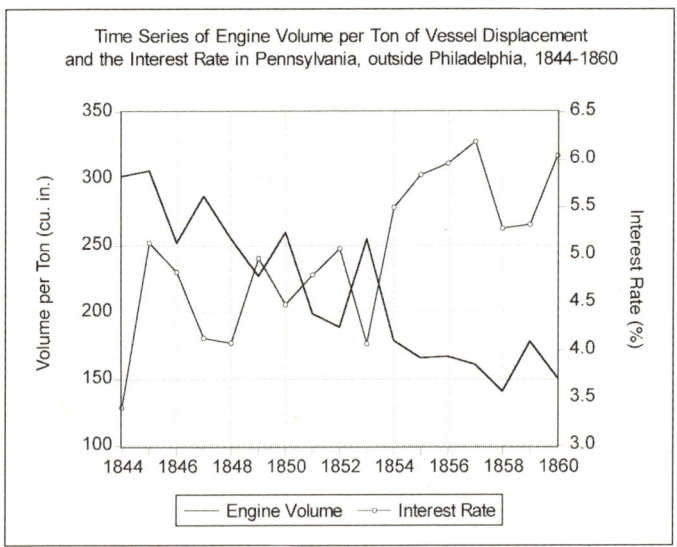

Figure 31. *Sources:* See notes to text.
Note: The interest rate is lagged one year. For example, the interest rate in 1843 appears here in 1844.

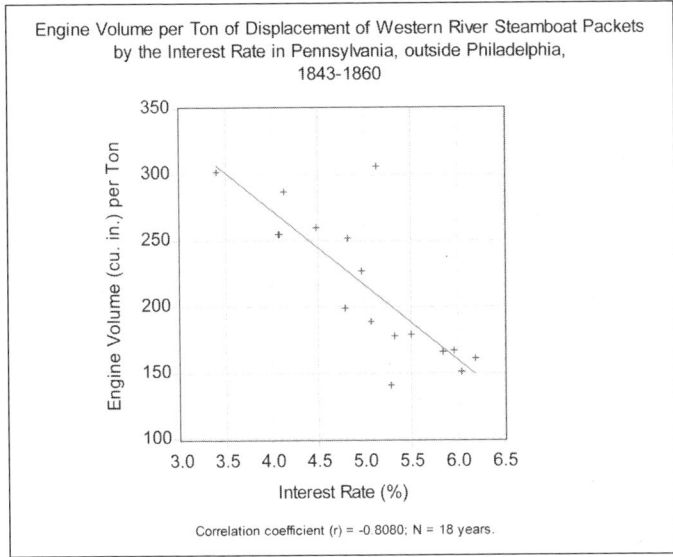

Figure 32. *Sources:* See notes to text.
Note: The interest rate is lagged one year.

years later, the War Department identified this same stretch of the river as being "very dangerous."[25] Shallower water presented the greatest danger from snags, and these, coupled with bars, often restricted easy passage to only a small part of the channel's width. In these narrowed channels, the risk of collision was heightened and attempts to avoid another vessel increased the dangers posed by bars, rocks, snags, and submerged steamboat wrecks. Under such circumstances, operations to remove these hazards call to mind an effort to sweep back the ocean with a broom. Still, the work went forward, though, for political reasons, often at a pace and with an intensity considered inadequate by those charged with doing it. That problem, however, was not the extent of their difficulties. By 1850, proponents of the river improvements program had to contend with another factor, the very nature of which was to work against their interests: the railroad had come into its own.

Even before the Mexican War, the railroad, as the newest transportation technology, had begun to divert the capital of investors and the interest of politicians at all levels of government away from the western rivers. The allure of the railroad is certainly understandable, and its story is familiar enough not to

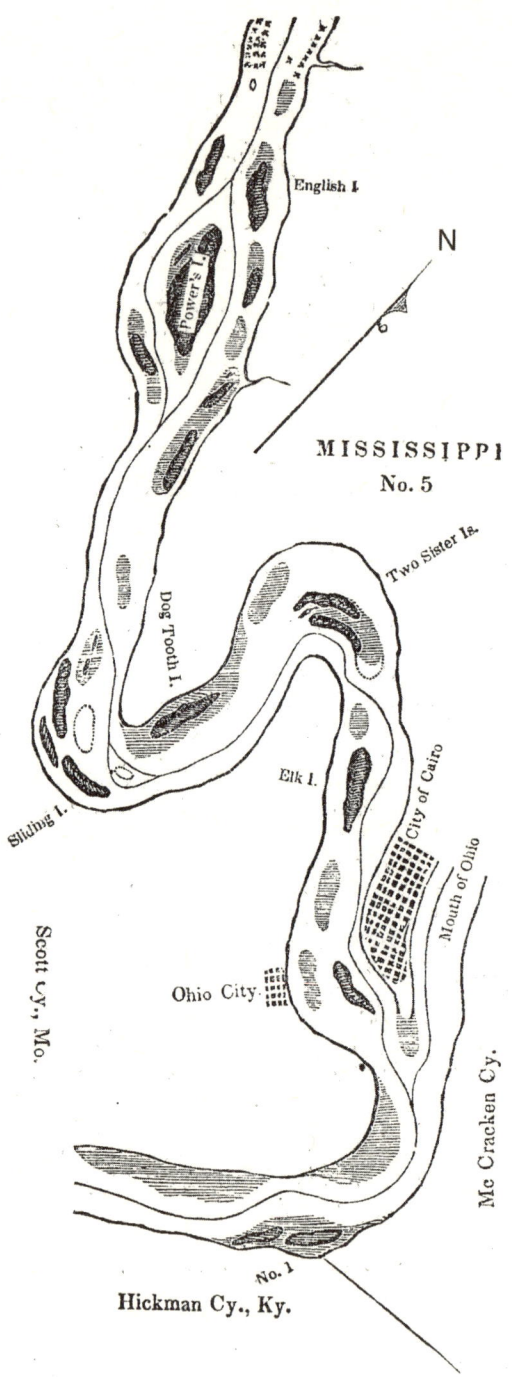

Map 2. The southern end of the "Graveyard." Adapted from Conclin, *Conclin's New River Guide*, 89.

require a complete retelling here.[26] There are, however, facets of that story that bear directly on what happened to the river improvements program and, therefore, on the level of hazards encountered on the western rivers.

It is probably not an exaggeration to describe the affection felt for the railroad by state legislatures and Congresses as something akin to infatuation, and there were many reasons for it. First, the railroad had the great virtue of comparative novelty—in 1850, the railroad as an innovation was slightly more than twenty years old; as an industry it was even younger—and it was modern and exciting. Second, railroad corporations projected their lines through congressional districts in a way that heretofore only turnpike corporations had been able to do. Whether a rail corporation selected a town to be a stop or a county for the right-of-way could determine whether that town or county prospered or languished. Finally, the railroad corporations were magnets for investors, drawing capital from this country and from abroad in amounts the likes of which had never been seen before.

Observers of the railroad's early decades of rapid growth and technological progress often described it as a Promethean force at work over the landscape and in society, an agent of sweeping changes on a vast geographic scale, the embodiment of the ultimate source of power, steam. No place was beyond its reach, nothing could prevent its growth, and nothing could withstand it, including the steamboat. A brief account in *Scientific American* of the competition in 1850 between a new rail line and an established steamboat service in Massachusetts, each party charging the same fare, ended with the verdict: "The conclusion was obvious, the steamboat cannot run against railroads."[27] Three years earlier, before an audience in Savannah, Georgia, Daniel Webster had invoked the mystery and majesty of the railroad as he described "the passage of the long train of cars through the dense and gloomy pine forests of your interior, self-moved by an inner power which gave no visible signs of its existence and left no trace behind it, cleaving those solitudes as a bird cuts the air, but urged by a power that could know no weariness and whose energies never flagged."[28]

But, for Webster, the steam railroad was not only a form of motive power; it was, along with the telegraph, an almost providential means of cementing the states together, of sustaining the Union. Seeking common ground with his southern audience, with whom the Mexican War was far more popular than in New England, Webster praised Georgia for its patriotism during the Revolution and for its embrace of internal improvements. Extolling the virtues of the Constitution and the Union, Webster described a country being reshaped by steam, one in which the laborer's toil is lessened, the lot of the poor is improved, and "a harmony of interest and feeling" is created among Americans who, living in "the most remote regions," are brought "face to face" because of the railroad. Not only that: the railroad "limits all distinctions. The poor and the rich, the prince and the peasant, enjoy now equal facilities of travel, . . . and when they travel, they sit side by side in the same rail-car. The individual is sinking, and the mass is rising up in the majesty of a common manhood."[29] He could have said the same thing about the steamboats on the western rivers, but he never did. For Webster and for many other Americans, the steamboat was transportation, but the railroad was a wonder.

Railroad enthusiasm assumed more concrete forms, as well, of which investment in railroad corporations was, of course, the ultimate expression. Another was the legislative activity of a succession of Congresses during the 1850s to aid railroad construction, often through grants of public land to states or directly to railroad corporations or, less commonly, by passing special legislation to advance railroad interests, as in the controversy over the Wheeling Bridge. The magnitude of congressional largesse on behalf of railroad construction was remarkable, and, as table 8 indicates, unprecedented. Federal land grants to promote other forms of transportation paled before those lavished on the railroads, and the railroad grants probably came at the expense of the alternatives to the railroad, including hazard removal on the western rivers. In fact, once federal aid to railroad construction got under way on a serious footing after 1850, public land grants to fund river improvements ceased altogether, as did such aid for canal construction with one relatively small, isolated exception in 1853. Land sales in support of wagon roads and canals resumed briefly on a modest scale during and after the Civil War until 1872, but then, as in the decade before the war, the overwhelming share of federal sales of public lands was dedicated to the railroad.[30]

This record of lavish federal support given for railroad construction is somewhat at odds with the insistence by champions of laissez-faire that an efficient allocation of transportation services "emerged fairly quickly once the government left the business of internal improvement to the workings of the market."[31] The case for market triumphalism is further undermined when federal land grants to railroads are considered in light of total capital formation by the railroad industry. In fact, the brave words uttered by laissez-fairists, including the railroad boosters among them, about entrepreneurial independence and self-sufficiency were often belied by the facts. As table 9 indicates, federal grants of public lands to promote railroad construction in 1853—an average year for such grants during the first half of the decade—were worth $6,572,500, which amounted to about only 7 percent of the railroad industry's $93.4 million in gross capital formation that year. Three years later, however, land grants for that purpose were equivalent to $35,212,500, or just under 46 percent of the industry's gross capital formation of $77 million.[32] Had the railroads been required to purchase the acreage that they received as grants, and had that purchase amount been added to the industry's capitalization, the more than $35 million in land costs would have represented more than 31 percent of the industry's total gross capital formation of $112.2 million. Any industry dependent on federal largesse for the equivalent of nearly one-third of its capital formation can hardly be considered the avatar of undiluted laissez-faire capitalism. Instead, the railroad industry, like its competitors in the transportation field, was a beneficiary of the vigorous involvement of the public sector in the antebellum American economy.

State governments were also energetic in promoting and subsidizing railroads, though the greater part of such support dried up in the wake of the depression that began in 1839 and lasted into the early 1840s, after which many state constitutions explicitly prohibited such participation.[33] But state-government support for railroads also took the forms of favorable terms of incorporation and provisions for lenient regulation, and the economic crisis of the late 1830s and early 1840s did not affect this sort of support at all.[34] One of the most successful of the state efforts was that of Georgia, and it was noted by Webster when he spoke at Savannah. Georgians, whose attitude toward federally funded internal improvements was at best ambivalent, embraced the railroad. The wisdom of their state government's decision to finance the construction of the Western and Atlantic Railroad to extend and further articulate a thin network of privately financed rail lines was confirmed in 1854 when the Western and Atlantic proved sufficiently profitable to enable it to contribute to the state's treasury.[35]

One might expect that a state with a sometimes lukewarm and, more often, hostile stand on a federal program of improvements would be drawn to the railroad as an alternative to river transportation. But the same sort of railroad-tropism was increasingly evident during the 1850s in the states of the western

Table 8
Public Land Grants by the Federal Government in Support of Transportation, 1823–1857 (thousands of acres)

Year	River Improvements	Wagon Roads	Canals	Railroads	Total Grants	Railroads' % of Total
1823	0	49	0	—	49	—
1827	0	202	2,071	—	2,273	—
1828	400	0	938	—	1,338	—
1838	0	0	139	0	139	—
1847	1,005	0	0	840	1,845	46
1851	0	0	0	3,752	3,752	100
1852	0	0	0	1,773	1,773	100
1853	0	0	750	2,629	3,379	78
1856	0	0	0	14,085	14,085	100
1857	0	0	0	6,689	6,689	100
1823–1857	1,405	251	3,898	29,768	35,322	84

Source: U.S. Bureau of the Census, *Historical Statistics of the United States, Colonial Times to 1970*, electronic ed., ed. Susan Carter et al. [machine-readable data file] (Cambridge: Cambridge University Press, 1997), Chapter J: Land and Water Utilization Series J 21–25, "Public Land Grants by United States to Aid in Construction of Railroads, Wagon Roads, Canals, etc.: 1823 to 1871." Figures in the column headed "Railroads' % of Total" and those in the row labeled "1823–1857" are derived from source.

Table 9
Federal Land Grants to Railroads and Railroad Capital Formation, 1847–1857 (selected years)

Year	Acres Granted (1000's)	Value of Acres Granted to RR's at $2.50 per Acre	Gross Capital Formation by All RR's	Column (2) ÷ Column (3) × 100 (%)
1847	840	2,100,000	27,300,000	7.7
1851	3,752	9,380,000	50,900,000	18.4
1852	1,773	4,432,500	66,400,000	6.7
1853	2,629	6,572,500	93,400,000	7.0
1856	14,085	35,212,500	77,000,000	45.7
1857	6,689	16,722,500	84,200,000	19.9

Sources: For acres granted, see table 8; the price per acre of $2.50 used in the second column is from Hibbard, *History of the Public Land Policies*, 245; the figures on gross capital formation are from Albert Fishlow, *American Railroads and the Transformation of the Ante-Bellum Economy* (Cambridge: Harvard University Press, 1965), table 52.

river valleys.[36] Perhaps the most telling symbol of that profound shift in sympathies was the city seal of Memphis, Tennessee, reproduced in a city directory of 1860. An outer ring of lettering proclaims the seal to be that of the "Corporation of the City of Memphis—Tenn." The seal's circular interior field is bisected horizontally by a line on which sits a steam locomotive, facing west. Arrayed alongside the line, which is really a track rail, are a barrel and two bales of cotton.[37] The Mississippi River and its steamboats, to which Memphis owed its existence and prosperity, are nowhere to be seen.

6

THE SUCCESS OF PUBLIC POLICY

The figurative turning away by Memphis from the river and the steamboats that had made the city was a symptom of a set of problems confronting advocates of river improvements. The federal government's efforts to remove hazards on the western rivers occurred in an environment of inconsistent funding and occasional hostility from Congresses and presidents. During the last two decades before the Civil War, additional difficulties beset the program as the siren call of the railroad made erstwhile supporters of men who had once championed river improvements and as national unity was strained by regional jealousies and animosities over the constitutional legitimacy and distributional equity of the improvements program. Often lost amid this clamor of competition for political and regional advantage, however, was the prosaic question of whether the program, as an expression of public policy, was in fact succeeding in reducing the dangers of domestic waterborne commerce by removing hazards to inland and coastal navigation.

The practical test of any public policy is whether it achieves its objectives. Champions of the rivers and harbors improvements program generally advanced two arguments in its favor: first, that it, along with the construction of roads and canals, would promote national economic and political integration, and, second, that it would make domestic waterborne commerce safer and more profitable by reducing hazards to shipping. For many, the Civil War dramatically demonstrated the program's failure to accomplish completely the first of these goals, a failure that has tended to obscure the record of the program's performance in achieving the second goal of reducing or eliminating barriers and dangers to commerce on the nation's rivers, lakes, and coastal waters.

Before the war, and even afterwards, opponents of the program, in and out of Congress, assailed it for what they insisted was its inherent unconstitutionality, the alleged ineptitude and inefficiency with which it was executed, the corruption with which it was suffused, and its aggravation of already rampant jealousies between states and regions over the division of federal revenues. For such critics, the program itself was perhaps as serious a problem as the hazards to navigation that it was supposed to eliminate. Some recommended that river and harbor improvements be confined to customs ports of entry, while others, notably Stephen A. Douglas, urged the replacement of the federally funded, or general, program of internal improvements with a decentralized system of federally sanctioned, but state-directed improvements, financed by "tonnage duties" to be charged by a particular state or city on a river or lake.

Even the program's advocates and defenders criticized it, not for doing too much, but for doing too little. They insisted that the funds appropriated and expended for the program's far-flung projects often were inadequate to their great tasks. Moreover, they asserted, an initial inadequacy of funding was all too often compounded by Congress's failure to sustain an appropriation made in one year with sufficient funds in succeeding years. The unfortunate result, they said, was to leave a desirable, even vital, improvement project uncompleted, thereby making the problem that the improvement was intended to solve not somewhat better but, in fact, worse than it had been before the work had commenced.[1]

Historians, writing long afterward, have turned in a mixed verdict on the river improvements program. The consensus is that, while the program enjoyed some early success prior to 1840, its effectiveness in lowering the risks involved in waterborne commerce was largely undermined during the 1840s and 1850s by a combination of declining federal revenues, a series of strict constructionist presidents who were hostile to the idea of such improvements, and an increasingly acrimonious sectional dispute in and out of Congress.[2] The problem with this judgment is that it has rested largely on a foundation of anecdote and conjecture and not on one of hard data. Any determination of the success or failure of the river improvements program should be based on an analysis of the quantitative record of appropriations and expenditures under the program and domestic waterborne shipping and steamboat wrecks. Fortunately, there are sufficient data available to permit a close examination of these matters and, consequently, a determination of the efficacy of the improvements program.[3]

There are two standards—the political and the operational—against which the program may usefully be judged. The first of these, the political, has to do with the geographical distribution of appropriations to remove or reduce hazards to shipping and essentially bears on the program's integrity. Did states with the greatest shipping losses, and therefore in the greatest need of assistance, receive the greatest shares of congressionally appropriated funds? If they did, then we might reasonably conclude that the process by which appropriations and expenditures were made was, at least in economic terms, a rational one. If, however, states with relatively low shipping losses were systematically the major recipients of the program's funds, while states with high loss rates went begging, then the sensible conclusion to draw would be that economic politics and not political economy drove the river improvements program.

The other standard, the operational, consists of three individual performance criteria for assessing the river improvements program's success or failure in reducing the incidence of wrecks on the nation's natural inland waterways. The first of these criteria is a determination that greater expenditures for hazard reduction or removal lowered the rate of steamboat losses, that is, the percentages of vessels and tonnages in service that were lost to natural hazards. A second test is evidence of a demonstrable increase in the life expectancy of western river steamboats during the four decades of the program's operation before 1861. The third performance criterion by which to judge the rivers and harbors improvements program is its economic impact, specifically its contribution to increasing productivity of the steamboats that plied the western rivers. A larger but ultimately less helpful question to ask is whether the cost of the hazard-removal program was at least offset by the social and economic benefits that accrued because of that program. This is the sort of question for which economists and economic historians have become notorious by asking and, even more so, by then answering.

Advocates of greater spending for river improvements frequently offered their own calculations of the costs of steamboat losses and argued that the value of the boats and cargoes to be saved by additional expenditures would far exceed what was spent to make the rivers less perilous. The figures offered to prove this contention varied greatly in magnitude with respect to their respective dates of origin and authors and usually were exaggerations. A tabulation of steamboat losses on the western rivers in 1843 put the number of vessels destroyed at 51, worth $789,378. Five years later, the *Cairo Delta* reported that there had been 251 "steamboat disasters on the Mississippi river" alone, for an aggregate loss of about $5 million. In 1855, *De Bow's Review* reprinted a report, originally published earlier that year in the *Cincinnati Gazette,* that claimed 39 steamboats worth about $573,700 had been "totally lost." According to that account, which probably pertained only to the Ohio River, another $1,229,800 in cargo and thirty-one lives were also lost. Even more alarming was a report of losses suffered in the first six months of 1860, during which time 125 steamboats, 127 coal boats, and 23 flatboats and barges worth about $1,732,500 were said to have been destroyed, along with 136 lives. These assertions of losses on the western rivers for 1843, 1848, 1855, and 1860 are very much at odds with the recoverable record of such losses, that is, the steamboat database compiled for this study. The figures from the latter are: for 1843, 19 steamboats lost; for 1848, 50 steamboats lost; for 1855, 65 steamboats lost; and for all of 1860, 66 steamboats lost.[4]

Despite any ostensible precision in the matter of tabulating losses, there is no reliable estimate of the aggregate dollar value of vessels and cargoes lost on the western rivers, and it is unlikely that one can be arrived at.[5] Besides, ultimately, in the opinion of many of those who lived along the rivers, *the* essential variable was missing from such benefit-cost calculations, as the memorial to Congress by some prominent citizens of Cincinnati pointed out: "And who," they asked, "shall count the value of their lives that float on the Ohio and the Mississippi? Shall the lives of free citizens of an enlightened Nation, be weighed against the amount of an appropriation in money, which would insure their safety?"[6]

Even if, in principle, the question posed in the memorial has an actuarial answer, it nevertheless has no meaningful practical one. No reliable, comprehensive tabulation of fatalities due to steamboat accidents exists, and none can be computed. Most accidents were due to causes other than fire or explosion. Although steamboat explosions or fires, with their attendant deaths and maimings, made riveting newspaper copy, such incidents were not especially frequent. Far more common were the more prosaic snaggings and founderings that claimed many boats but comparatively few lives, and it was mainly these that drove congressional appropriations and expenditures of money to improve the rivers. That is still another reason for not trying to strike some balance between the value of human life and the cost of hazard reduction. More in keeping with the concerns of Congress and also more useful here as a set of criteria by which to judge the performance of the river improvements program would be a determination of the program's political and operational efficacy, and that goal is well within reach.

CUI BONO?

There is reason to believe that on occasion Congress designated improvements appropriations earmarked for specific states in proportion to each state's population, that is, the size of its delegation in the House of Representatives. Such behavior would, after all, have been consistent with what the political process was about: the catering to local, and therefore state, interests in the division of federal funds. It was also what Lewis Campbell, William Ward, and some other partisans of the river improvements program in the House of Representatives insisted upon for their districts, and they complained bitterly when they believed that their constituents were not getting their proportionate shares of the program's outlays. Their complaints on such occasions only served to give additional ammunition to the program's opponents, who had been saying all along that its chief source of propulsion through Congress was log-rolling and the consequent corruption of the political integrity of members of the House.

When opponents of river and harbor bills assailed them for being sinks of corruption, they did not necessarily mean that individual members of Congress had enriched themselves through graft. No one in Congress ever made such an allegation and supported it with evidence. More often, what they meant was that such legislation had compromised the political process as it was supposed to work, a process in which a congressman's or senator's vote was a disinterested one, determined solely by the merits of a proposed piece of legislation and by no other consideration. Such purity demanded that there be no thought of trading one's vote for money in the form of an appropriation for one's district or state through log-rolling and pork-barreling legislation, for such considerations were the essence of systemic political corruption. In that sense, and largely only in that sense, can the rivers and harbors program be said to have been corrupt. Log-rolling and pork-barrel legislation were intrinsic to the drafting and passing of the program's legislation. Such practices were not, of course, confined to Congress but were also standard operating procedure in state legislatures, where, as Stanley L. Engerman and Kenneth L. Sokoloff have noted, "[l]og- rolling along geographic lines was a common feature of crafted political alignments throughout the antebellum period, wherever there was legislative consideration of internal improvements."[7]

Over the course of forty years of river and harbor improvements, there was unlikely to have been a substantial general appropriations bill directed at projects within particular states that was not crafted with an eye to making sure that the political arithmetic of congressional delegations added up. Figure 33 illustrates this point by plotting an individual state's total appropriations during the 1821–60 period for projects stipulated to be conducted within its territory against its corresponding share of the national population in 1850.[8] The relationship between the amount of appropriations for state-specific projects received by each state and their respective shares of the national population is clear and lends support to the charges hurled at the improvements program by its critics.[9]

Such considerations aside, however, a succession of Congresses also made miscellaneous appropriations for improvements to a particular river or to a part or all of an entire river system, on a where-needed basis. In some years, such appropriations were the major component of river improvement appropriations.[10] The general relationship between the magnitude of vessel losses due to natural causes and the level of congressional funding for river and harbor improvements is shown in figure 34, which illustrates the essentially direct relationship between states' steamboat losses and the levels of congressional funding for river improvements that they received.[11] While this connection was true of the twenty years before 1840, when spending for the program was at its highest sustained levels, it was also true of the last two decades of the antebellum period, when increasing federal deficits and political ideology combined to limit spending. No doubt because some of the most treacherous reaches of the Mississippi River lay within or adjacent to Louisiana, it was foremost among all states in steamboat losses (160 vessels) due to natural causes, especially snags.[12] The Mississippi River and its mouth, as well as the great importance of New Orleans as both a river and a customs port, were also responsible for making Louisiana one of the three largest recipients of congressionally appropriated funds for river and harbor improvements; only New York and Delaware got more.

Delaware Bay and New York Harbor were the two exceptions to this rule of funding in proportion to losses. Together the two ports accounted for a substantial share of the nation's foreign commerce. That Congress would appropriate to these areas sums that even strict constructionist presidents would spend and that were well in excess of the amounts warranted by vessel losses is certainly understandable in light of the importance of tariff revenues to the national treasury.[13] Allowing for these two exceptions and, of course, the log-rolling that was built into almost any money bill that Congress considered, much of the money appropriated for the river and harbor improvements went where it had the chance of doing the most good—a notable achievement in any era—and, at least on that basis, the program achieved a measure of success.

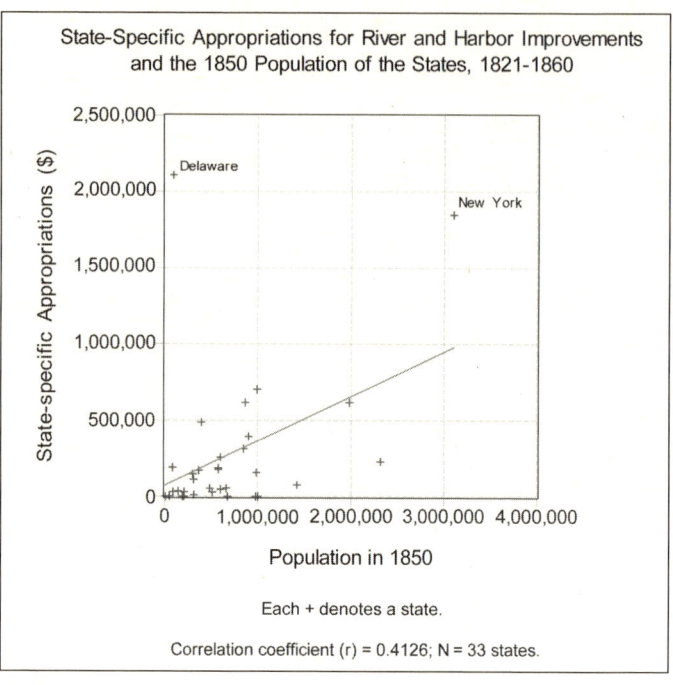

Figure 33. *Sources:* See notes to text.

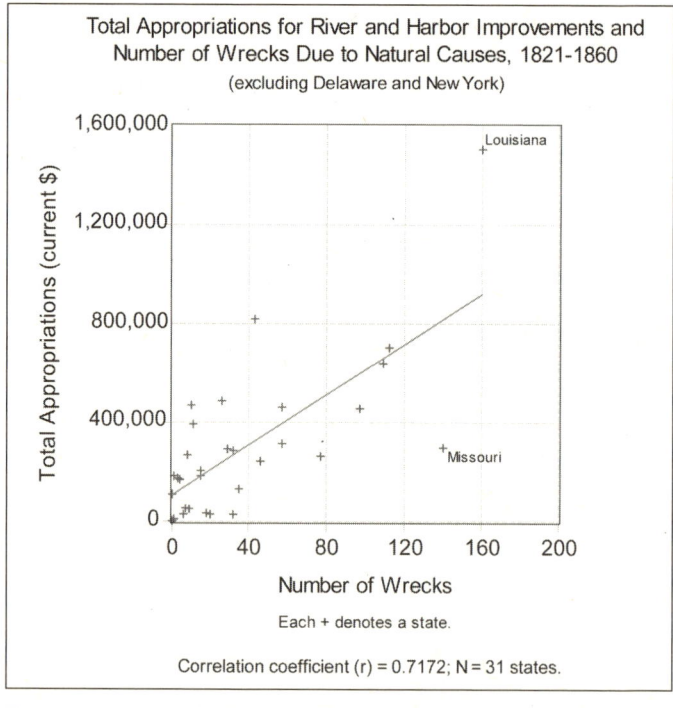

Figure 34. *Sources:* See notes to text.

REDUCED RATES OF LOSSES

A second indication of the program's success was the reduction, or at least suppression, of the incidence rate of wrecks due to natural hazards, especially snags. As figures 35 and 36 indicate, the proportion of vessels in operation lost to snags and the proportion of all naturally caused losses attributable to snags fell dramatically from 1821 to about 1840 and, with the exception of an upward spike in 1841 and 1842, only slightly increased from then until 1861.[14] Over the entire period under consideration, the rate of steamboat losses due to snags generally declined as the level of federal expenditures for river improvements increased. The relationship between the two was fairly strong, as figure 37 shows, suggesting that the reduction in losses was at least partly a consequence of spending for hazard removal. There was, however, another influence at work to lower the rate of steamboat losses on the western rivers.

During the decades before the Civil War, steamboat builders constructed western river craft of ever larger average tonnage. The larger vessels, especially the packets, generally plied the major rivers, most of which had long been objects of attention from the army Topographical Corp's snag boats. These main channels were therefore less likely to be plagued by snags than were their narrower, shallower tributaries, which smaller steamboats were able to navigate because of the progressive reduction in vessel drafts. In this way, the increasing average size of the western river steamboat acted to reduce somewhat the rate of loss due to snags. So much is clear from figure 38, which shows that, after the mid-1830s, the average size of snagged vessels in any given year tended to be smaller than the average size of those in operation that year. This was especially the case after 1845, when the average size of snagged steamboats was about 82 percent of the average size of those still in operation.

Also evident in figure 38 is the effect of decades of steamboat evolution and hazard removal on the western rivers. Before the concerted effort to pull snags got under way during the late 1820s and early 1830s, the average steamboat that fell prey to a snag was generally quite a bit larger than the average steamboat in operation, in many years in that early period as much as 10 to 15 percent larger. That observation makes sense when we recall that the first decades of steamboating on the western rivers were a time when steamboats mainly stuck to the major rivers, entire stretches of which, such as the infamous "Graveyard" on the Mississippi, were all but infested with snags. Smaller vessels, with their

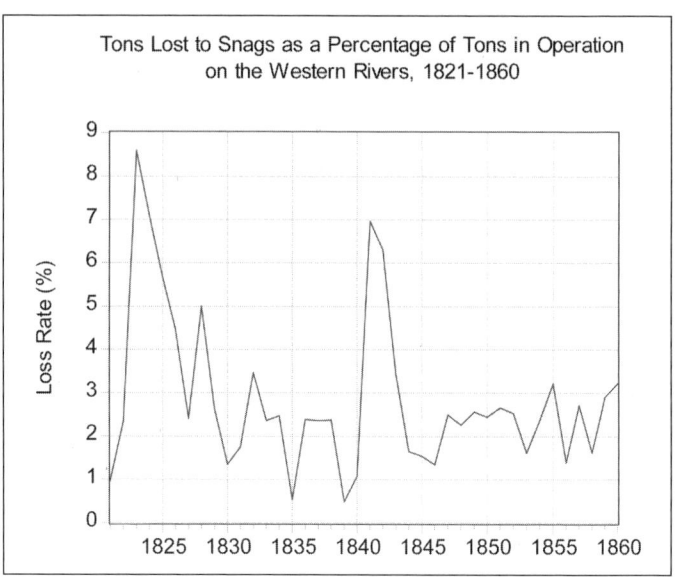

Figure 35. *Source:* See notes to text.

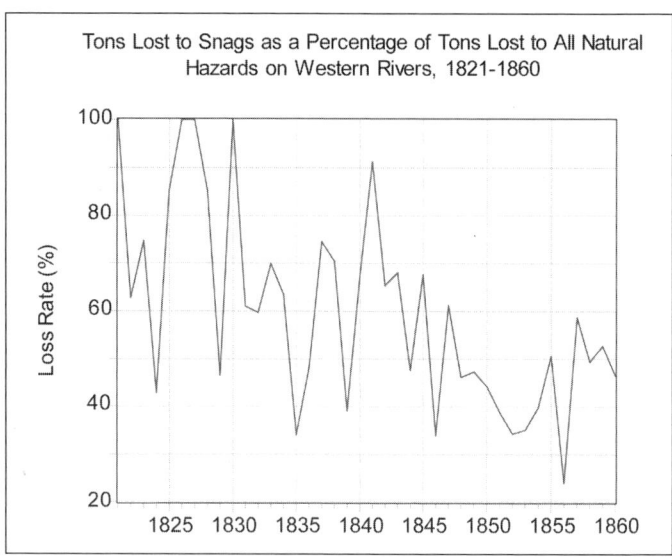

Figure 36. *Source:* See notes to text.

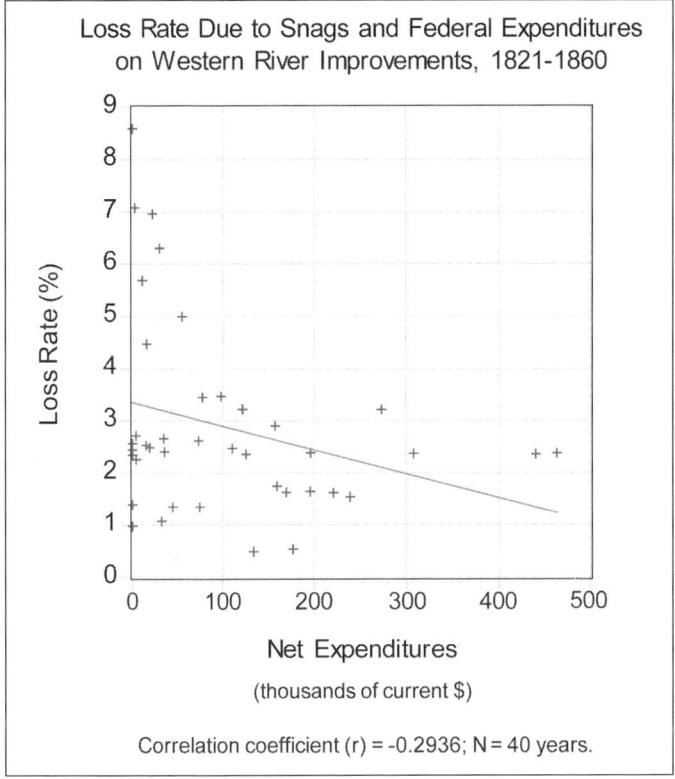

Figure 37. *Sources:* See notes to text.

shallower drafts, had a better chance of avoiding snags in the main channels of the larger rivers than did the much larger steamboats. Once the major concentrations of snags had been addressed by the Topographical Corps and steamboats began plying the smaller rivers and the larger rivers' shallower channels and tributaries, the balance swung away from the smaller craft and in favor of the larger ones.[15]

LONGEVITY

Natural hazards, human error, and technological considerations combined to claim a large and, as the volume of river traffic increased, growing number

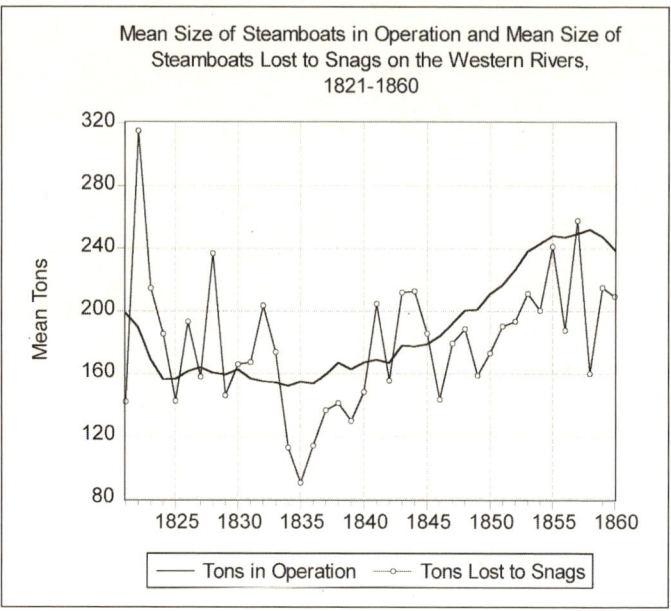

Figure 38. *Sources:* See notes to text.

of steamboats. Because snags were the single-greatest cause of steamboat disasters on the western rivers, a reduction in the rate of loss due to snags necessarily meant an extension of the average life expectancy of steamboats. Over the four decades preceding the Civil War, the average service life of a western river steamboat was between three and five years. A steamboat's longevity depended primarily on its tonnage, design, and when it had been built—all of which more or less defined it technologically—and on which rivers it navigated.

The relationship between size and vintage to longevity is presented in table 10, which makes clear the advantage that larger vessels generally enjoyed over smaller ones in terms of survival, a matter that has already been discussed in some detail. More significant is the fact that, for each tonnage grouping, steamboat longevity generally increased over time, suggesting that the rivers were becoming safer during the last three decades before the Civil War. That there was no uniformity to this improvement of conditions is also evident in table 10. Just how discontinuous the improvement actually was is indicated in figure 39. The inconsistency from one year to the next with which steamboat longevity

Table 10
Average Age in Years of Western River Steamboats, 1832–1860 (for major tonnage groupings)

Period	Tonnage Range		
	Under 101	101–200	201–400
1832–37	2.64 (11)	3.32 (22)	3.33 (3)
1837–42	3.00 (12)	2.90 (30)	2.70 (10)
1842–47	2.25 (12)	3.10 (40)	4.72 (25)
1847–52	2.84 (19)	3.78 (54)	4.38 (45)
1852–57	3.75 (16)	4.62 (37)	4.28 (46)
1857–60	4.72 (18)	4.00 (44)	4.78 (27)

Source: Steamboat Database.

Note: A vessel's age equals the difference between the year when it was wrecked and the year when it was built; the figures in parentheses indicate the number of steamboats for which those data, as well as tonnage data, are available.

increased or decreased resulted from the ever-variable conditions on the rivers, conditions created by nature and also by the captains, pilots, crews, and builders of the vessels.

Western river steamboats came in two types: side-wheelers and stern-wheelers. Each design had its advantages over the other. Stern-wheelers were small and light and, unlike the larger and heavier side-wheelers, could operate under conditions of extreme low water.[16] The two great advantages of side-wheelers were their greater size, which meant greater carrying capacity, and their greater maneuverability, which resulted from the presence of two independently controllable paddle wheels. Greater maneuverability translated into

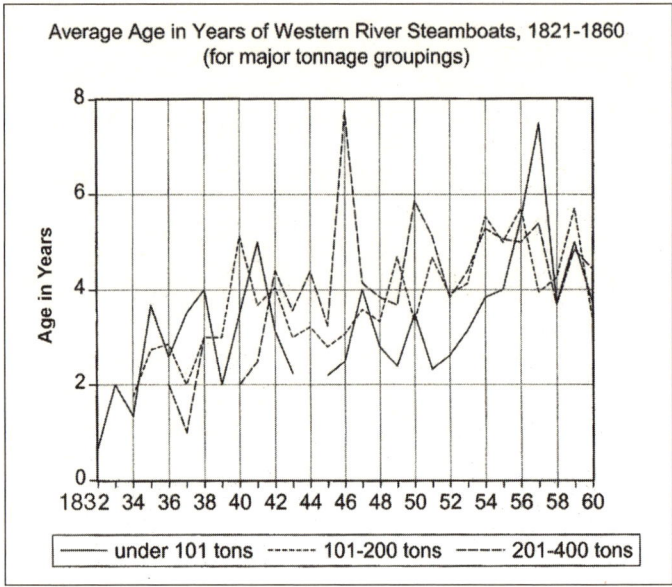

Figure 39. *Source:* Steamboat Database.

a greater ability to avoid such river hazards as snags, bars, rocks, and wrecks, and therefore into a better chance at survival.[17] The truth of this last observation is illustrated in figure 40, which charts the average longevity of 841 western river steamboats—676 sidewheelers and 165 sternwheelers.[18]

Steamboat longevity also varied with respect to the rivers on which vessels worked, and, as table 11 indicates, the variation could be substantial, especially before about 1850. Improved conditions on each of the rivers were reflected in the increased life spans of the steam packets that plied them. While packets on the Mississippi enjoyed the greatest increase in longevity, no doubt because that river was the major object of the Topographical Bureau's efforts to remove snags and bars, those on the other major rivers, as well as on their tributaries, also lasted longer.

The overall impact of the army's campaign to clear the western rivers of their hazards to navigation can perhaps best be appreciated by comparing the survival rates of steamboats built in each of six successive five-year periods from 1831 through 1860, as in figure 41, which presents data for 846 steamboats. As figure 41 indicates, a steamboat's chance of surviving to any age up to and including six years was appreciably better for a vessel built during the 1850s than

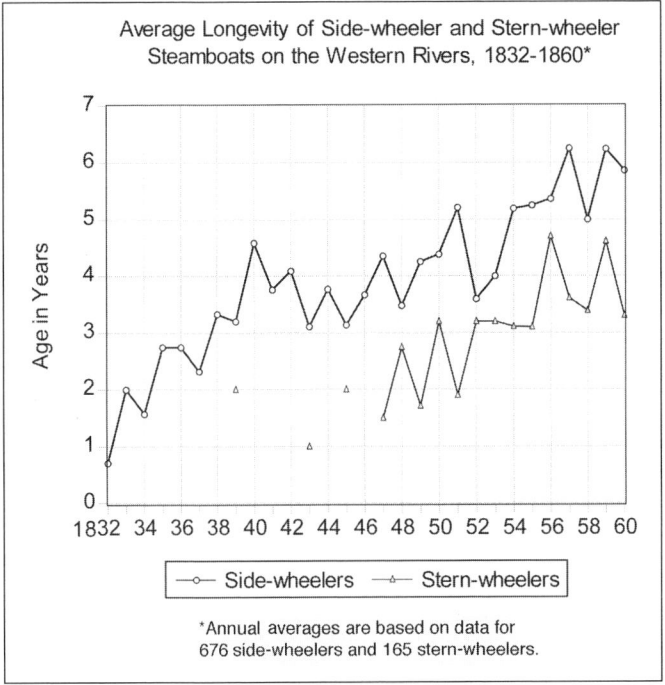

Figure 40. *Source:* See notes to text.

during the 1840s or 1830s. Similarly, steamboats constructed during the 1840s had a better chance of surviving to a particular age than did those built the decade before. The demonstrable improvement in steamboat survival rates not only resulted in greater safety for passengers and cargoes but also in increases in the efficiency and productivity of western river steamboat operation, which benefited steamboat owners and customers alike.

PRODUCTIVITY

In 1975, Erik F. Haites, James Mak and Gary M. Walton published the first rigorous analysis of productivity growth in western river transportation. They found evidence of significant increases in steamboat productivity and concluded, emphatically, that these increases were due almost entirely to the actions of private enterprise, including efficiency gains from learning-by-doing, routinization, and technological advances in steamboat design of the sort ex-

Table 11
Average Age of Steam Packets Lost by All Causes on Western Rivers by Five-Year Period and River, 1841–1860

River System	1841 1845	1846 1850	1851 1855	1856 1860
Missouri	3.0*	a	2.5	3.4
	(1)		(4)	(17)
Mississippi		3.8	5.0	5.7
		(32)	(46)	(62)
Ohio	5.0*	4.2	4.8	5.4
	(1)	(29)	(51)	(60)
Major Tributaries	3.7	4.4	4.6	5.2
	(3)	(9)	(8)	(20)
All	2.9	4.1	4.7	5.3
	(5)	(70)	(109)	(159)

Source: Steam Packet Database.

Note: Age at time of wreck is given by the number appearing on the same line as the name of the river on which the wreck occurred; the number in parentheses just below the age is the corresponding number of wrecked steamboats from which the average age is computed.

[a] Blanks indicates insufficient data from which to compute average ages.

* One value, and not an average.

plored in chapter 5.[19] For them, the role of government, that is, the public sector, in spurring productivity was relatively insignificant, and they offered their findings as an overdue corrective to a view of nineteenth-century U.S. economic history that had emphasized the importance of public investment in transportation development, especially improvements on the western rivers, but also government assistance in the construction of canals and railroads.[20] Their interpretation was an early instance among economic historians of what, for many, would become a retreat from a neo-Keynesian, demand-side conception of economic development and the adoption in its place of the so-called "new classical," supply-side view.

The basis for the revision offered by Haites, Mak and Walton was a calculation of a set of time series of indexes of inputs and outputs involved in steam-

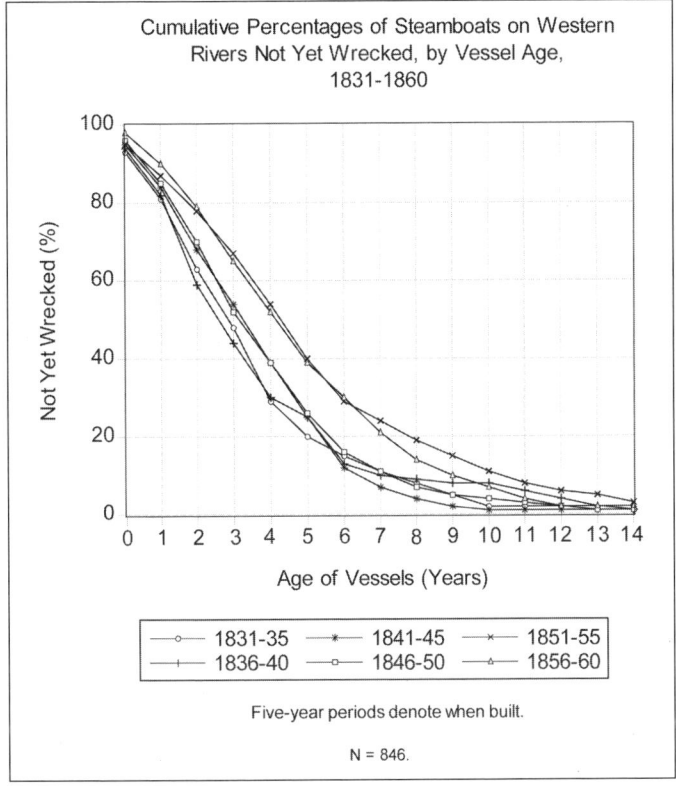

Figure 41. *Source:* Steamboat Database.

boat operation for the period 1815–60. Inputs included such items as fuel, labor, and insurance costs; outputs were measured principally in terms of upriver and downriver freight charges per ton of cargo. These weighted indexes are charted for the period 1821–60 in figures 42 and 43. Using these individual annual indexes, Haites, Mak, and Walton computed an index of total factor productivity, P_I/P_O, which was the ratio of the prices of inputs to those of outputs. This ratio's value, charted in figure 44, increased over time, a result consistent with the increasing efficiency of steamboat operation and one that reflected only moderately increasing input costs and dramatically falling freight rates. We can use these indexes to determine the impact of river improvements on steamboat productivity.

The variables that determined the productivity of western river steamboats fall into two categories: market-related influences, including technological change, and public-sector activity in the form of river improvements. In looking for their effects on input prices, output prices, and total factor productivity, we should bear in mind that, because of the way in which Haites, Mak, and Walton defined these indexes, steamboat tonnage played an integral part in the calculations of all three. Consequently, between any of the indexes and a tonnage measurement or any tonnage-related dimensional measurement of a steamboat, there is inevitably something of an autocorrelation. That is to say, a statistical measurement of the relationship between, for example, the mean size of steamboats in operation, measured in tons, and the index of input prices will suggest a stronger relationship than that which actually existed because each index is, itself, to a significant degree, based on vessel tonnage.

Figures 45 and 46 lay out the essential facts of the relationships between the size of western river steamboats, given in tons, and each of the two measures of productivity. Figure 45 illustrates the inverse relationship for the period 1821–60, between the weighted index of output prices and the mean tonnage of steamboats in operation each year, which for our purposes may serve as a measure of the state of steamboat technology. The point clearly made in the graph is that larger vessels were more efficient than smaller ones in terms of freight

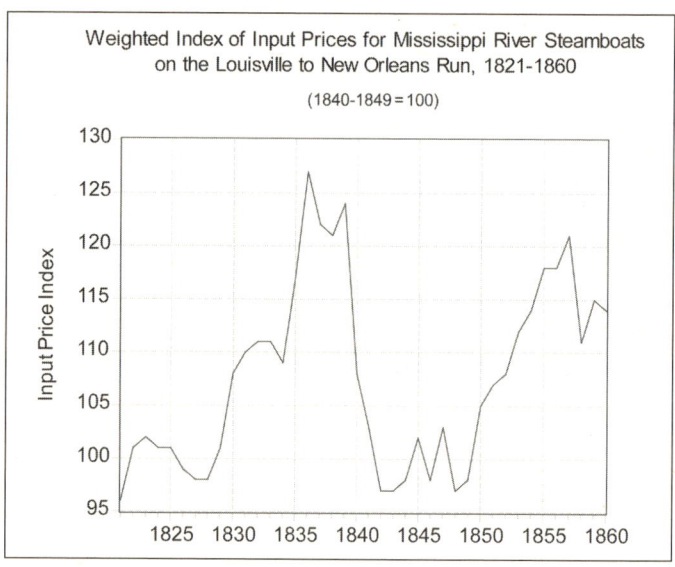

Figure 42. *Source:* See notes to text.

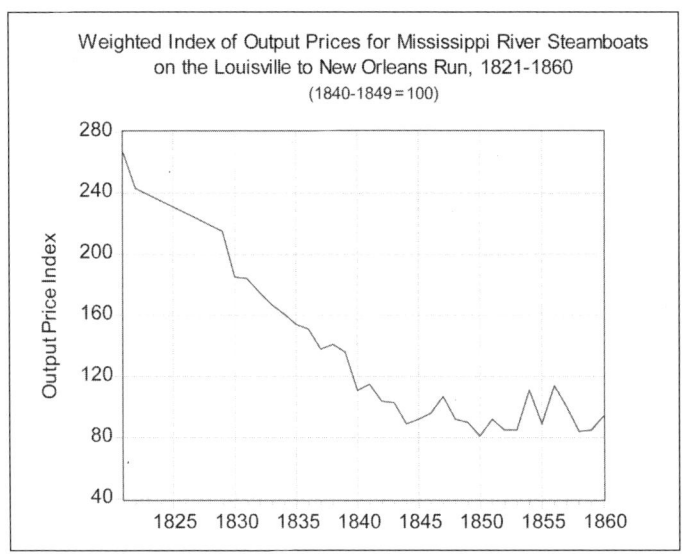

Figure 43. *Source:* See notes to text.

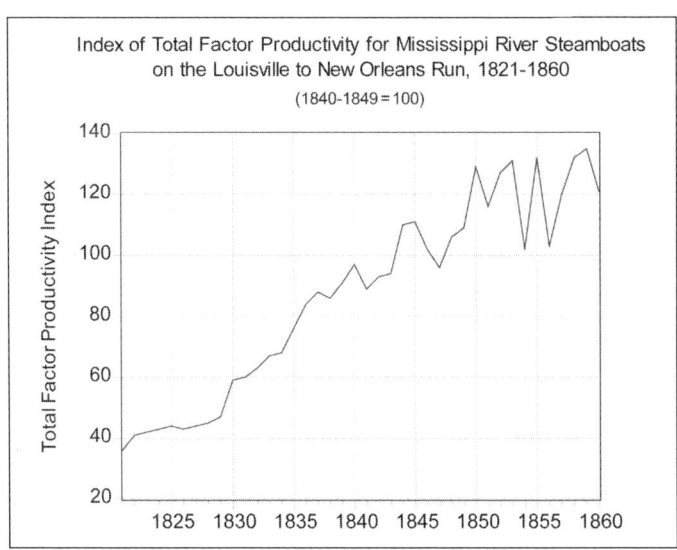

Figure 44. *Source:* See notes to text.

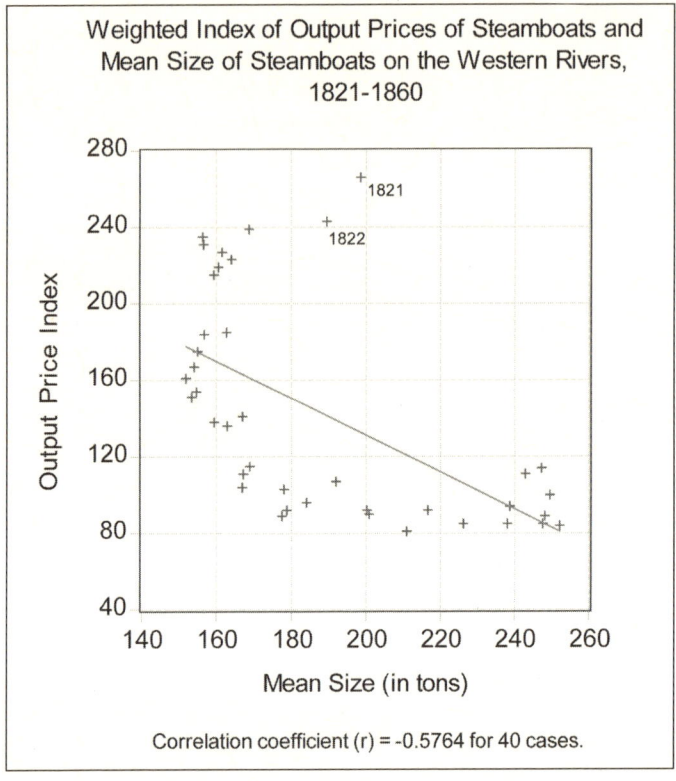

Figure 45. *Sources:* See notes to text.

and passenger rates charged. The straight oblique line of best fit between the points plotted in figure 45 only approximates the true nature of the relationship between the two measures, which was that, beyond about 180 tons, increases in steamboat size were ever less availing of further reductions in rates.[21] This point would be clearer still if the data for the first two years of the period covered by the graph—1821 and 1822—were excluded. These years were near the beginning of the steamboat era on the western rivers, and, as would be true of the price structure of any innovation in its infancy, freight and passenger shipping rates were then at their highest levels.

The relationship between total factor productivity and average steamboat size was also a strong one and is illustrated in figure 46. As one would expect, larger vessels were considerably more productive than smaller ones because they were able to realize size-related economies of labor and fuel.[22] Although

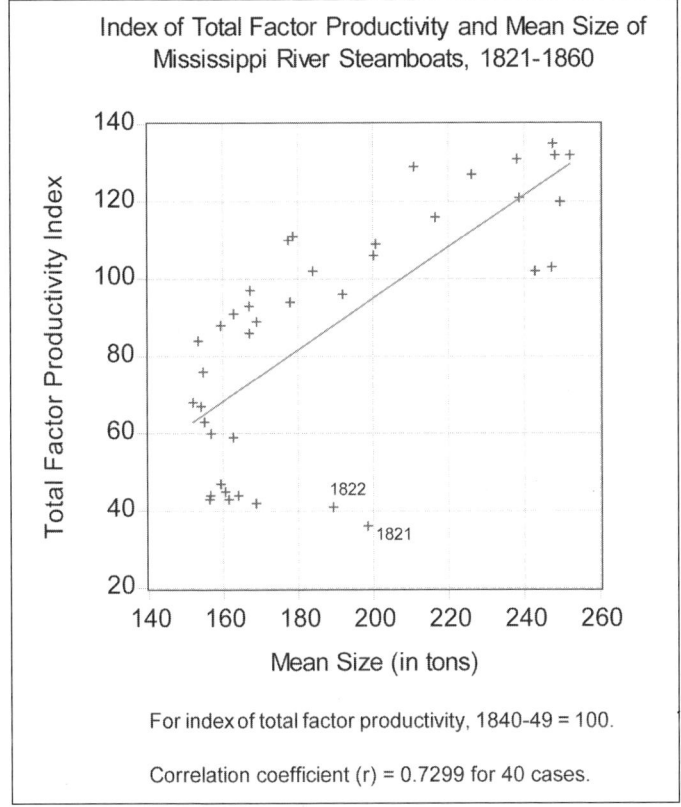

Figure 46. *Sources:* See notes to text.

this result and the one plotted in figure 45 are impressive and of some value, we should take them with more than one grain of salt because of the inherent bias in favor of finding a strong statistical relationship between the indexes and average steamboat size. Again, the problem of autocorrelation exists because each of the indexes is, in part, a function of steamboat size. Fortunately, this problem does not arise with respect to the relationship between river improvements and freight rates and productivity.[23]

It is hardly surprising to find that the removal of hazards to navigation on the western rivers and the resulting reduction in the rate of loss for steamboats on those rivers had a salutary effect on steamboat shipping rates and productivity. Haites, Mak, and Walton supposed as much, at least for the years before

1840, and Americans who called for river improvements assumed that a reduction of the dangers on the rivers would decrease the cost of doing business on them.[24] As figure 47 suggests, they were right, a point of no small significance. The relationship charted in figure 47 is one in which higher values of the index of output prices, that is, shipping rates, are strongly associated with higher rates of loss for steamboats on the western rivers. Similarly, figure 48 shows that higher values of total factor productivity for western river steamboats were tied to lower rates of steamboat losses (the anomalous character of the two pairs of years, 1821 and 1822 and 1841 and 1842, already noted above, is evident here as well). These results confirm the opinions of contemporaries and of the historians who have second-guessed them. More to the point, the results presented in figures 47 and 48 are further evidence of the success of the federal government's river improvements program.

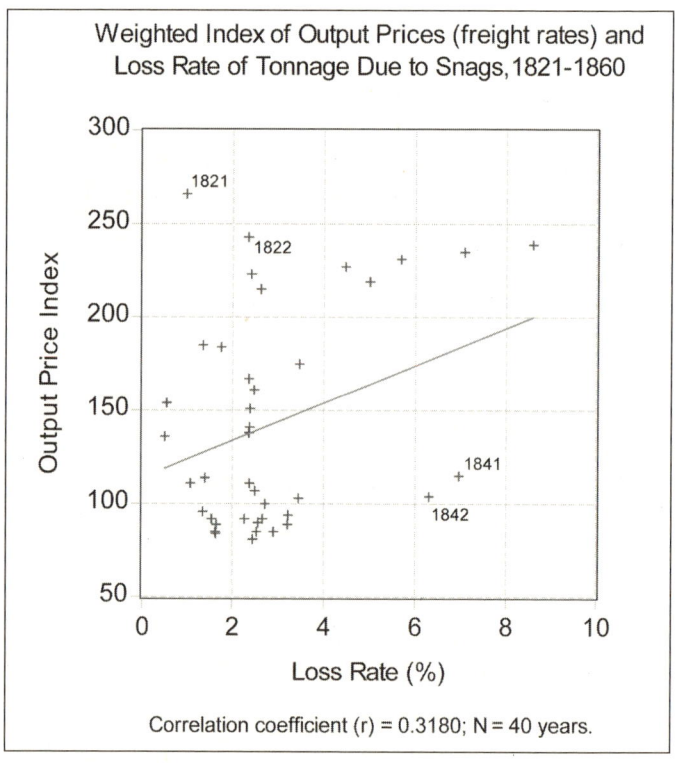

Figure 47. *Sources:* See notes to text.

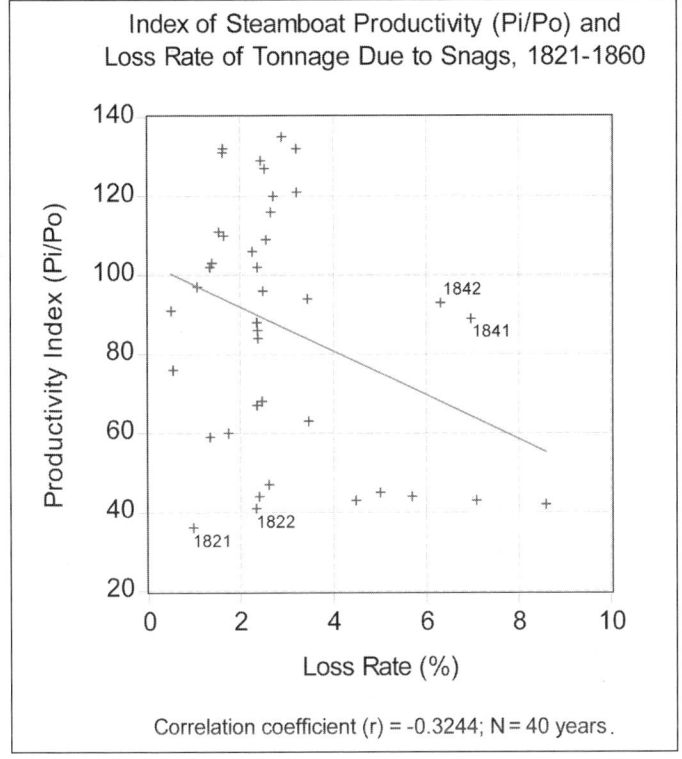

Figure 48. *Source:* See notes to text.

Even during the last two decades before the Civil War, when the river improvement program's funding and activity were more sporadic, the program enjoyed a fair measure of success. As figure 37 showed, supporters of the program who justified their calls for increased funding with claims that spending more money on snag removal would lower the rate of loss on the western rivers were essentially correct.[25] There is, however, another conclusion to be drawn from the evidence presented in figure 37, and that is that expenditures of more than $100,000 to $200,000 beyond the large initial expenditures became ever less availing of further significant reductions in the rate of loss. This observation makes sense, after the fashion of the law of diminishing marginal returns (that is, beyond a certain level of investment, there is a progressively smaller return from each additional dollar invested), and it undermines the strong argument of the program's friends that substantial expenditures would significantly reduce losses on the rivers. One reason that this was not so, and why the dramatic reductions in losses were achieved with the first few hundred thousand or so dollars, was that many of the most accessible and dangerous snag concentrations were tackled first, funded by that initial expenditure. Moreover, the clearing of snags from a stretch of river did not have the long-term impact that the removal of rocks from a channel had; that is, the removal of snags was, of necessity, a repetitive operation, akin to weeding a garden, because new snags could, and often did, form in the same part of the river.[26] Another, related reason was that the additional amounts of expenditures proposed by advocates of a more ambitious improvements program were, in fact, rather picayune in comparison with the amounts that were ultimately to prove necessary to accomplish the task.

Notwithstanding the program's limited capacity to clear the western rivers of their hazards, it was successful enough. Steamboats that otherwise might have been wrecked worked the rivers without serious mishap. Just how much steamboat tonnage would have fallen prey to the rivers' natural hazards had there been no concerted federal effort to remove them is, of course, impossible to say with certainty. We can, however, arrive at a rough but useful estimate of this aspect of the river improvement program's impact. The initial step in this process is to compute the average of the annual rates of steamboat losses due to snags from 1821 through 1825, the first five years of the river program. Using this average of 4.93 percent as the loss rate that would have prevailed from 1826 through 1860 in the absence of a snag-removal program, we can estimate the additional annual steamboat tonnage that would have been lost. The results of this procedure are presented in figure 49 and indicate that, in terms of losses prevented, the river improvement program more than paid for itself by the mid-1830s.[27]

The only manner in which that program may be said to have failed was that the actual, as opposed to the relative, number and tonnage of steamboats lost to river hazards rose throughout the four decades of the program's antebellum existence. Despite this shortcoming, the federal government's river improvements program may still be judged to have been an overall success. While Congress often appropriated large sums of money for hazard removal on the basis of log-rolling, it also appropriated substantial amounts on the basis of where the money was needed and would do the most good. A consequence of this effort was the reduction and subsequent stabilization of the rate of loss of steam-

THE SUCCESS OF PUBLIC POLICY

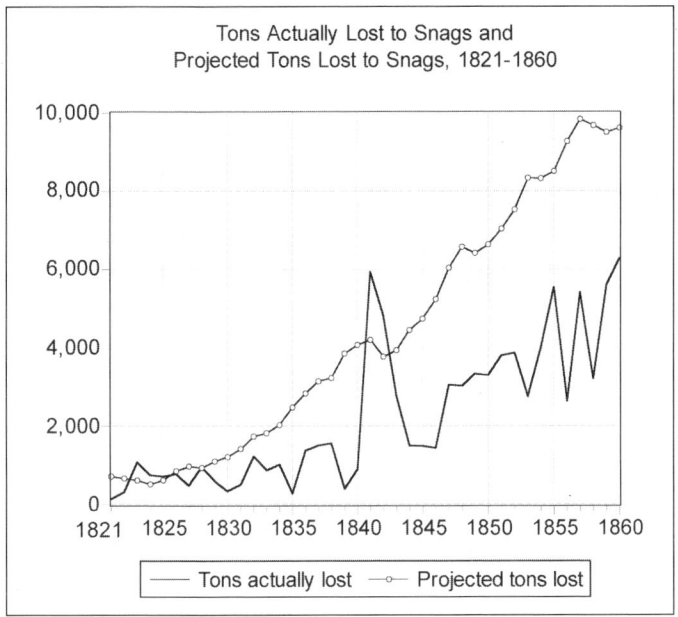

Figure 49. *Source:* See notes to text.

boats on the most hazardous stretches of the western rivers, even as the volume of traffic on the rivers increased. As the rate of loss fell, the life expectancy of steamboats increased, which, in the face of a more or less stable vessel technology, meant an increase in net returns for steamboat owners. Greater steamboat longevity was one consequence of reductions in the rate of loss on the western rivers; increased productivity of steamboat operations was another.

Together, gains in steamboat longevity and productivity meant greater efficiency not only for river transportation but for the general economy as well. The success of the federal government's distribution of public lands was another important, if somewhat indirect, effect of the river improvements program. Considered in light of these achievements of the program and in terms of its specific accomplishments, the antebellum federal river improvements program was a remarkable undertaking. At a time when the institutional arrangements of the market economy were taking shape and the power of private capital was growing in strength and influence in American society, the improvements program was the principal means by which the interests and resources of the public sector were projected in the national economy.[28] And, all things considered, the program accomplished quite a lot, not least being the provision of a vital means for the settlement and development of the lands of the public domain. That achievement and the significant reduction in the rate at which steamboats succumbed to the natural hazards of the western rivers helped to transform the economic life of the United States before the Civil War. This transformation, while not complete by 1861, was well advanced and was one of the two grand objectives set for the policy of federally funded improvements by its advocates. As for the other goal—the welding together of distant and disparate regions into an ever-stronger federal union—the secession winter of 1860–61 and the four years of war that followed would show that there were limits to what even a successful public policy could do.

EPILOGUE

The Civil War was the great divide in the development of the program to improve the nation's rivers and harbors. Before the war, partisan bickering, sectional rivalries, and constitutional objections had combined to curtail and, often, eliminate appropriations. The program's friends and enemies had fought over spending levels that only a few years after the war seemed insignificant. Before the war, annual appropriations had exceeded $1 million only three times and in most years were far lower. The actual amounts spent on improvements were generally lower still, surpassing the million-dollar level only twice and the $500,000 level only fourteen times during the forty years from 1821 through 1860.

Spending levels before the war might have been greater had critics and even friends of the program grasped its sheer geographical extent. Even excluding the entire Atlantic and Gulf coasts, and all the rivers east of the Appalachians, the magnitude of the program at its antebellum peak in the early 1850s was remarkable. The federal government's topographical engineers were responsible for surveying and removing hazards along more than 2,100 miles of the Mississippi, 549 miles of the Missouri just to St. Joseph, more than 1,000 miles of the Ohio, and more than 3,300 miles of the Wabash, White, Black, Red, Cumberland, Tennessee, and other rivers—some 7,000 miles of river in all. Along with this task went the responsibility for improving or maintaining numerous harbors along the lengthy shoreline of the Great Lakes, and the dredging and clearing of East Coast harbor entrances at Charleston, Savannah, Wilmington (North Carolina), Richmond, Norfolk, Baltimore, Philadelphia, New York, New Haven, Boston, and Portsmouth, New Hampshire.

That proponents of the program thought that sustained annual expenditures along the lines of the early 1830s, that is, at an annual level of two or three hundred thousand dollars, would be sufficient to eliminate the hazards posed by shoals, bars, and snags on the western rivers was only in part an indication of their naiveté. It was also a function of programmatic thinking, a type of conceptualization that was already becoming established in scientific, business, and, increasingly, government circles. A problem, once identified, described, and appraised necessarily would yield before the determined application of machinery and manpower, both guided by a complete faith in the triumph of the instruments of progress over nature.

Opponents of a federal improvements program were even more unrealistic in their insistence that the clearing of the rivers and harbors was within the fiscal capacity and political will of individual states or municipalities. If nothing else, the antebellum record of state spending for river and harbor improvements should have discouraged such thinking. But, having already made up their minds on the matter, they did not wish to be confused by facts.

This propensity to underestimate the difficulty and ultimate cost of these ambitious civil works projects was not peculiar to the debate over clearing river channels and building new harbors and improving existing ones. A similar lack of realism was characteristic of the increasingly acrimonious controversy over proposals to expand levee construction along the Mississippi River. Just months before the destructive flood of 1851, Albert Stein, a Mobile, Alabama, engineer, advocated a sustained effort to monitor and deepen the river's channel, beginning at its mouth, to increase its rate of flow as a means of forestalling flooding and, more important, the abandonment of its channel for a new course.[1] After the flood of 1851, Charles Ellett—one of the founders of the profession of civil engineering—conducted a study, as a paid consultant for the federal government, of what steps should be taken to avert damage from floods of that magnitude in the future. He recommended in 1861 that the existing levees "from Red River to New Orleans" be raised by two feet.[2] For his pains, Ellett became a target of ridicule and abuse from the government's engineers and bureaucrats, who made sure that his recommendations went unheeded.[3]

In the decades that followed, catastrophic flooding by the Mississippi River converted many of the skeptics and scoffers into true believers. As Herman Haupt, a close student of the river, noted: "The prophecy of Mr. Ellett has been more than realized. Instead of 2 feet [as proposed by Ellett], levees are now 5 feet higher than in 1851."[4] Just as wide of the mark was the $26 million in costs projected by federal engineers in 1861 to construct a levee system from Cairo to the mouth of the Mississippi River. By 1897, the system, which was not then yet finished, had cost "more than $50,000,000, and the cost of the levee system when completed will not be less than $100,000,000, if built at once, but more, if built gradually."[5] Such was the gap between perception and reality.

Stephen A. Douglas's proposal for "tonnage duties" to be levied by cities or states along the western rivers and Great Lakes was much discussed in 1852 and was, in its own way, no more realistic. Had Congress implemented it, it would have pitted each port against all others nearby in a beggar-thy-neighbor competition. Although free-market purists might have applauded such contests, confident that ultimately the most efficient arrangement of river and lake ports would emerge, the playing out of such a game would necessarily have entailed politically unacceptable consequences and probably would have raised commodity prices to consumers for some time. Also, by increasing transaction costs involved in river and lake transportation, Douglas's plan might very well have accelerated the growth of the railroad throughout the Midwest and, perhaps, the lower Mississippi Valley as an alternative to transport by water. However desirable in the long run, such an outcome still would not have made the rivers and lakes appreciably safer.

What was required to do that was a level of annual spending that, in 1882, exceeded $11.6 million, or more than twenty times the expenditure of a typical antebellum year. Spending on anything approaching that scale had seldom even been envisioned before the war. Still, as figure 50 indicates, the prewar share of the federal government's total expenditures, exclusive of debt servicing, that went for river and harbor improvements compared favorably with the level of commitment reached after the war. In fact, not until 1879 did that ratio surpass the highest prewar level, reached in 1831. The average value for the antebellum period, about 2 percent, was not reached in the postwar years until 1871. Given the limitations of resources and understanding during the prewar decades, the striking thing about the antebellum rivers and harbors program is not that it was small, but that it was as large as it was, not that it failed of its ultimate objective, but that it came as close to achieving it as it did.

The failure of antebellum political leaders to grasp the magnitude of the civil works projects before them is understandable, if only because nothing in their experience had prepared them for thinking on that scale. Apart from funding the nation's wars, the Mexican War having been the latest and, at $120 million (in current dollars),[6] the most expensive per year of conflict since the Revolution, the federal government's role in the economy as a mobilizer of capital and

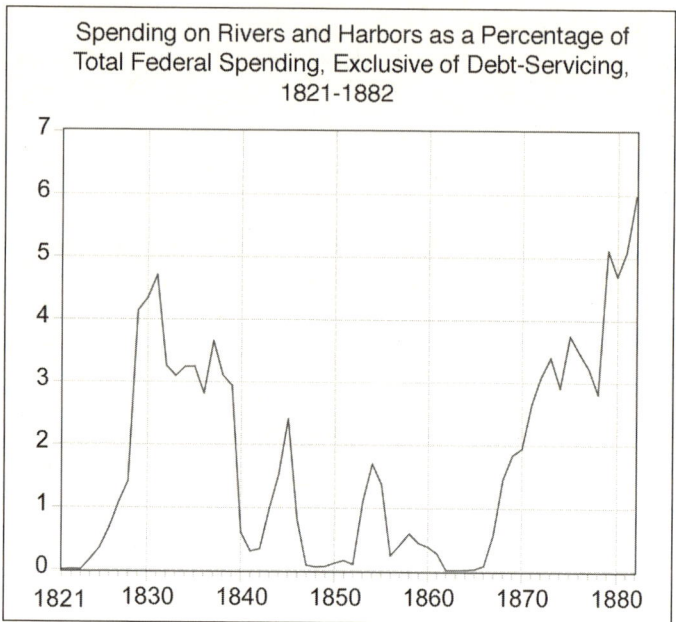

Figure 50. *Sources:* See notes to text.

projector of great plans had always been relatively constrained. The ambitious undertakings of the American System of John Quincy Adams's one-term presidency, especially the canals and roads projects, had been the exceptions. And, although spending levels during Andrew Jackson's two terms had been even larger, the money had gone primarily for interstate rivers and harbors improvements. Jackson's constitutional and political aversion to intrastate projects was profound and became the touchstone of internal improvements policy for almost all of his successors before Lincoln.

It was just such thinking that, in concert with the growing sectional dispute during the late 1840s and 1850s over the selection of a route for the much-discussed transcontinental railroad, had guaranteed that that project would go nowhere. The war, of course, changed all that, and not simply because secession had removed southern Democrats as a political impediment to northern railroad and mining interests. The war had, in a very real sense, been about the concentration, direction, and application of power, military and political of course, but also economic and technological power. That the exercise had achieved its primary objective of crushing secession was beyond dispute. That achievement was convincing evidence that the present and the future belonged to those who dreamed large and acted boldly.

After the war, with the Republican Party in control of national policy and committed to a nationalist program of economic development, the scale and scope of river improvements expanded dramatically. Beginning with 1868, net annual expenditures over the next decade and a half never fell below $3 million and averaged well above $5 million. Of course, the revenue base out of which such expenditures came was much larger after the war than before, thanks in large part to the Morrill tariffs and a greatly expanded volume of foreign trade. But there was a difference, a significant difference, between the things that made such growth in spending possible and those that made it desirable, even imperative. The war, as John Sherman had observed in a letter to his brother the general at the war's end, had enlarged capitalists' conceptions of the scale on which they should operate and what they could accomplish.[7] That change in thinking was necessary in order for the improvements program to achieve its long-sought objective, that is, the elimination of natural hazards to river navigation. Just how much thinking had changed from before the war first became clear in the Senate debate over the river and harbor bill of 1872.

The river and harbor bill sent to the Senate by the House of Representatives in 1872 was unprecedented by virtue of its contemplated level of spending—more than $5 million. That fact alone was enough to generate opposition along lines that would have been familiar to anyone who had participated in the debates over improvements before the war. According to critics, the proposed appropriations were evidence of a staggering profligacy. Even a senator friendly to the bill, Zachariah Chandler, a hard-core Michigan Republican, warned against the ever-present danger of new appropriations from the Senate being added to an already swollen House bill.[8] The result would be, predicted Republican Henry B. Anthony of Rhode Island, that "it will end in such log-rolling that we will break the bill down by putting on fifteen or twenty millions. That is the way we have done before."[9]

Chandler and Anthony were of course correct, just as they would have been in 1852 or 1836. For that matter, had they wished to indulge in prophecy, they could have said that the same would be true in 1880. Log-rolling was intrinsic to the appropriation process, as was pork. And so, for that matter, was a style of congressional disputation that by 1872 had become almost scholastic in its attention to obscurely small, even picayune, bits of language. The focus of the debate at one point resolved itself into a difference between two Republicans, Lot M. Morrill of Michigan and New York's Roscoe Conkling, over whether the appropriations bill was a "general," that is, national, bill or was, instead, one with a narrower, more parochial purpose. If it was the latter, Conkling argued, then it could not be dealt with on an annual basis because the long-familiar annual practice of passing appropriations measures was unconstitutional, appropriations being, by their constitutional nature, for two-year periods. Although inherently a matter of some consequence, it was, at that point and in Conkling's hands, little more than a delaying tactic, and Morrill quickly disposed of it.[10]

A more immediately significant point emerged in the debate when Lyman Trumbull, an Illinois Republican, offered an amendment requiring the secretary of war to determine not only the cost to be incurred "to make [a] river navigable" but also "the necessity for navigating the river."[11] Here, at last, was an issue at the core of the entire question of internal improvements: just because a proposed improvement was practicable, did that fact make it desirable? Essentially, Trumbull was calling for the formulation of an economic impact statement as a justification for a river project. Although his amendment passed, the reaction from some of his colleagues was indicative of the cynicism that had long characterized discussion of internal improvements projects.

Allen Thurman, a Democrat from Ohio, was especially sarcastic in his opposition to Trumbull's amendment, predicting that the secretary of war's determination of the necessity for any proposed project would inevitably and invariably be flawed. The problem was not so much with the secretary as it was with the proponents of an improvement project for a particular place. These people, Thurman warned, would mislead the secretary, who would then forward to Congress "a parcel of statistics . . . that will make out that that is the greatest place that ever was seen, that there is more commerce, that there are more people who trade there, more business done there or ever will be done there if the improvement be made than was ever done at Tyre or Venice."[12] Roscoe Conkling agreed and added his own objection that any information collected by the secretary of war for the Senate's consideration under Trumbull's amendment would be redundant because it was already contained in the census and other sources readily available to interested senators at no additional cost to the national treasury.[13]

After an exchange between Trumbull and Conkling, the Senate debate on the river and harbor bill proceeded smoothly to its conclusion, almost every reasonable and specious ground for objection or amendment having been exhausted, and, on June 10, the measure passed. The bill that eventually emerged from the Congress appropriated more than $11,554,000 for 1873, about twice the bill's original size when it had come over to the Senate from the House of Representatives.[14] Zachariah Chandler's warning at the beginning of the Senate's consideration of the House bill had been prescient; it had also been futile. Pressure for the passage of a large improvements bill in 1872 had been building for some time and had become acute with the meeting that year of a convention of river interests in St. Louis.[15] The pressure became more intense the next year when a "Congressional Convention" was held in St. Louis to mobilize business and political support, particularly among members of Congress, for river improvements.[16]

Addressing the opening session of the convention, Missouri's Democratic governor, Silas Woodson, stood beneath a banner proclaiming "'Cheap Transit Benefits All.'"[17] He noted that there had been "some unkind remarks . . . concerning the assembling of this Convention. I have heard it intimated here and there, that in all probability there was some political purpose or scheme to be subserved by calling this Convention." Nothing, the governor assured his audience, many members of which were congressmen and governors from the river states, could be further from the truth. Rather, the objective of those who had organized the convention and those who were in attendance was "to see this great country of ours become united. . . . We want to see no sectionalism in it. . . . In other words, gentlemen, we want, no North, no East, no South, no West, but a great, a free, a prosperous country, enjoying equal rights under the Constitution in every part of it." The governor, addressing an audience beyond St. Louis, insisted that those assembled in convention in St. Louis "wanted to see equal rights and equal privileges extended to every part of our country" and so he had made one of the traditional and most highly charged demands of the champions of improvements: equality. Applause punctuated his remarks.[18]

The appropriation for 1873 was larger than any earlier appropriation for rivers and harbors, but it would be surpassed only two years later when Congress appropriated over $12,382,000. By 1880, even that amount would seem modest. The days of hardscrabble internal improvements had long since come to an end. The war had done that. When, in May 1880, the House of Representatives took up that hardy perennial, the annual bill to fund internal improvements, the ensuing debate was rich in ironies and juxtapositions. Samuel Cox, a Democrat from the Sixth District in New York City and formerly of Ohio's Seventh District, asserted the primacy of states' rights and warned of an unholy alliance between protectionists and advocates of intrusive, federally funded improvements and their joint assault on strict constitutional constructionism. These views were essentially the ones that he had held as a congressman during the late 1850s, and the war had done nothing to change them.[19]

Cox had been, at best, a lukewarm supporter of the Union cause during the Civil War and had rallied to the defense of Clement Vallandigham when the Ohio Copperhead had been sentenced to expulsion and exile to rebel territory.[20] Cox himself had walked a fine line between loyal opposition to the Lincoln administration's policies and giving aid and comfort to the Confederacy—and he knew it. In an apologia published in 1865, he sought to cast his words and actions during the war in the best possible light, asserting that "whatever may have seemed to a superficial observer an unpatriotic opposition, was only and truly an opposition to the arbitrary proceedings with which the war was accompanied."[21] Now, fifteen years after Appomattox, he read approvingly from a speech against the nationalizing of local interests, originally given in 1846 "by a southern statesman," the late George S. Houston of Alabama.[22]

Houston had been a widely respected and long-serving member of the House when he resigned in January 1861 in support of Alabama's secession. He returned to Washington, D.C., in February 1866 as senator-elect from Alabama but was denied his seat by the Republican-dominated Senate. After serving as governor of his state, he was elected once again to the Senate in 1878 and took his seat in March 1879, only to die on the last day of that year.[23] Cox had served in the House with Houston and knew him as a fellow Democrat, and so it was perhaps not surprising that he would invoke the memory of his late friend, "a member of irreproachable honesty, courage, and sense."[24] Even so, his resurrection of the very arguments used by Houston more than thirty years before, arguments presumably refuted by the war, was more than a little strange.

James R. Chalmers, a Mississippi Democrat who had been wounded while serving the Confederacy as a general under Braxton Bragg, also objected to the bill, but for very different reasons.[25] Unlike Cox, who was distressed by all that the bill was designed to do, Chalmers was aggrieved by the fact that it did not propose to do enough for his district: "I look in this bill for appropriations for the great Mississippi River, filled with snags and sand-bars from Cairo to its mouth; and I find a pitiful appropriation of $100,000 to remove these obstructions in twelve hundred miles of the greatest channel of internal commerce known in the world." That the "little" Raritan River in New Jersey was also the object of a $100,000 appropriation was especially galling, and Chalmers noted that one of the members of the Committee on Commerce, where the bill had originated, was from the Garden State.[26] As many had been before him, Chalmers was shocked to find log-rolling and pork-barreling in Congress. Still, one indication of just how much the political landscape had changed from before the war was that a Mississippi Democrat now objected to a proposed river and harbor bill because it was not sufficiently ambitious.

The carping prompted John H. Reagan, a Democrat from Palestine, Texas, and the chairman of the Committee on Commerce, to take issue with the bill's critics. The committee, Reagan said, had not intended to and, indeed, had not slighted the Mississippi River. Quite the contrary, the committee had "made proportionately larger appropriations for the Mississippi River and its tributaries than for any other waters of the Union."[27] As for Cox's echoing of the departed Houston's objections to such bills, Reagan directed Cox's attention to decisions by the U.S. Supreme Court that had affirmed the federal government's responsibility for undertaking river improvements regardless of "what the length of the river is, whether it be in one or more than one State, if its commerce passes into other States or into foreign countries by navigable waters and adds to the commerce of the whole country."[28]

The casting in this bit of political dramaturgy was at least novel: a New Yorker who objected to the corrupt and sweepingly unconstitutional character of a river and harbor bill, and who bolstered his arguments by invoking those of a dead Alabama secessionist; a Mississippian who complained about the bill's inadequate spending provisions for the Mississippi River; and a Texan who rose to rebut and correct them both in the name of the larger interests of the Union and the constitutional powers of the federal government to conduct such a program. That they were all Democrats and that two were former Confederates (Reagan had been the Confederacy's postmaster general) made their exchange all the more remarkable.[29]

The rehearsing of the same old lines—in Cox's case, the quoting of lines spoken decades earlier—suggested that all participants understood that the outcome was never in doubt. Cox, for his part, had said as much at the outset, admitting, "I know very well that all opposition to this bill will be unavailing."[30] But each player had said his piece. Even Reagan, who had helped to write the bill, came close to admitting, in answer to a direct question from Cox, that he did not like it.[31] The many Democrats and the few Republicans who were against the bill were like the Bourbons: they had forgotten nothing, and they had learned nothing. In any event, Cox was correct; their opposition was to no avail, and the bill, which appropriated more than $8 million for an ambitious program of rivers and harbors improvements, passed overwhelmingly by a vote of 179 to 48, with 65 listed as not voting.[32]

The river and harbor bill of 1880 was a notable achievement for the champions of the improvements program. The large amount of money appropriated by it confirmed the end of a retrenchment in the program from 1873 through 1878, induced by the falling off of revenue due to the depression of those years. Moreover, the large appropriation made in 1880, coming on the heels of the even larger ($10,321,034) appropriation in 1879, signaled the disposal of the awkward question of whether such appropriations were to be made on an annual or biennial basis. Even more important was the creation by Congress of the Mississippi River Commission at the end of June 1879, which was the decisive step in changing fundamentally how Mississippi River improvements were conceived and funded. Thereafter, engineers began to approach the problems of removing hazards to navigation and preventing floods along the great river on the scale of the river as a whole and not on a piecemeal and sporadic basis, as had so often been the case before.[33] In 1881, Congress appropriated a still larger amount, $11,824,985, and, in 1882, a smaller but still impressive $9,741,853. These appropriations fueled a heightened rate of spending on river and harbor improvements.[34]

In those years, 1879 through 1882, as in years past, steamboats traveled up and down the Mississippi River and its major and minor tributaries. But in those years, unlike the all-too-eventful ones before the war when so many steamboats had come to grief, only seventeen steamboats were lost—ten by fire and seven in collisions. In fact, during the first half of the 1880s the loss rate for steamboat tonnage on the western rivers was well under 3 percent, perhaps as low as below 2 percent.[35] The federal government's massive program of improvements had done what the states and the private sector had been unable to do: it had made the great rivers of the interior safe for commerce. In those five years, only one steamboat was lost to a snag. By 1883, the year Mark Twain published *Life on the Mississippi*, his memoir of younger days, the river that had claimed so many boats and lives was largely a thing of the past, as was the dispute over whether river improvements were a federal responsibility.

APPENDIX A

MAJOR SOURCES OF QUANTITATIVE DATA

The quantitative data for this study come primarily from: Bruce D. Berman, *Encyclopedia of American Shipwrecks* (Boston: Mariners Press, 1972); William M. Lytle and Forrest R. Holdcamper, *Merchant Steam Vessels of the United States, 1790–1868: The Lytle-Holdcamper List, Initially Compiled from Official Merchant Marine Documents of the United States and Other Sources*, revised and edited by C. Bradford Mitchell, with the assistance of Kenneth R. Hall (Staten Island, NY: Steamship Historical Society of America; Baltimore: distributed by University of Baltimore Press, 1975); Frederick Way Jr., comp., *Way's Packet Directory, 1848–1994: Passenger Steamboats of the Mississippi River System since the Advent of Photography in Mid-Continent America*, rev. ed. (Athens: Ohio University Press, 1994); and the published congressional *Statement of Appropriations and Expenditures for Public Buildings, Rivers and Harbors, Forts, Arsenals, Armories, and Other Public Works from March 4, 1789 to June 30, 1882*, more familiarly known as serial 1992. As their titles suggest, the first three of these sources, but especially Berman's *Encyclopedia* and *Way's Packet Directory*, provided the information from which was compiled a database of 1,848 documented shipwrecks that occurred between 1800 and 1861 on American waters. More than 1,200 of these wrecks involved steamboats on the western rivers and Great Lakes and in the harbors along the Gulf coast. Berman's *Encyclopedia* provided data for each of these vessels, including its tonnage, type of steamship (that is, side-wheeler or stern-wheeler), rigging (if a sailing craft), year and place of construction, month and year when it was wrecked, and the location and cause of the wreck.

The data from Berman's Encyclopedia were scanned by a Kurzweil Optical Reader and then written onto disks, with the generous support of the Text Processing Center of the College of Arts and Sciences at Louisiana State University. The information on the disks was then run through a SPITBOL program written by Mr. Randy Hebert, then of the Office of Research of Louisiana State University. The SPITBOL program transposed the database into Dbase III format. I then edited the transposed database to produce the "Steamboat Database," which I then analyzed using a variety of statistical programs. *Way's Packet Directory* was an invaluable source of data on steamboat hull dimensions and engines. Data from it were abstracted and put into computerized form as the "Steamboat Packet Database" by Ms. Catherine P. Chang. Lytle's work, usually referred to simply as "Lytle's List," served here primarily as a check on the completeness of Berman's compilation. The *Statement of Appropriations and Expenditures* presented a comprehensive listing of every appropriations measure passed by Congress for the broadly described purpose of constructing public works, including those intended to contribute to the national program of internal improvements. These measures constitute the records of an appropriations and expenditures database that includes 2,266 specific spending projects, as well as 108 miscellaneous projects. While working to compile this database, I learned of the work of Laurence Joseph Malone, who compiled a similar congressional funding database for his 1991 dissertation, "Opening the West: Federal Internal Improvements before 1860," done at the New School for Social Research and subsequently published as *Opening the West: Federal Internal Improvements before 1860* (New York: Greenwood Press, 1998).

The steamboat databases rest on a firm foundation of work laid down by several historians, but the work of Louis C. Hunter has been exceptionally important to this study. His magisterial *Steamboats on the Western Rivers: An Economic and Social History*, published in 1949, has been the jumping-off point for two generations of scholarship on almost every aspect of the large subject of western river transportation. The more recent (1975) book by Erik F. Haites, James Mak, and Gary M. Walton, *Western River Transportation: The Era of Early Internal Development, 1810–1860*, examined many of Hunter's findings from a more rigorous economic perspective and broader base of data on flatboat and steamboat construction and operation. An innovative historical perspective on steamboat design, construction, and operation is *The Western River Steamboat* by Adam I. Kane, a nautical archaeologist. An equally novel approach to steamboat history is taken by Carl A. Brasseaux and Keith P. Fontenot in their *Steamboats on Louisiana's Bayous: A History and Directory*, which presents a comprehensive analysis and listing of steamboat traffic on the waterway's of the New Orleans hinterland, and by Harry P. Owens in *Steamboats and the Cotton Economy: River Trade in the Yazoo-Mississippi Delta*, which examines steamboat traffic on an important smaller tributary of the Lower Mississippi River and also provides a detailed listing of steamboats in service there during the nineteenth century.

APPENDIX A

Jacques D. Bagur's *A History of Navigation on Cypress Bayou and the Lakes* provides comparable information concerning the steamboat traffic near Shreveport, Louisiana. Many other sources, especially government documents, including the annual reports of the army's Topographical Bureau and several congressional executive documents, have provided essential data and commentary. The most valuable repositories for research in the primary sources concerning river improvements have been the libraries of the U.S. Army Corps of Engineers in New Orleans and, especially, in Washington, D.C.

APPENDIX B
APPORTIONMENT OF MISCELLANEOUS APPROPRIATIONS

Assigning to each state its proper share of the "miscellaneous" congressional appropriations for river improvements is not as straightforward as it might at first seem. Typically, a particular "miscellaneous" appropriation was aimed at improvements to navigation along a river that flowed through or along two or more states. Consider, for example, the single-largest object of such appropriations, the Mississippi River, with its headwaters in Minnesota, its head of navigation also in Minnesota, at the Falls of St. Anthony, and its mouth on the Gulf of Mexico. Of the more than $2 million appropriated by Congress from 1821 through 1860, how much went to each of the states with interests along the river? Apart from the desirable but also impractical method of tracing the destination of each dollar of miscellaneous appropriations for the Mississippi, four different means of apportioning the miscellaneous appropriations are available.

First, we might divide the $2-million Mississippi River appropriation evenly among the nine states along or through which the river flowed. Although this is certainly the simplest method, it does a good deal of violence to the facts by its implication that Congress might have been as concerned with improving navigation along Minnesota's short stretch of a small Mississippi River as with the giant river through Louisiana. An alternative approach would be to assign to each state a share of the appropriations proportional to its share of the total number of wrecks on the river. The chief objection to this approach is that it rests on the assumption that Congress appropriated funds solely on the basis of need and that the common politics of the log-roll and the pork barrel did not enter significantly into members' deliberations. As was demonstrated in chapters 2 and 3 with respect to state-specific appropriations, such political practices were integral to that appropriations process. Moreover, this assumption is one of the hypotheses tested here and, as such, cannot very well serve as an initial assumption concerning congressional behavior. The third approach would be to distribute the appropriations to the states in proportion to their respective populations, the assumption being that political calculations and not the

APPENDIX B

facts of physical geography and the hazards confronting waterborne commerce determined the voting in Congress on miscellaneous appropriations. Again, this sort of behavior was clearly in evidence when members considered state-specific appropriations, and the possibility that it occurred when Congress considered miscellaneous appropriations is certainly within the bounds of reason. It was therefore considered in some detail in chapter 2.

A fourth apportionment method, and the one that is used here, assigns miscellaneous appropriations for improvements along a particular river to states on the basis of their respective percentage shares of the river's total mileage. Although this method is necessarily somewhat ad hoc and, therefore, inevitably somewhat inaccurate, it has a number of virtues. It takes into account the physical reality of river conditions; it imposes the least number of a priori behavioral assumptions on the actions of Congress; and, most important, it is consistent with the language of the appropriations bills themselves and corresponds to the instructions given to those charged with the responsibility of hazard removal. On the latter point, the language of the appropriations acts of 1824 and 1827 for improvements along the Mississippi and Ohio rivers is instructive and worth quoting at length. The act of May 24, 1824, provides:

> That for the purpose of improving the navigation of the Mississippi River from the mouth of the Missouri to New Orleans, and of the Ohio River from Pittsburgh to its junction with the Mississippi, the President of the United States is hereby authorized to take prompt and effectual measures for the removal of all trees which may be fixed in the bed of said river; and, for this purpose, he is authorized to procure and provide, in that way which his discretion may be most eligible, the requisite water-craft, machinery, implements, and force, to raise all such trees, commonly called "planters, sawyers, or snags," as may be found in the current of the said rivers at the lowest stage of water, and to saw or cut them off, as near as practicable, to the bottom of the stream; and where trees are found upon sand-bars, upon the points of islands, or near the bank of the river, which may, at the lowest stage of the water, endanger the safety of navigating said rivers, they shall in like manner be cut, removed, or sawed off; and all roots or limbs belonging to those parts of said trees, which are fastened in the earth, shall be carefully cut away.[1]

The wording of the 1827 act was similar to that used in 1824, but the sense is even clearer that the method of hazard removal would entail searching out targets of opportunity along the Ohio River's course along "Ohio, Pennsylvania, West Virginia [then part of Virginia], Kentucky, Indiana, and Illinois." The act of March 3, 1827, states:

> That all snags, sawyers, stumps, logs, and obstructions of every description, which tend to endanger the steamboat navigation of the Ohio River at any navigable stage of the water, *and which present themselves, and are to be found on the banks and sides of the river, shall be removed* so that the navigation of said river may be rendered at all times safe; and the same shall be done under the supervision and direction of the Secretary of War, and through the aid of some practical agent acquainted with the situation of the river, its respective bars, islands, and dangerous places and parts; and he shall likewise cause the channel of said river, at a part usually called the Grand Chain, near its mouth, so to be deepened by a proper channel formed, that at the usual state of the water steamboats may be enabled safely to pass and repass the same.[2]

The mechanics of this method of apportioning congressional river and harbor appropriations can perhaps better be understood through the application of the method to a practical example, the division and assignment of $30,000 in miscellaneous appropriations for the Ohio River in 1827. Six states—Pennsylvania, Ohio, Indiana, Illinois, Virginia, and Kentucky—had interests along the Ohio River's 967-mile length. The states' respective mileages along the river and their corresponding shares of the $30,000 in miscellaneous appropriations are indicated in table B.1.

The chief defect in this method of apportioning miscellaneous appropriations is that it rests, in part, on the implicit assumption that hazards to waterborne commerce were distributed uniformly along each mile of water course. Obviously, that was not the case. Offsetting this problem, however, is the likelihood that, on average, longer stretches of rivers contained more water hazards than did shorter stretches. This was, in fact, the experience of the Office of Topographical Engineers, which directed the hazard removal on the rivers.[3] Using this approach as the basis for apportioning miscellaneous appropriations means that a state having the greatest mileage along a particular river is deemed likely to have had the greatest number of hazards to river navigation, and so is allocated the greatest share of the appropriations for hazard removal on that river.

APPENDIX B

Table B.1
Shares of Ohio River Mileage and the 1827 Miscellaneous Appropriation

State	Miles on Ohio River	% of Miles	Share of $30,000
PA	60 x 2 = 120[a]	6	1,800
OH	427	22	6,600
IN	351	18	5,400
IL	129	7	2,100
VA	253	13	3,900
KY	654	34	10,200
All	1,934	100%	$30,000

Source: Mileage figures for the Ohio River are from U.S. Congress, House, Department of the Treasury, *Report on the Internal Commerce of the United States 1887,* William F. Switzler, part 2 of *Report on the Internal Commerce of the United States,* 50th Cong., 1st sess., 1888, H. Exec. Doc. 6, serial 2552, 586. The source of appropriations amounts is serial 1992.

[a] The Ohio River's first 60 miles are entirely within Pennsylvania. Consequently, the total length of the state's banks along the river is equal to twice 60 miles, or 120 miles.

APPENDIX C

PRICE DEFLATORS AND CONGRESSIONAL FUNDING

The sometimes violent price swings for a wide range of commodities during the period 1821–60 raise the question of whether a price deflator should be applied to appropriations and expenditures to correct for inflationary and deflationary pressures on the dollar. The question is more readily posed than answered, and, even assuming that a deflator should be applied, there are only a few available candidates from which to choose the most suitable one, and each of these is far from ideal.

The deflators considered here are the Warren-Pearson Wholesale Price Index for the United States, the wholesale price index for Cincinnati, and the wholesale price index for New Orleans, presented in table C.1.[1] There are two distinct types of problems associated with the use of these price indexes. One is that of geographic bias and is especially acute with respect to the Warren and Pearson index, which is based almost exclusively on New York City prices. There is little likelihood that the latter would have prevailed in the states of the interior where the provisions, fuel, and supplies needed for the removal of western river hazards were purchased. In fact, as the sometimes sharp divergences of Cincinnati and New Orleans price indexes from the Warren and Pearson index indicate (figure C.1), eastern metropolitan prices and those along the western rivers often were quite different in level and direction of movement. Just such a circumstance argues against the use of the Warren and Pearson index as a deflator with which to adjust nominal appropriations and expenditures levels. But there are also problems associated with the use of the Cincinnati and New Orleans indexes for that purpose.

Although the Warren and Pearson index's usefulness is diminished by the fact that its constituent prices are essentially those that prevailed in New York City, that index at least has the virtue of being broadly based, representing the weighted conflation of more than 100 individual commodity series across a wide range of commodity classes. That is not at all true of either the Cincinnati or New Orleans index, each of which rests on a much narrower base of primarily agricultural commodities—21 to 50 in the case of Cincinnati, and 49 to 50 in that of New Orleans. Thus, these indexes probably have a somewhat tenuous connection to actual expenditures for hazard removal on the Ohio and Mississippi rivers.

The conflicting and confusing situation that emerges from the application of the three price indexes is evident in figure C.2, which plots the annual dollar difference between nominal net expenditures (that is, in current dollars) and the adjusted annual net expenditures computed by applying each of the price index deflators to the annual nominal amounts. Immediately apparent is the fact that, with the exception of three periods—1836–38, 1843–45, and 1854–55—the nominal and the three adjusted levels of net expenditures are nearly identical. The three periods when they are markedly different are notable because of the unusual economic conditions that prevailed at those times.

The first period, 1836–38, embraces the gyrations in public land sales and wholesale commodity prices that preceded, accompanied, and immediately followed the Panic of 1837. The second period, 1843–45, was the tail end of the depression that began in 1839 and persisted well into 1844. The behavior of prices during the third anomalous period, 1854–55, is less readily explained. Not years of panic or depression, 1854 and 1855 were nevertheless tinged with political uncertainty by the political tensions associated with the controversy over the Kansas-Nebraska Act, and such uncertainty may have resonated in the price system. Perhaps of greater immediate significance was the dramatic rise of public land sales in those two years to consecutive post–1836 highs of $9,285,534 and $11,485,385, respectively, up from only $1,804,653 in 1853.[2] The effect of these sharp increases may well have been to reduce the money supply, by withdrawing currency from circulation, and therefore to depress prices, however slightly.

How significant, then, are the differences between appropriations and expenditures levels when expressed in current dollars and constant dollars? As figure C.3 illustrates, the disparity between the current and constant valuations of the two was sometimes significant, especially during the 1836–38 and 1843–45 periods, when nominal and adjusted expenditures varied over a range of more than $150,000 and about $130,000, respectively. But here, again, the problems inherent in using these price indexes as deflators argue against placing very much reliance on them to adjust the annual nominal amounts of appropriations and expenditures for improvements.

Even were it possible to overcome these problems and construct a proper price index, there would probably be little benefit in doing so. Although the differences between current and constant dollars that result from the application of the various price indices as deflators to the total federal net expenditures for internal improvements to rivers, lakes, and harbors vary with the particular index, and the differences between the current and constant dollar amounts are sometimes marked, the overall effect in each instance is not decisive. Moreover, the largest-single items in many appropriations measures passed by Congress for the removal of river hazards were the snag boats and dredges themselves, the costs of which can scarcely have been reflected in any of the wholesale commodity indexes used here. In the end, therefore, the use of constant dollars, rather than current dollars, would not materially alter the weight or substance of the conclusions reached here.

Table C.1
Wholesale Price Indices, 1821–1860

Year	Cincinnati[a]	New Orleans[b]	Warren & Pearson[c]
1821	86	115	102
1822	98	124	106
1823	101	105	103
1824	98	110	98
1825	100	130	103
1826	93	95	99
1827	91	90	98
1828	92	91	97
1829	98	90	96
1830	93	86	91
1831	99	80	94
1832	101	88	95
1833	102	99	95
1834	95	96	90
1835	117	123	100
1836	145	132	114
1837	131	108	115
1838	129	107	110
1839	138	116	112
1840	104	91	95
1841	89	93	92

(Table C.1, continued)

Year	Cincinnati[a]	New Orleans[b]	Warren & Pearson[c]
1842	72	75	82
1843	72	70	75
1844	77	75	77
1845	**87**	**74**	**83**
1846	76	78	83
1847	90	93	90
1848	75	68	82
1849	77	80	82
1850	**86**	**103**	**84**
1851	90	89	83
1852	93	85	88
1853	104	91	97
1854	110	90	108
1855	**123**	**103**	**110**
1856	121	114	105
1857	128	136	111
1858	102	104	93
1859	114	107	95
1860	**110**	**105**	**93**

Source: U.S. Bureau of the Census, *Historical Statistics of the United States, Colonial Times to 1957* (Washington, D.C.: U.S. Government Printing Office, 1960), "Series E 90–95, Wholesale Price Indexes (Berry), for Cincinnati, 1816 to 1861, and Ohio River Valley, 1788 to 1817," 121; "Series E 96–100, Wholesale Price Indexes (Taylor), for New Orleans: 1800 to 1861," 122; and "Series E 1–12, Wholesale Price Indexes (Warren and Pearson), by Major Product Groups: 1749 to 1890," 115.

[a] Cincinnati: 1824–46=100.

[b] New Orleans: for the 1821–42 period, 1824–42=100; for the 1840–60 period, 1843–61=100.

[c] Warren & Pearson: 1910–14=100.

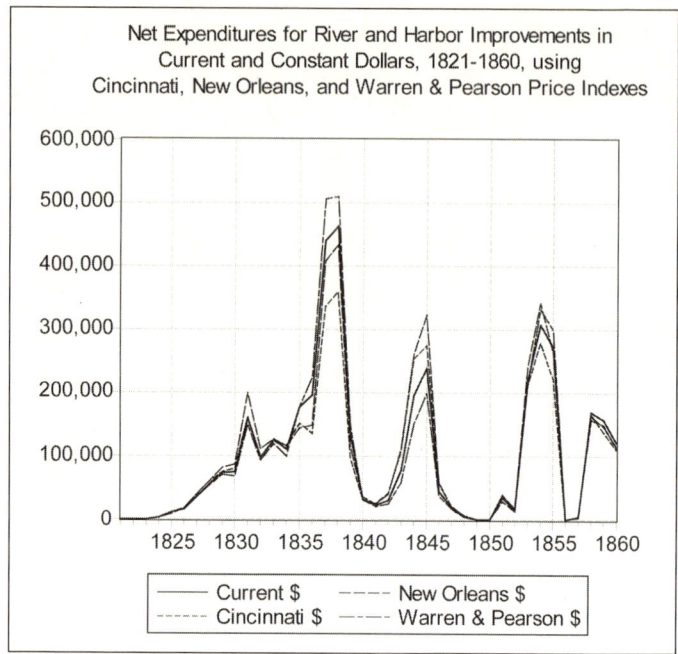

Figure C.1. *Sources:* See table C.1 and notes to text.

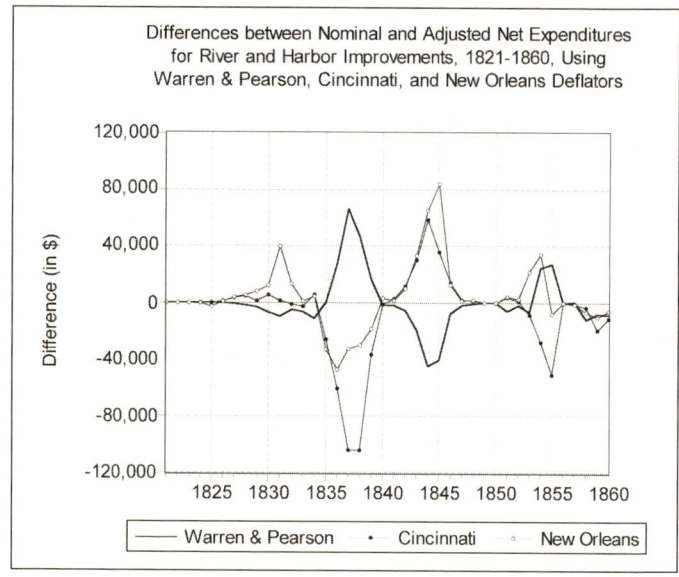

Figure C.2. *Source:* Derived from figure C.1.

APPENDIX C

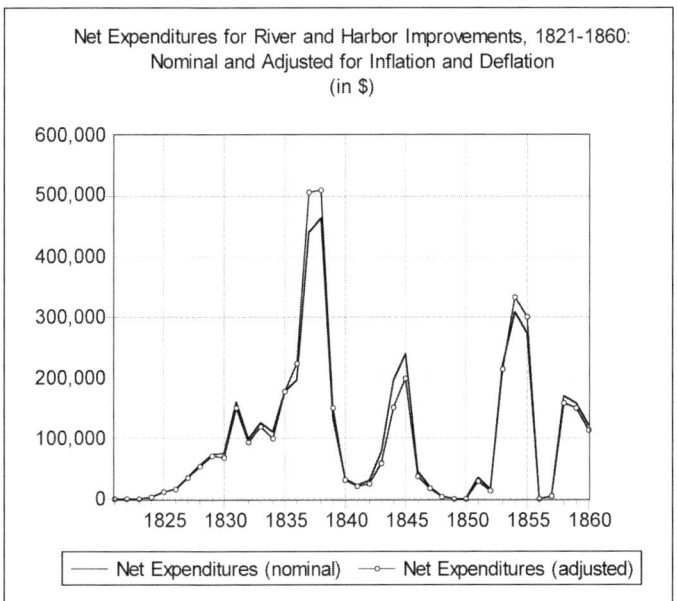

Figure C.3. *Source:* Derived from figure C.1.

APPENDIX D

Table D.1
Number of Wrecks on U.S. Waters, by Cause and Five-Year Period, 1821–1899

Cause	21–25	26–30	31–35	36–40	41–45	46–50	51–55	56–60	66–70[a]	71–75	76–80	81–85	86–90	91–95	96–99	1821–99 Total
Burned	2	9	18	20	30	85	78	78	160	45	17	8	14	12	6	582
Capsized	0	0	1	0	0	1	1	0	0	0	0	0	1	0	0	4
Collided	2	2	10	9	14	21	28	23	21	9	7	2	39	12	0	199
Exploded	1	3	4	9	12	16	20	16	23	8	0	0	1	0	0	113
Foundered	0	0	1	4	1	7	10	14	28	16	8	8	38	23	12	170
Ice	1	0	2	1	0	4	5	20	12	17	5	0	2	0	0	69
Snagged	16	17	27	43	86	79	93	102	106	17	6	1	0	0	0	593
Stranded	5	3	3	11	16	24	28	24	58	17	9	5	21	6	0	230
Unknown[b]	0	0	0	0	0	2	1	3	42	49	64	129	456	254	205	1,205
Total	27	34	66	97	159	239	264	280	450	178	116	153	572	307	223	3,165

Source: Steamboat Database; see appendix A.

[a] The number of steamboats lost during the Civil War is not included here, but the war's impact on vessel losses is suggested by the figures for the five-year period 1866–70, during which the incidence of wrecks due to almost every cause sharply increased. Much of the beneficial effect of prewar hazard removal was probably undone through neglect and the hazard presented by numerous vessels sunk or wrecked during the fighting on the rivers and coastal waters.

[b] Includes wrecks listed as "lost," "sank," due to "storm," due to "unknown" cause, and simply "wrecked."

Table D.2
Tonnage of Wrecks on U.S. Waters, by Cause and Five-Year Period, 1821–1899

Cause	21–25	26–30	31–35	36–40	41–45	46–50	51–55	56–60	66–70[a]	71–75	76–80	81–85	86–90	91–95	96–99	1821–99 Total
Burned	418	1,886	3,246	3,692	5,178	21,642	24,962	23,826	57,793	16,064	6,856	2,356	3,480	3,851	2,332	177,762
Capsized	0	0	0	0	0	65	183	0	0	0	0	0	0	0	0	248
Collided	341	261	1,783	1,424	2,697	5,234	6,960	5,402	5,188	4114	3,133	0	8,420	0	0	44,957
Exploded	295	588	450	1,219	2,064	3,626	4,881	7,305	5,733	1,314	0	0	563	0	0	28,038
Foundered	0	0	87	764	194	658	2,078	3,213	10,910	5,775	2,204	2,091	2,824	2,447	6,158	39,403
Ice	269	0	119	129	0	568	1,063	4,447	3,292	4,764	777	0	457	0	0	15,885
Snagged	2,984	3,110	3,887	5,699	16,218	13,557	19,248	21,649	23,965	5,537	1,352	148	0	0	0	117,344
Stranded	1,054	575	515	1,300	2,734	5,858	9,374	6,426	28,965	5,828	5,228	1,634	4,801	808	2,150	77,250
Unknown[b]	0	0	0	0	237	784	121	359	3,714	5,646	9,476	10,155	19,123	19,856	28,768	98,239
Total	5,361	6,420	10,077	14,227	29,322	51,992	68,870	72,627	139,560	49,042	29,026	16,564	39,668	26,962	39,408	599,126

Source: Steamboat Database; see appendix A.

[a] The tonnage of steamboats lost during the Civil War is not included here, but the war's impact on vessel losses is suggested by the figures for the five-year period 1866–70, during which the incidence of wrecks due to almost every cause sharply increased. Much of the beneficial effect of prewar hazard removal was probably undone through neglect and the hazard presented by numerous vessels sunk or wrecked during the fighting on the rivers and coastal waters.

[b] Includes wrecks listed as "lost," "sank," due to "storm," due to "unknown" cause, and simply "wrecked."

Table D.3
Side-wheel and Stern-wheel Steam Packets on the Western Rivers, 1841–1860[a] (by five-year periods)

	1841 1845	1846 1850	1851 1855	1856 1860
Number in use				
STW	7	8	75	96
SW	80	150	180	188
Total	87	158	255	284
% by rig				
STW	8	5	29	34
SW	92	95	71	66
Tonnage in use				
STW	947	944	14,713	14,781
SW	21,154	42,014	64,138	70,371
Total	22,101	42,958	78,851	85,152
Tonnage % by rig				
STW	4	2	19	17
SW	96	98	81	83
Mean tonnage in use				
STW	135	118	196	154
SW	264	280	356	374
Total	254	272	309	300
Tons Snagged				
STW	0	115	303	2,054
SW	341	1,979	2,652	5,502
Total	341	2,094	2,955	7,556
% Snagged				
STW	0	12	2	14
SW	2	5	4	8
Total	2	5	4	9

Source: Steamboat Packet Database, derived from Way, *Way's Packet Directory, 1848–1994.*

[a] Figures refer only to packets for which tonnage data are available in source.

APPENDIX D

Table D.4
Data on Steam Packet Engines

Year	Number in File[a]	Mean Tons	Mean C.I.D.[b]	Mean Stroke[c]	Mean Total Engine Volume[d]	Mean Engine Volume/Ton[e]
1834	1	98	27.0	5.5	37,769	385
1836	1	194	25.5	7.0	85,755	442
1840	1	473	25.0	8.0	94,200	199
1843	1	306	19.0	8.0	54,410	178
1844	5	297	23.4	8.1	92,751	302
1845	12	287	22.9	8.2	86,082	306
1846	7	244	20.9	7.3	65,273	252
1847	11	297	22.0	7.7	74,487	287
1848	13	319	22.7	7.7	78,197	255
1849	17	311	20.9	7.4	67,337	227
1850	9	392	25.6	7.5	123,807	260
1851	7	273	20.3	6.4	57,523	199
1852	20	420	22.6	7.3	87,943	189
1853	17	479	27.0	8.0	135,604	255
1854	18	429	23.0	6.9	81,224	179
1855	18	359	21.2	6.5	61,463	166
1856	24	370	21.2	6.7	61,873	167
1857	36	434	22.4	6.9	72,164	161
1858	13	404	20.3	6.4	56,972	141
1859	15	390	22.0	6.8	70,628	178
1860	21	386	20.9	6.4	61,717	151

Source: Steamboat Packet Database, derived from Way, *Way's Packet Directory, 1848–1994*. Although the title of Way's *Directory* indicates a beginning date of 1848, there are several entries for steam packets that were built prior to 1848.

[a] Number of steam packets for which engine data are given in the source.

[b] C.I.D. is the inside diameter, in inches, of the steam packet engine's cylinder.

[c] Stroke is the length of the piston stroke, given in feet.

[d] Mean total engine cylinder volume, in cubic inches, $= n[(\pi(C.I.D.2/)^2)(12(stroke))]$, where n = number of engines aboard.

[e] In cubic inches of engine volume per ton of vessel displacement.

APPENDICES

Table D.5
Public Land Sales Receipts in Selected States, 1814–1860 (in thousands of dollars)

Year	AL	AR	FL	IA	IL	IN	LA	MI	MO	MS	OH	WI	All States
1814	87				168						1,620[a]		1,889
1815	157				53	369				175	1,656[a]		2,410
1816	398				207	1,202				501	1,332		3,640
1817	1,718				572	1,080				298	1,416		5,084
1818	8,676				1,491	1,272		119	793	387	881		13,619
1819	4,148				661	458		52	2,461	293	958		8,981
1820	1,067				87	272		11	128	29	142		1,736
1821	378	4			64	363	140	9	123	42	155		1,278
1822	202	30			35	326	108	26	40	13	236		1,016
1823	181	2			76	203	4	38	105	41	158		808
1824	169	1			58	724	0	94	98	113	243		1,500
1825	318	10	91		82	223	5	136	108	115	194		1,282
1826	230	17	66		110	250	33	75	73	106	169		1,129
1827	163	5	189		81	263	143	55	200	90	215		1,404
1828	215	5	45		121	315	5	33	185	87	208		1,219
1829	597	3	68		282	490	35	90	192	127	277		2,161
1830	447	3	79		402	604	85	185	268	136	199		2,408
1831	842	17	36		420	713	72	403	372	49	442		3,366
1832	519	13	12		261	685	91	323	314	42	544		2,804
1833	565	52	15		381	692	111	563	297	801	695		4,172
1834	1,444	213	20		440	843	106	623	281	1,473	599	21	6,063
1835	1,986	700	65		2,688	2,078	572	2,272	824	3,836	828	317	16,166
1836	2,378	1,205	111		4,003	4,063	944	5,242	2,072	2,532	1,665	720	24,935
1837	478	354	126		1,271	1,571	289	969	820	321	590	153	6,942

218

APPENDIX D

1838	210	231	87	344	983	756	220	122	645	69	305	40	4,012
1839	159	193	72	373	1,421	778	848	175	1,307	25	316	822	6,489
1840	74	141	33	661	492	150	222	33	713	27	41	161	2,748
1841	70	70	8	93	440	120	120	28	339	31	64	128	1,511
1842	152	32	7	64	544	73	65	36	201	59	54	165	1,452
1843	226	60	10	179	520	64	139	17	555	46	19	214	2,049
1844	115	69	19	139	616	137	126	29	570	39	48	333	2,240
1845	98	46	25	264	611	99	94	34	315	36	289	551	2,462
1846	98	71	47	330	600	146	135	36	238	145	151	885	2,882
1847	187	126	35	344	615	348	140	67	303	113	195	800	3,273
1848	119	120	23	195	374	708	146	94	257	37	115	345	2,533
1849	181	78	18	123	319	277	92	63	209	44	77	260	1,741
1850	369	83	33	141	313	135	125	93	292	54	53	97	1,788
1851	361	194	44	250	421	150	133	193	610	44	81	92	2,573
1852	144	77	27	42	492	49	65	49	304	44	58	37	1,388
1853	352	217	69	1,015	1,218	104	182	331	803	212	106	393	5,002
1854	573	212	126	4,605	1,562	78	166	668	1,234	286	95	1,529	10,934
1855	369	137	93	4,310	897	6	171	410	1,539	157	14	2,039	10,142
1856	228	257	36	735	473	18	92	160	1,304	72	3	760	4,138
1857	62	547	22	588	155	11	95	45	973	82	2	121	2,703
1858	68	417	28	97	12	1	277	24	567	72	2	62	1,627
1859	176	492	32	18	12	1	266	37	325	198	5	52	1,614
1860	178	450	27	11	0	31	298	46	117	136	2	41	1,337
Total													192,650

Source: Arthur H. Cole, "Cyclical and Sectional Variations in the Sale of Public Lands, 1816–1860," *Review of Economic Statistics* 9, no. 1 (January 1927), table 2.

[a] Approximate figure, described in the source as "In a small degree estimated" (52n).

Table D.6
State-Specific Federal Appropriations and Expenditures for Rivers and Harbors: 1790–1860, 1861–1882, and 1790–1882

State	Appropriations			Expenditures		
	1790–1860	1861–1882	1790–1882	1790–1860	1861–1882	1790–1882
Alabama	275,752	816,000	1,091,752	235,143	720,999	956,142
Arkansas	0	316,500	316,500	0	315,000	315,000
California	30,000	1,687,000	1,717,000	30,000	1,463,429	1,493,429
Connecticut	167,714	1,418,213	1,585,927	153,824	1,373,625	1,527,449
Delaware	2,107,261	1,219,904	3,327,165	2,154,133	889,503	3,043,636
Florida	169,350	591,000	760,350	138,352	542,001	680,353
Georgia	393,964	1,018,633	1,412,597	349,774	1,014,290	1,364,064
Idaho Territory	0	10,000	10,000	0	10,000	10,000
Illinois	311,801	2,360,504	2,672,305	312,601	2,039,704	2,352,305
Indiana	156,204	630,000	786,204	156,199	630,000	786,199
Iowa	2,500	0	2,500	2,499	0	2,499
Kentucky	0	457,000	1,731,818	0	367,500	367,500
Louisiana	27,700	217,700	245,400	24,928	122,882	147,810
Maine	175,273	1,300,611	1,475,884	172,428	1,232,461	1,404,889
Maryland	185,500	1,546,318	1,731,818	185,486	1,300,284	1,485,770
Massachusetts	774,931	2,344,568	3,119,499	652,768	2,276,012	2,928,780
Michigan	472,916	7,511,961	7,511,961	471,195	7,357,161	7,828,356
Minnesota	0	447,500	447,500	0	447,500	447,500
Mississippi	46,000	331,900	377,900	25,675	269,500	295,175
Missouri	0	22,000	22,000	0	22,000	22,000

APPENDIX D

New Hampshire	10,000		202,500	10,000	165,500	175,500
New Jersey	52,063	192,500	1,099,063	43,430	944,067	987,497
New York	1,847,675	1,047,000	10,237,611	1,833,119	7,706,855	9,539,974
North Carolina	617,059	8,389,936	2,399,059	600,203	1,661,000	2,261,203
Ohio	620,252	1,782,000	3,116,147	618,028	2,239,003	2,857,031
Oregon	0	2,495,895	654,000	0	649,306	649,306
Pennsylvania	228,081	654,000	1,158,042	227,490	839,611	1,067,101
Rhode Island	34,700	929,961	733,700	34,636	698,977	733,613
South Carolina	54,000	699,000	963,000	53,000	878,342	931,342
Tennessee	0	909,000	85,000	0	85,500	85,000
Texas	29,500	85,500	2,566,200	27,528	2,138,606	2,138,606
Vermont	111,000	2,536,700	551,980	111,000	434,311	545,311
Virginia	48,580	440,980	1,734,880	48,075	1,635,300	1,683,375
Washington Territory	0	1,686,300	5,500	0	5,500	5,500
West Virginia	0	5,500	1,531,300	0	1,387,587	1,387,587
Wisconsin	146,564	1,531,300	4,659,542	145,719	4,470,777	4,616,496
District of Columbia	1,500	4,512,978	291,500	1,209	251,993	253,202
Grand Total	9,097,840	290,000	61,533,702	8,818,442	45,296,193	54,114,635
		52,435,862				

Source: Serial 1992.

Table D.7
Specific Federal Appropriations and Expenditures for Rivers and Harbors, 1790–1860: Specified by State (in current dollars)

State	River or Harbor	Appropriations ($)	Expenditures ($)
Alabama	Mobile Harbor	231,702	221,642
	Pass au Heron	24,050	13,501
	Total	**275,752**	**235,143**
California	San Diego Harbor	30,000	30,000
	Total	**30,000**	**30,000**
Connecticut	Bridgeport Harbor	20,000	19,857
	Cedar Point Beach	1,000	1,000
	Connecticut River	130	61
	Mill River	10,587	10,587
	New Haven Harbor	6,000	6,000
	Saybrook Harbor	39,182	23,421
	Southport Harbor	2,500	2,500
	Stonington Harbor	34,777	36,954
	Thames River	40,150	40,000
	Westport Harbor	13,388	13,444
	Total	**167,714**	**153,824**
Delaware	Delaware Breakwater	1,874,700	1,952,408
	New Castle Harbor	104,469	86,184
	Reedy Island Harbor	95,736	95,596
	Wilmington Harbor	32,356	29,945
	Total	**2,107,261**	**2,154,133**
Florida	Appalachicola River	42,250	32,192
	Chipola River	9,000	3,000
	Indian River	6,500	4,995
	Ochlawaha River	10,000	3,912
	Ochlochney River	5,000	5,000
	St. Augustine Harbor	33,570	33,473
	St. John's River	10,000	9,173
	St. Mark's River	37,530	36,011
	Suwannee River	15,000	10,154
	Yellow River	500	442
	Total	**169,350**	**138,352**
Georgia	Brunswick Harbor	10,000	44
	Savannah River	373,964	340,359
	Survey of rivers	10,000	9,371
	Total	**393,964**	**349,774**

APPENDIX D

State	River or Harbor	Appropriations ($)	Expenditures ($)
Illinois	Chicago Harbor	266,801	267,601
	Illinois River	30,000	30,000
	Waukegan Harbor	15,000	15,000
	Total	311,801	312,601
Indiana	Michigan City Harbor	156,204	156,199
	Total	156,204	156,199
Iowa	Des Moines and Iowa rivers, survey	1,000	999
	Red Cedar River, survey	1,500	1,500
	Total	2,500	2,499
Louisiana	Bayou La Fourche, survey	2,500	486
	Bayou Teche, survey	200	0
	Lake Pontchartrain Harbor	25,000	24,442
	Total	27,700	24,928
Maine	Belfast Harbor	1,200	1,034
	Cobscook Bay	5,300	4,174
	Kennebec River	18,520	18,152
	Kennebunk River	44,175	44,092
	Matincus Island breakwater survey	1,000	162
	Owl's Head Harbor	17,902	17,897
	Piscataqua River	8,510	8,500
	Penobscot River	300	297
	Portland Harbor	61,366	61,276
	Richmond Island Harbor	10,000	9,844
	Saco Harbor	7,000	7,000
	Total	175,273	172,428
Maryland	Baltimore Harbor	55,000	55,000
	Chesapeake Bay, survey of headwaters	500	486
	Patapsco River	120,000	120,000
	Susquehanna River	10,000	10,000
	Total	185,500	185,486
Massachusetts	Bass River	20,150	20,150
	Boston Harbor	411,526	277,970
	Edgartown Harbor	3,000	3,000
	Hyannis Harbor	75,932	75,858
	Marblehead Harbor	900	749
	Merrimac River	60,367	58,467
	Nantucket Harbor	45,835	40,924

(Table D.7, continued)

State	River or Harbor	Appropriations ($)	Expenditures ($)
	New Bedford Harbor	17,691	10,035
	Plymouth Beach and Harbor	57,267	57,177
	Provincetown Harbor	36,350	32,685
	Sandy Bay Breakwater	39,233	69,229
	Scituate Harbor	1,180	1,091
	Taunton River	3,000	2,950
	Wood's Hole Harbor	2,500	2,483
	Total	**774,931**	**652,768**
Michigan	Black Lake Harbor	8,000	7,999
	Clinton River	5,000	5,000
	Grand Haven Harbor	2,000	1,378
	La Plaisance Bay	19,803	19,694
	Monroe Harbor	124,000	123,980
	New Buffalo Harbor	8,000	7,752
	St. Clair Flats & Canal	65,000	65,000
	St. Joseph's Harbor	141,113	141,112
	St. Mary's River & St. Mary's Falls Canal	100,000	100,000
	Total	**472,916**	**471,195**
Mississippi	Pascagoula River	46,400	25,675
	Total	**46,000**	**25,675**
New Hampshire	Cocheco River	10,000	10,000
	Total	**10,000**	**10,000**
New Jersey	Cranberry Inlet	1,000	1,000
	Flat Beach	100	71
	Little Egg Harbor	23,500	15,048
	Newark Bay	12,000	11,876
	New Brunswick Harbor	13,963	13,941
	Shrewsbury River	1,500	1,494
	Total	**52,063**	**43,430**
New York	Black River	37,401	37,071
	Black Rock Harbor	52,098	52,095
	Buffalo Harbor	262,895	253,214
	Cattaraugus Creek	57,410	57,410
	Charlotte Harbor	178,871	177,637
	Dunkirk Harbor	123,446	123,288
	East River & Hell Gate	20,000	20,000
	Great Sodus Bay	163,620	163,620
	Hudson River	420,000	420,000
	Oak Orchard Harbor	30,500	30,500

APPENDIX D

State	River or Harbor	Appropriations ($)	Expenditures ($)
	Ogdensburg Harbor	3,000	2,085
	Oswego Harbor	271,488	271,472
	Plattsburg Harbor	60,180	60,180
	Port Jefferson Harbor	1,200	1,200
	Portland Harbor	56,616	56,442
	Port Ontario Harbor	50,000	49,663
	Sackett's Harbor	6,000	5,477
	Sag Harbor, survey	150	134
	Sandy Creek, survey	300	127
	Staten Island, ice-breaker construction	19,500	18,504
	Whitehall Harbor	33,000	13,000
	Total	1,847,675	1,833,119
North Carolina	Beaufort Harbor	5,000	5,000
	Cape Fear River	363,229	359,501
	Croatan Sound	50,000	38,089
	New River	50,000	48,834
	Ocracoke Inlet	133,750	133,732
	Pamlico & Tar rivers	10,000	10,000
	Pascotank River	80	47
	Washington Harbor	5,000	5,000
	Total	617,059	600,203
Ohio	Ashtabula Harbor	80,002	79,590
	Black River Harbor	68,205	68,179
	Cunningham Creek	19,781	19,779
	Cleveland Harbor	179,559	178,247
	Conneaut Harbor	58,306	58,175
	Grand River Harbor	74,598	74,595
	Huron River & Harbor	55,774	55,771
	Sandusky City Harbor	30,400	30,069
	Vermillion River & Harbor	53,627	53,623
	Total	620,252	618,028
Pennsylvania	Chester Harbor	10,100	9,921
	Erie Harbor	212,981	212,815
	Marcus Hook Harbor	5,000	4,754
	Total	228,081	227,490
Rhode Island	Church's Cove Harbor	28,200	28,195
	Providence River & Harbor	6,500	6,441
	Total	34,700	34,636

(Table D.7, continued)

State	River or Harbor	Appropriations ($)	Expenditures ($)
South Carolina	Charleston Harbor	50,000	50,000
	Georgetown Harbor, survey	4,000	3,000
	Total	**54,000**	**53,000**
Texas	Colorado River	20,000	19,776
	San Antonio River	1,500	1,455
	Survey of rivers and harbors	5,000	4,914
	Trinity River	3,000	1,383
	Total	**29,500**	**27,528**
Vermont	Burlington Harbor	90,000	90,000
	Hero Islands Channel	21,000	21,000
	Total	**111,000**	**111,000**
Virginia	Elizabeth River, survey	80	24
	James & Appomattox rivers	45,500	45,232
	Rappahannock River survey	3,000	2,819
	Total	**48,580**	**48,075**
Wisconsin	Kenosha Harbor	37,500	37,494
	Manitowoc Harbor	8,000	8,000
	Milwaukee Harbor	65,564	65,000
	Neenah River	2,500	2,345
	Racine Harbor	22,500	22,498
	Sheboygan Harbor	10,000	10,000
	Winnebago Lake	500	382
	Total	**146,564**	**145,719**
District of Columbia	Potomac River	1,500	1,209
	Total	**1,500**	**1,209**
Grand Total	**All rivers and harbors**	**$9,097,840**	**$8,818,442**

Source: Serial 1992.

Table D.8
Appropriations and Net Expenditures for Miscellaneous Improvements of Rivers and Harbors, prior to 1861 (in current dollars)

River System or Harbor	States Affected	Appropriations ($)	Expenditures ($)
Arkansas River	Arkansas Indian Territory Kansas	160,000	158,577
Chattahoochee and Flint rivers	Alabama Florida Georgia	7,000	0
Choctawhatchee River	Alabama Florida	15,000	12,877
Cumberland River	Kentucky Tennessee	155,000	153,980
Delaware River	Delaware New Jersey New York Pennsylvania	15,000	14
Dismal Swamp Canal entrance	North Carolina Virginia	35,000	34,962
Escambia River	Alabama Florida	10,500	5,000
Dubuque Harbor	Iowa	29,500	29,500
Des Moines Rapids	Iowa	300,000	257,328
Below the Rapids	Iowa	90,000	89,994
St. Louis	Missouri	75,000	51,665
Mississippi River, mouth	Louisiana	690,000	651,119
Mississippi Delta, survey	Louisiana	100,000	98,940
Mississippi and Missouri rivers, above mouth of Ohio	Illinois Iowa Kansas Minnesota Missouri Wisconsin	100,000	98,541
Mississippi and Ohio rivers, from mouth of Missouri River to New Orleans, and from Pittsburgh to Cairo	Arkansas Illinois Indiana Kentucky Louisiana	677,712	631,500

(Table D.8, continued)

River System or Harbor	States Affected	Appropriations ($)	Expenditures ($)
Mississippi, Missouri, Ohio rivers	Mississippi Ohio Pennsylvania Tennessee Virginia Arkansas Illinois Indiana Iowa Kansas Kentucky Louisiana Minnesota Mississippi Nebraska Ohio Pennsylvania Tennessee Virginia Wisconsin	223,000	222,924
Mississippi, Missouri, Ohio, Arkansas rivers	Arkansas Illinois Indiana Iowa Kansas Kentucky Louisiana Minnesota Mississippi Nebraska Ohio Pennsylvania Tennessee Virginia Wisconsin	430,000	429,937
Missouri River	Iowa Kansas Missouri Nebraska	40,000	40,000

APPENDIX D

River System or Harbor	States Affected	Appropriations ($)	Expenditures ($)
Falls of Ohio River and Louisville Canal Ohio River	Kentucky Ohio	5,000	4,261
	Illinois Indiana Kentucky Ohio Pennsylvania Virginia	406,479	401,456
Red River of the South	Arkansas Louisiana Texas	535,766	532,220
Rock River	Illinois Wisconsin	1,000	935
St. John's and St. Mary's rivers	Florida Georgia	78,000	69,958
Tennessee River	Alabama Kentucky Tennessee	52,957	52,762
Wabash River	Illinois Indiana	500	500
White, Black, and St. Francis rivers	Arkansas Missouri	2,500	1,623
Great Lakes harbors, repairs	Generic	20,000	19,929
Harbor and River improvements, preservation, repair	Generic	85,000	84,269
Atlantic Coast, repairs and contingencies	Generic	10,000	8,658
Transportation, fuel, etc.	Generic	12,127	11,278
Great Lakes, hydrographic surveys	Generic	640,000	640,820
Great Lakes and Bays dredging machines	Generic	122,683	120,356
Great Lakes, build iron steamer	Generic	50,000	50,000
Great Lakes, charts, etc.	Generic	48,000	29,995
Western Rivers, snag and dredge boats	Generic	150,000	149,494
All rivers and harbors		5,372,724	5,144,372

Source: Serial 1992.

Note: Dollar amounts refer to the entire river system or harbor listed.

Table D.9
Appropriations and Net Expenditures for Miscellaneous Improvements of Rivers and Harbors, 1861–1882 (in current dollars)

River System or Harbor	States Affected	Appropriations ($)	Expenditures ($)
Arkansas River	Arkansas Indian Territory Kansas	107,000	101,000
Chattahoochee and Flint rivers	Alabama Florida Georgia	138,000	133,000
Choctawhatchee River	Alabama Florida	37,000	37,000
Cumberland River	Kentucky Tennessee	466,000	466,500
Delaware River	Delaware New Jersey New York Pennsylvania	1,051,000	963,000
Dismal Swamp Canal entrance	North Carolina Virginia	0	0
Escambia River	Alabama Florida	13,000	13,000
Dubuque Harbor	Iowa	41,000	33,000
Des Moines Rapids	Iowa	4,088,500	4,099,700
Below the Rapids	Iowa	0	0
St. Louis	Missouri	89,600	29,600
Mississippi River, mouth	Louisiana	1,525,000	1,559,874
Mississippi Delta, survey	Louisiana	25,000	25,576
Mississippi and Missouri rivers, above mouth of Ohio	Illinois Iowa Kansas Minnesota Missouri Wisconsin	0	0
Mississippi and Ohio rivers, from mouth of Missouri River to New Orleans, and from Pittsburgh to Cairo	Arkansas Illinois Indiana Kentucky Louisiana	0	0

APPENDIX D

River System or Harbor	States Affected	Appropriations ($)	Expenditures ($)
Mississippi, Missouri, Ohio rivers	Mississippi Ohio Pennsylvania Tennessee Virginia Arkansas Illinois Indiana Iowa Kansas Kentucky Louisiana Minnesota Mississippi Nebraska Ohio Pennsylvania Tennessee Virginia Wisconsin	0	0
Mississippi, Missouri, Ohio, Arkansas rivers	Arkansas Illinois Indiana Iowa Kansas Kentucky Louisiana Minnesota Mississippi Nebraska Ohio Pennsylvania Tennessee Virginia Wisconsin	2,135,000	2,055,000
Missouri River	Iowa Kansas Missouri Nebraska	1,261,500	1,260,500

(Table D.9, continued)

River System or Harbor	States Affected	Appropriations ($)	Expenditures ($)
Falls of Ohio River and Louisville Canal	Kentucky Ohio	1,286,563	1,286,563
Ohio River	Illinois Indiana Kentucky Ohio Pennsylvania Virginia	2,475,000	2,340,000
Red River of the South	Arkansas Louisiana Texas	762,500	677,500
Rock River	Illinois Wisconsin	0	0
St. John's and St. Mary's rivers	Florida Georgia	0	0
Tennessee River	Alabama Kentucky Tennessee	2,178,500	2,178,695
Wabash River	Illinois Indiana	380,000	380,000
White, Black, and St. Francis rivers	Arkansas Missouri	0	0
Great Lakes harbors, repairs	Generic	250,000	250,000
Harbor and River improvements, preservation, repair	Generic	0	0
Atlantic Coast, repairs and contingencies	Generic	100,000	101,342
Transportation, fuel, etc.	Generic	0	546
Great Lakes, hydrographic surveys	Generic	2,290,379	2,267,030
Great Lakes and Bays, dredging machines	Generic	0	0
Great Lakes, build iron steamer	Generic	0	0
Great Lakes, charts, etc.	Generic	60,000	61,676
Western Rivers, snag and dredge boats	Generic	646,000	645,960
All rivers and harbors		5,372,724	5,144,372

Source: Serial 1992.

Note: Dollar amounts refer to the entire river system or harbor listed.

Table D.10
Appropriations and Expenditures for Lighthouses, Light Stations, Beacons, and Buoys in Each State, 1791–1860 (current dollars)

State	Appropriations	Expenditures
Alabama	108,520	91,382
California	388,789	132,138
Connecticut	281,672	240,339
Delaware	520,456	355,792
Florida	1,243,140	1,026,946
Georgia	275,293	141,985
Illinois	183,992	82,202
Indiana	2,300	8,512
Louisiana	730,235	443,661
Maine	1,302,721	600,193
Maryland	280,665	191,485
Massachusetts	715,845	932,943
Michigan	527,544	337,773
Minnesota	15,000	12,202
Mississippi	147,126	73,709
New Hampshire	97,578	49,623
New Jersey	573,708	320,158
New York	1,018,771	723,530
North Carolina	729,785	486,876
Ohio	189,998	145,668
Oregon	32,000	26,997
Pennsylvania	52,238	40,986
Rhode Island	231,508	201,504
South Carolina	393,666	270,585
Texas	217,000	141,430
Vermont	14,800	14,014
Virginia	459,588	277,730
Washington Territory	106,000	68,245
Wisconsin	188,100	69,314
All states	11,048,737	7,507,922

Source: Serial 1992.

Table D.11
Number of Memorials Concerning Internal Improvements Sent by Each State to the 19th–36th Congresses, 1825–1861

State	\multicolumn{19}{c}{Congress Number}

State	19	20	21	22	23	24	25	26	27	28	29	30	31	32	33	34	35	36	Total
AL												1			1				2
AR						1					3	2							6
CA														1				1	2
DC				1															1
FL											2								2
GA				1															1
IA							1												1
IL							1	3		1	2	4							11
IN	1					4	3	2	7	1	1								19
KY				1															1
LA								1						1					2
MA														1		1			2
MD							3	1		1	1								6
ME									1										1
MI						1	2	3						1		1	3		11
MN																		1	1
MO					1			1		1		1							4
MS								1			3	1		1		1			7
NC						1								2		2			5
NE																	1		1

APPENDIX D

State																			All
NH				1															1
NJ				1					1	1						4			7
NM															1				1
NY						3	1	1											5
OH						1	1			1	1					2			6
PA			3			1		1			1		1			1			7
TN								1		1									2
TX								1											1
VA									1										1
WI						4	1		2		3								10
Other	1						1	2	2				1			2		1	9
All	2	0	0	5	4	9	19	15	12	9	14	13	7	7	3	12	4	2	137

Sources: House and Senate Documents, serial set; Gales and Seaton's *Register of Debates in Congress*; and *Congressional Globe and Appendix*.

Note: "NM" denotes the New Mexico Territory; "Other" includes one memorial from Charles Ellet and eight of unknown origin.

Table D.12
Maverick Members of the House of Representatives in the 29th–32nd Congresses, 1845–1853

Congress	Congressman	Party	State	District	Water Interest[a]	Scaling Score[b]
29th	Smith	D	IL	1	R	2
	Wentworth	D	IL	4	L	2
29th	Cathcart	D	IN	9	R	3
	Henley	D	IN	2	R	2
	Wick	D	IN	4	R/N	4
29th	Bell	W	KY	4	N	18
	Tibbatts	D	KY	10	R	2
	Young	D	KY	5	R	0
29th	Dunlap	D	ME	2	R/N	8
29th	Constable	D	MD	5	R	5
	Giles	D	MD	4	R	3
29th	Chipman	D	MI	2	L/R	5
	Hunt	D	MI	3	L/R	7
	McClelland	D	MI	1	L/R	3
29th	Bowlin	D	MO	n/a	GT	2
	Price	D	MO	n/a	GT	12
	Relfe	D	MO	n/a	GT	2
29th	Sykes	D	NJ	2	R	3
29th	De Mott	D	NY	27	L	5
	Ellsworth	D	NY	26	N	3
29th	Goodyear	D	NY	21	C	5
	Hough	D	NY	23	L	7
	Hungerford	D	NY	19	L	10
	Jenkins	D	NY	20	C	5
	King	D	NY	8	L	5
	Lawrence	D	NY	1	LIS	9
	Niven	D	NY	9	R/N	6
	Rathbun	D	NY	25	C	5
	Strong	D	NY	22	N	4
	Wood	D	NY	13	C/N	8
	Woodruff	D	NY	5	R	2
	Woodworth	D	NY	8	R	3
29th	Baringer	W	NC	2	N	19
	Dockery	W	NC	4	N	19
	Graham	W	NC	1	R/N	19

APPENDIX D

Congress	Congressman	Party	State	District	Water Interest[a]	Scaling Score[b]
29th	Brinkerhoff	D	OH	11	R/N	2
	Cunningham	D	OH	2[c]	N	0
	Faran	D	OH	1	R	2
	Fries	D	OH	17	R	3
	McDowell	D	OH	7	R	2
	Morris	D	OH	15	R	2
	Parrish	D	OH	13	R	3
	St. John	D	OH	6	L	6
	Sawyer	D	OH	5	L	4
	Starkweather	D	OH	18	C	2
	Thurman	D	OH	8	R	2
29th	Brodhead	D	PA	10	R/N	4
	Foster	D	PA	19	C	4
	Ingersoll	D	PA	4[d]	R	0
	Leib	D	PA	11	C	5
	Thompson	D	PA	23	L	3
29th	Brown	W	TN	11	R/N	14
	Crozier	W	TN	3	N	17
	Stanton	W	TN	10	R	11
30th	Smith	D	IL	1	R	5
	Turner	D	IL	6	R	2
	Wentworth	D	IL	4	R	0
30th	Cathcart	D	IN	9	L	4
	Henley	D	IN	2	R	5
	Pettit	D	IN	8	N	6
	Rockhill	D	IN	10	N	6
30th	Leffler	D	IA	2	R	2
30th	Bingham	D	MI	3	L	2
	McClelland	D	MI	1	L	2
	Stuart	D	MI	2	L	2
30th	Bowlin	D	MO	1	R	2
30th	Edsall	D	NJ	3	R	4
30th	Birdsall	D	NY	22	N	3
	Collins	D	NY	18	R/L	2
	Jenkins	D	NY	20	C	2
	Lawrence	D	NY	15	N	2
	Maclay	D	NY	4	R/H	3
	Nicoll	D	NY	3	R/H	3

(Table D.12, continued)

Congress	Congressman	Party	State	District	Water Interest[a]	Scaling Score[b]
	Lord	W	NY	1	LIS	10
	St. John	D	NY	9	R	1
	Starkweather	D	NY	2	C	2
30th	Dickinson	D	OH	6	L	2
	Faran	D	OH	1	R	4
	Lahm	D	OH	18	C	6
	Morris	D	OH	7	R	2
	Ritchey	D	OH	13	R	6
30th	Ingersoll, C.	D	PA	4	R	3
	Ingersoll, J.	D	PA	2	R	0
	Thompson	D	PA	23	L	2
30th	Thurston	D	RI	2	H	2
30th	Cocke	D	TN	2	N	0
30th	Peck	D	VT	4	N	2
30th	Darling	D	WI	1	L	2
	Lynde	D	WI	2	L	2
31st	Bowden	D	AL	7	R	0, 9
	Cobb	D	AL	6	R	0, 8
	Harris	D	AL	3	N	0, _
	Hubbard	D	AL	5	R	0, _
	Inge	D	AL	4	N	1, 8
31st	Johnson	D	AR	1	R	0, 1
31st	Gilbert	D	CA	1	H	0, 5
31st	Cleveland	D	CT	3	LIS	4, 8
31st	Stephens	W	GA	7	N	3, 9
	Jackson	D	GA	1	R/O	0, 8
	Toombs	W	GA	8	R	_, 8
31st	Bissell	D	IL	1	R	0, _
	Harris	D	IL	7	N	0, 4
	McClernand	D	IL	2	R	0, _
	Richardson	D	IL	5	R	0, 4
	Wentworth	D	IL	4	L	1, 1
	Young	D	IL	3	R	0, 8
31st	Albertson	D	IN	1	R	0, 9
	Brown	D	IN	5	N	1, 9
	Dunham	D	IN	2	R	2, 8
	Fitch	D	IN	9	L	0, 1
	Gurman	D	IN	6	R	0, 8

APPENDIX D

Congress	Congressman	Party	State	District	Water Interest[a]	Scaling Score[b]
	Harlan	D	IN	10	N	0, _
	Robinson	D	IN	3	R	0, 8
31st	Leffler	D	IA	2	R	0, 0
31st	Marshall	W	KY	7	R	9, 2
	Mason	D	KY	9	R	2, _
	Stanton	D	KY	10	R	0, 9
31st	La Sere	D	LA	1	R/G	0, 9
	Morse	D	LA	4	R/G	1, 9
31st	Evans	W	MD	5	B	7, 0
	Kerr	W	MD	6	B	7, 0
	McLane	D	MD	4	B	0, 0
31st	Rockwell	W	MA	7	N	7, 0
31st	Bingham	D	MI	3	L	0, 1
	Buel	D	MI	1	L	0, 0
31st	Brown	D	MS	4	R/G	0, 8
	Featherston	D	MS	2	R	0, 8
	McWillie	D	MS	3	R	3, 8
31st	Bay	D	MO	2	R	0, _
	Bowlin	D	MO	1	R	0, 4
	Green	D	MO	3	R	0, _
	Hall	D	MO	4	R	0, 8
	Phelps	D	MO	5	R	0, 7
31st	Alexander	W	NY	17	N	7, 0
	Clarke	W	NY	19	L	7, 1
	Jackson	W	NY	26	C	9, 0
	Nelson	W	NY	7	R	7, 0
	Reynolds	W	NY	12	R	7, 1
	Rumsey	W	NY	30	C	7, 0
	Walden	D	NY	21	C	1, 1
31st	Clingman	D	NC	1	N	3, 0
	Deberry	W	NC	3	N	9, 8
	Outlaw	W	NC	9	O	7, 1
31st	Campbell	W	OH	2	N	9, 1
	Cartter	D	OH	18	C	9, 2
	Evans	W	OH	14	N	7, 1
	Hoagland	D	OH	16	N	0, 8
	Hunter	W	OH	15	R	8, 2
	Morris	D	OH	7	R	9, 2

(Table D.12, continued)

Congress	Congressman	Party	State	District	Water Interest[a]	Scaling Score[b]
	Potter	D	OH	5	L	0, _
	Root	W	OH	21	L	7, 1
	Sweetser	D	OH	10	N	_, 1
	Vinton	W	OH	12	R	7, 0
	Whittlesey	D	OH	13	R	0, 7
	Wood	D	OH	6	L	3, _
31st	Butler	W	PA	11	N	7, _
	Danner	W	PA	15	R	3, 8
	Dickey	W	PA	7	R	7, 0
	Freedley	W	PA	5	R	4, 9
	Moore	W	PA	3	R	7, 0
	Ogle	W	PA	18	N	7, 1
	Reed	W	PA	20	R	7, 1
	Stevens	W	PA	8	R	7, _
31st	Dixon	W	RI	2	B	7, _
31st	McQueen	D	SC	4	O	0, 8
	Orr	D	SC	2	R	1, 8
31st	Ewing	D	TN	8	R	1, 8
	Stanton	D	TN	10	R	0, 0
	Watkins	D	TN	2	N	7, 1
31st	Howard	D	TX	2	R/G	1, 0
	Kaufman	D	TX	1	R/G	3, _
31st	Hebard	W	VT	2	R	7, _
	Henry	W	VT	1	R	7, 1
	Meacham	W	VT	3	L	7, 1
31st	Morton	W	VA	9	R	5, 9
31st	Doty	D	WI	3	L	0, 1
32nd	Harris	D	AL	3	N	1
32nd	Johnson	D	AR	1	R	0
32nd	McCorkle	D	CA	2	O	0
32nd	Riddle	D	DE	1	B/R	11
32nd	Allen	D	IL	2	R	0
	Bissell	D	IL	1	R	1
	Campbell	D	IL	6	R	6
	Ficklin	D	IL	3	N	0
	Molony	D	IL	4	L	0
	Richardson	D	IL	5	R	0
32nd	Davis	D	IN	7	N	8

APPENDIX D

Congress	Congressman	Party	State	District	Water Interest[a]	Scaling Score[b]
	Fitch	D	IN	9	L	1
	Gorman	D	IN	6	R	5
	Hendricks	D	IN	5	N	2
	Lockhart	D	IN	1	R	0
	Mace	D	IN	8	N	0
32nd	Clark	D	IA	2	R	0
	Henn	D	IA	1	R	0
32nd	Breckinridge	D	KY	8	R	11
	Ewing	W	KY	3	N	19
	Stanton	D	KY	10	R	1
	Stone	D	KY	5	R	3
32nd	Penn	D	LA	3	R	1
	St. Martin	D	LA	1	R/G	1
32nd	Appleton	D	ME	2	R/O	3
	Washburn	W	ME	7	O	16
32nd	Rantoul	D	MA	2	O	0
32nd	Stuart	D	MI	2	L	2
32nd	Brown	D	MS	4	O	0
32nd	Hall	D	MO	4	R	10
	Phelps	D	MO	5	R	0
32nd	Price	D	NJ	5	R	4
32nd	Bennett	W	NY	22	N	17
	Boyd	W	NY	14	N	16
	Hascell	W	NY	33	C	14
	Martin	W	NY	31	L	18
	Schoonmaker	W	NY	10	N	16
	Seymour	D	NY	12	N	10
	Stephens	D	NY	7	R	1
32nd	Caldwell	W	NC	2	N	19
	Clingman	D	NC	1	N	5
	Morehead	W	NC	4	N	16
32nd	Cable	D	OH	17	R	7
	Cartter	D	OH	18	C	4
	Disney	D	OH	1	R	2
	Edgerton	D	OH	5	L	4
	Gaylord	D	OH	13	R	11
	Green	D	OH	6	L	2
	Hunter	W	OH	15	R	14

(Table D.12, continued)

Congress	Congressman	Party	State	District	Water Interest[a]	Scaling Score[b]
	Olds	D	OH	9	N	1
	Sweetser	D	OH	10	N	11
	Townshend	D	OH	21	L	4
32nd	Allison	W	PA	20	R	16
	Curtis	D	PA	23	L	13
	Florence	D	PA	1	R	1
	Gilmore	D	PA	24	C	10
	McNair	D	PA	5	R	10
32nd	Thurston	D	RI	2	B	13
32nd	Polk	D	TN	6	R	8
	Stanton	D	TN	10	R	3
32nd	Howard	D	TX	2	R/G	0
	Scurry	D	TX	1	R/G	0
32nd	Doty	D	WI	3	L	0
	Eastman	D	WI	2	R/L	0

Sources: All columns, except "District" and "Water Interest," are from Joel Silbey, *The Shrine of Party: Congressional Voting Behavior, 1841–1852* (Pittsburgh: University of Pittsburgh Press, 1967), appendix 2, "Congressmen's Scale Positions on All Issues, 1841–1852," 29th–32d Congresses. Districts are from Kenneth C. Martis, *The Historical Atlas of Political Parties in the United States Congress, 1789–1989* (New York and London: Macmillan, 1989). I have derived "Water Interest" from a variety of topographical and geophysical maps and from Martis, *Historical Atlas of Political Parties,* passim.

[a] "Water Interest" symbols for 29th–32nd Congresses are B, Bay; C, canal; G, Gulf of Mexico; GT, general territory; H, ocean harbor; L, one of the Great Lakes; LIS, Long Island Sound; N, no significant water body or course; O, ocean coast; R, river, generally one of those slated for improvement under the federal program.

[b] Scores are those given by Silbey on the internal improvements roll-call votes:

Congress	Pro-	Moderate support	Anti-
29th	0–5	6–12	13–19
30th	0–2	3–6	7–10
31st	0–2	3–6	7–9
32nd	0–5	6–8, 10–13	14–19

[c] District was near Cincinnati.

[d] District embraced Philadelphia.

Table D.13
Maverick Members of the House of Representatives in the 29th Congress, 1845–1847, on the Issues of the Tariff and Internal Improvements

Congressman	Party	State	District	Water Interest[a]	Improve Score[b]	Tariff Score[c]
Smith	D	IL	1	R	2	3
Wentworth	D	IL	4	L	2	—
Cathcart	D	IN	9	R	3	0
Henley	D	IN	2	R	2	0
Wick	D	IN	4	R/N	4	0
Bell	W	KY	4	N	18	7
Tibbatts	D	KY	10	R	2	3
Young	D	KY	5	R	0	7
Dunlap	D	ME	2	R/N	8	2
Constable	D	MD	5	R	5	1
Giles	D	MD	4	R	3	4
Chipman	D	MI	2	L/R	5	1
Hunt	D	MI	3	L/R	7	1
McClelland	D	MI	1	L/R	3	2
Bowlin	D	MO	n/a	GT	2	2
Price	D	MO	n/a	GT	12	—
Relfe	D	MO	n/a	GT	2	2
Sykes	D	NJ	2	R	3	7
De Mott	D	NY	27	L	5	1
Ellsworth	D	NY	26	N	3	1
Goodyear	D	NY	21	C	5	1
Hough	D	NY	23	L	7	1
Hungerford	D	NY	19	L	10	7
Jenkins	D	NY	20	C	5	7
King	D	NY	8	L	5	1
Lawrence	D	NY	1	LIS	9	1
Niven	D	NY	9	R/N	6	—
Rathbun	D	NY	25	C	5	6
Strong	D	NY	22	N	4	1
Wood	D	NY	13	C/N	8	1
Woodruff	D	NY	5	R	2	7
Woodworth	D	NY	8	R	3	—
Baringer	W	NC	2	N	19	7
Dockery	W	NC	4	N	19	7
Graham	W	NC	1	R/N	19	7
Brinkerhoff	D	OH	11	R/N	2	5

(Table D.13, continued)

Congressman	Party	State	District	Water Interest[a]	Improve Score[b]	Tariff Score[c]
Cunningham	D	OH	2[d]	N	0	0
Faran	D	OH	1	R	2	2
Fries	D	OH	17	R	3	1
McDowell	D	OH	7	R	2	0
Morris	D	OH	15	R	2	0
Parrish	D	OH	13	R	3	1
St. John	D	OH	6	L	6	4
Sawyer	D	OH	5	L	4	—
Starkweather	D	OH	18	C	2	4
Thurman	D	OH	8	R	2	2
Brodhead	D	PA	10	R/N	4	7
Foster	D	PA	19	C	4	7
Ingersoll	D	PA	4[e]	R	0	7
Leib	D	PA	11	C	5	7
Thompson	D	PA	23	L	3	7
Brown	W	TN	11	R/N	14	7
Crozier	W	TN	3	N	17	7
Stanton	W	TN	10	R	11	0
Dillingham	D	VT	4	N	7	7

Sources: All columns, except "District" and "Water Interest," are from Joel Silbey, *The Shrine of Party: Congressional Voting Behavior, 1841–1852* (Pittsburgh: University of Pittsburgh Press, 1967), appendix 2, "Congressmen's Scale Positions on All Issues, 1841–1852," 29th Congress. Districts are from Kenneth C. Martis, *The Historical Atlas of Political Parties in the United States Congress, 1789–1989* (New York and London: Macmillan, 1989). I have derived "Water Interest" from a variety of topographical and geophysical maps and from Martis, *Historical Atlas of Political Parties,* passim.

Note: Bold, italic type denotes congressmen whose voting behavior on both issues was unorthodox, in terms of their respective parties' positions.

[a] "Water Interest" symbols are C, canal; GT, general territory; L, one of the Great Lakes; LIS, Long Island Sound; N, no significant water body or course; R, river, generally one of those slated for improvement under the federal program.

[b] Scores are those given by Silbey on the internal improvements roll-call votes: 0–5, pro-; 6–12; moderate support; and 13–19, anti-. See Silbey, *The Shrine of Party,* table 5.7.

[c] Scores are those given by Silbey on the tariff roll-call votes: 0–2 indicates support for a low tariff; 3–5, support for a moderate tariff; and 6–7, support for a high tariff. See Silbey, *The Shrine of Party,* table 5.1.

[d] District was near Cincinnati.

[e] District embraced Philadelphia.

APPENDIX D

Table D.14
Regional Origin of Memorials Concerning Improvements Received by the 19th–36th Congresses, 1825–1861

Congress and Years		New England	Middle Atlantic	South Atlantic	Gulf Coast	Miss. Valley	Far West	Misc.	Total
19th	1825–27	0	0	0	0	1	0	1	2
20th	1827–29	0	0	0	0	0	0	0	0
21st	1829–31	0	0	0	0	0	0	0	0
22nd	1831–33	2	0	1	0	1	0	1	5
23rd	1833–35	0	3	0	0	1	0	0	4
24th	1835–37	0	2	1	0	6	0	0	9
25th	1837–39	0	4	3	0	12	0	0	19
26th	1839–41	0	0	1	0	13	0	1	15
27th	1841–43	1	1	1	0	7	0	2	12
28th	1843–45	0	0	1	0	6	0	2	9
29th	1845–47	0	1	1	2	10	0	0	14
30th	1847–49	0	2	0	1	10	0	0	13
31st	1849–51	0	1	0	1	3	0	2	7
32nd	1851–53	1	0	2	0	3	1	0	7
33rd	1853–55	0	0	0	1	0	1	1	3
34th	1855–57	1	5	2	0	4	0	0	12
35th	1857–59	0	0	0	0	3	1	0	4
36th	1859–61	0	0	0	0	1	1	0	2
All	1825–61	5	19	13	5	81	4	10	137

Sources: House and Senate documents in the serial set; *Gales and Seaton's Register* and the *Congressional Globe*. For the number of memorials from each state to each Congress, see table D.11.

Note: Although the regional groupings of states used here generally conform to current practice, departures from that practice reflect the political and economic circumstances of the antebellum period. Thus, while Mississippi and Louisiana lie on the Gulf of Mexico, the greatest part of their shipping occurred along the Mississippi River and its tributaries. For that reason, both states are grouped within the "Mississippi Valley and Great Lakes" region, which embraces the states along the western Great Lakes, the Mississippi, Ohio, and Missouri rivers (lower reaches), and their major tributaries. The absence of a state from a regional grouping, that is, Rhode Island from "New England" and South Carolina from "South Atlantic," indicates only that no memorial from either state was recorded. States within each region are New England (Maine, New Hampshire, Vermont, Massachusetts, Connecticut); Middle Atlantic (New York, New Jersey, Pennsylvania, Delaware); South Atlantic (Maryland, North Carolina, Virginia, Georgia); Gulf South (Alabama, Florida, Texas); Mississippi Valley and Great Lakes (Minnesota, Wisconsin, Michigan, Iowa, Illinois, Indiana, Ohio, Kentucky, Tennessee, Missouri, Arkansas, Mississippi, Louisiana); Far West (Nebraska, New Mexico Territory, California); Miscellaneous (Washington, D.C., or a memorial from an individual or from an unknown source).

Table D.15
Maverick Voting on Internal Improvements in the House of Representatives in the 29th–32d Congresses, 1845–1853

Congress No. and Years		Maverick Dems.		Maverick Whigs		Maverick Dems. as % all Dems.	Maverick Whigs as % all Whigs
		Slave States	Free States	Slave States	Free States		
29th	1845–47	7/69 (10)	40/78 (51)	7/22 (32)	0/58 (0)	47/147 (32)	7/80 (9)
30th	1847–49	2/55 (4)	32/54 (59)	0/36 (0)	1/83 (1)	34/108 (31)	1/119 (1)
31st	1849–51 vote #1	28/62 (45)	23/51 (45)	6/29 (21)	22/85 (26)	51/112 (46)	28/109 (26)
31st	1849–51 vote #2	7/62 (11)	11/51 (22)	7/29 (24)	2/85 (2)		
32nd	1851–53	16/59 (27)	37/74 (50)	3/34 (9)	8/60 (13)	53/140 (38)	11/88 (13)

Sources: For each party's total number of members in each Congress, see Kenneth C. Martis, *The Historical Atlas of Political Parties in the United States Congress, 1789–1989* (New York and London: Macmillan, 1989), 98–102. For all other figures, and also for classifications of types of water interests used to determine maverick behavior, see table D.12.

Note: Ratios indicate the number of maverick members of a party to the total number of members of that party in a Congress. Figures in parentheses express these ratios as percentages.

Table D.16
Maverick Voting on Internal Improvements and a Protective Tariff in the House of Representatives in the 29th Congress, 1845–1847

	Political Party		
	Democrats	Whigs	Both
Maverick members	47	7	54
Total No. members	143	77	220
% mavericks	33	9	25
Mavericks with water interests	28	1	29
% mavericks with water interests	60	14	54
Mavericks from slave states	7	7	14
% mavericks from slave states	15	100	26
Mavericks on both issues	18	1	19
% mavericks on both issues	38	14	35
Mavericks on improvements only	29	6	35
% mavericks on improvements only	62	86	65

Sources: For each party's total number of members, see Harold W. Stanley and Richard G. Niemi, *Vital Statistics on American Politics, 1999–2000: A Comprehensive Reference of over 200 Tables and Figures* (Washington, D.C.: CQ Press, 2000), table 1–10. For all other figures, and also for classifications of types of water interests, see table D.13.

Table D.17
Major Sources of Federal Revenue in Each Presidential Administration[a]

President	Period	Average Annual Receipts ($)	% from Customs	% from Land Sales	Loans, Notes, etc. as a % of Avg. Annual Receipts
Washington	1789–1797	6,609,081	55	.01	38
Adams, John	1797–1801	10,510,158	75	.23	17
Jefferson	1801–1809	14,200,711	92	3.02	0.1
Madison	1809–1817	30,025,699	42	3.28	46
Monroe	1817–1825	23,180,338	79	7.78	7
Adams, J. Q	1825–1829	24,957,821	87	5.13	5
Jackson	1829–1837	31,507,671	74	23.07	0
Van Buren	1837–1841	31,454,168	51	16.08	20
Harrison/Tyler	1841–1845	29,318,547	56	4.83	37
Polk	1845–1849	41,996,955	65	6.31	28
Taylor/Fillmore	1849–1853	53,132,608	73	3.70	20
Pierce	1853–1857	68,698,750	87	11.10	< 2
Buchanan	1857–1861	53,336,200	93	4.45	< 3

Sources: Data for the administrations of Washington through Taylor and Fillmore, inclusive, are derived from a table, "Statement of the Receipts of the United States from 1789 to 1851, Inclusive. Annual Average during Each Administration," in the section "Journal of Banking, Currency, and Finance," *Hunt's Merchants' Magazine* 27, no. 4, October 1852, 480; data for the Pierce and Buchanan administrations are from *Historical Statistics of the United States from Colonial Times to 1970*, electronic ed., chapter Y: Government Series Y 352–357. "Federal Government Receipts—Administrative Budget, 1789–1939."

[a] All amounts are in current dollars, that is, unadjusted for inflation or deflation.

Table D.18
Major Objects of Federal Expenditures in Each Presidential Administration[a]

President	Period	Government, Foreign, Misc.	Army	Navy	Total Expenditures[b]	Public Debt Paid	Expenditures less Debt Paid	Debt as % of Total Expends.	Army + Navy as % of Net Expends.	Total Expends.
Washington	1789–1797	633,152	1,105,503	106,768	6,497,957	4,511,620	1,986,337	69.4	61.0	18.7
Adams, John	1797–1801	1,186,237	2,019,188	2,017,694	10,076,578	4,739,491	5,337,088	47.0	75.6	40.1
Jefferson	1801–1809	2,049,058	1,334,529	1,535,665	13,285,898	8,148,300	5,137,599	61.3	55.9	21.6
Madison	1809–1817	1,899,201	11,287,490	4,541,638	28,514,235	10,428,617	18,085,618	36.6	87.5	55.5
Monroe	1817–1825	3,295,304	4,596,848	3,181,997	25,716,203	12,670,764	13,045,439	49.3	59.6	30.2
Adams, J. Q	1825–1829	2,934,563	3,671,908	3,862,663	23,951,364	11,325,883	12,625,481	47.3	59.7	31.5
Jackson	1829–1837	4,259,584	6,263,460	3,986,375	27,585,896	9,361,800	18,224,096	33.9	56.2	37.2
Van Buren	1837–1841	7,193,859	10,648,054	6,268,622	35,640,486	5,208,036	30,432,450	14.6	55.6	47.5
Harrison/Tyler	1841–1845	5,528,601	5,884,751	6,156,058	27,623,628	7,007,430	20,616,199	25.4	58.4	43.6
Polk	1845–1849	6,175,533	20,122,220	7,504,468	40,966,537	7,189,315	33,777,222	17.5	81.8	67.4
Taylor/Fillmore	1849–1853	15,576,778	12,057,175	8,896,976	51,823,846	15,265,452[c]	36,557,595	29.5	57.3	40.4
Pierce	1853–1857	30,503,000	13,351,000	12,281,000	58,886,000	2,751,000	56,135,000	4.7	45.7	43.5
Buchanan	1857–1861	31,958,000	21,100,000	13,223,000	68,546,000	2,265,000	66,281,000	3.3	51.8	50.1

Sources: Data for the administrations of Washington through Taylor and Fillmore, inclusive, are derived from a table, "Statement of the Expenditures of the United States from 1789 to 1851, inclusive. Annual Average during Each Administration," in the section "Journal of Banking, Currency, and Finance," *Hunt's Merchants' Magazine* 27, no. 3, September 1852: 349; data for the Pierce and Buchanan administrations are from *Historical Statistics of the United States from Colonial Times to 1970*, electronic ed., chapter Y: Government Series Y 457–465, "Outlays of the Federal Government: 1789 to 1970," 1115.

Note: The figures and percents concerning the public debt paid during the Pierce and Buchanan administrations are probably not strictly comparable to the corresponding figures and percents during the preceding administrations. Whereas the figures for "Public Debt Paid" for the Pierce and Buchanan administrations are the amounts paid in interest on the debt, the figures for the preceding administrations represent principal and interest.

[a] All amounts are in terms of current dollars, that is, unadjusted for inflation or deflation, and represent annual average expenditures during each administration.

[b] The amount given under "Total Expenditures" for any administration is always greater than the sum of its individual spending categories because two categories of included expenditures, pensions and the Indian Department, are not listed.

[c] Included here is the average annual payment to Mexico of $5,896,000 in accordance with the Treaty of Guadalupe Hidalgo.

Table D.19
Incorporation of Transportation Companies in Ohio, 1821–1860

Year	Railroads	Canals	Turnpikes	Steamboats	Others	Year	Railroads	Canals	Turnpikes	Steamboats	Others
1821	0	0	0	0	0	1841	1	0	7	0	0
1822	0	0	0	0	0	1842	2	0	6	0	0
1823	0	0	0	0	0	1843	0	1	2	0	0
1824	0	0	1	0	1	1844	3	0	17	0	0
1825	0	0	1	0	0	1845	8	0	30	0	0
1826	0	1	2	0	0	1846	11	0	12	1	1
1827	0	2	1	0	0	1847	4	0	14	0	0
1828	0	2	3	1	0	1848	15	0	37	2	1
1829	0	0	3	0	0	1849	12	0	73	0	1
1830	0	2	1	2	1	1850	19	0	148	1	0
1831	1	0	2	1	0	1851	21	0	103	0	1
1832	11	0	9	2	0	1852	—	—	—	—	—
1833	0	0	6	1	0	1853	—	—	—	—	—
1834	5	1	4	2	0	1854	—	—	—	—	—
1835	7	2	5	0	1	1855	—	—	—	—	—
1836	32	5	11	0	1	1856	7	0	17	0	0
1837	10	8	19	4	1	1857	2	0	8	0	1
1838	2	1	20	1	0	1858	3	0	25	0	1
1839	4	0	34	0	0	1859	7	0	42	0	19
1840	2	0	4	1	0	1860	0	0	7	0	6

Source: Abstracted and adapted from George Heberton Evans Jr., *Business Incorporations in the United States, 1800–1943*, Publications of the National Bureau of Economic Research, no. 49 (New York: National Bureau of Economic Research, 1948), table 10-A, "Ohio Business Incorporations by Special Acts, 1803–1851," and table 10-B, "Under General Laws, 1856–1899."

Note: Figures for 1821–1851, inclusive, are the numbers of incorporations under special acts of the Ohio Legislature; figures for 1856–1860, inclusive, are the numbers of incorporations under the state's general law of incorporation; there are no figures in the source for 1852–1855, inclusive.

NOTES

INTRODUCTION

1. For tonnage of the *New Orleans* and the date, location, and cause of its destruction, see William M. Lytle and Forrest R. Holdcamper, *Merchant Steam Vessels of the United States, 1790–1868: The Lytle-Holdcamper List, Initially Compiled from Official Merchant Marine Documents of the United States and Other Sources,* revised and edited by C. Bradford Mitchell, with the assistance of Kenneth R. Hall (Staten Island, NY: Steamship Historical Society of America; Baltimore: distributed by University of Baltimore Press, 1975), 211. A succinct account of the vessel's abbreviated career is given in Louis C. Hunter, *Steamboats on the Western Rivers: An Economic and Technological History* (Cambridge: Harvard University Press, 1949; reprint, New York: Dover, 1993), 5, 12, 15–16, and 20.

2. Steamboats lost to collisions are included here among losses due to natural hazards, though one could reasonably exclude them, arguing that human (that is, pilot) error was responsible. In most cases, however, collisions occurred because of poor visibility due to darkness or fog or because a combination of poor visibility and strong currents precluded maneuvering out of the way of an oncoming steamboat in narrow channels. Numbers of steamboats lost and in operation are from, respectively, Paul F. Paskoff, Steamboat Database, abstracted and compiled by the author from Bruce D. Berman, *Encyclopedia of American Shipwrecks* (Boston: Mariners Press, 1972), passim.

3. For Yancey's remarks, see *Appendix to the Congressional Globe,* 29th Cong., 1st sess., House, March 10, 1846, 358; for Davis's remarks, see ibid., March 16, 1846, 437.

4. *Appendix to the Congressional Globe,* 29th Cong., 1st sess., House, March 10, 1846, 358.

5. John Lauritz Larson, *Internal Improvement: National Public Works and the Promise of Popular Government in the Early United States* (Chapel Hill: University of North Carolina Press, 2001), 5–6, 252–55.

6. Erik F. Haites, James Mak, and Gary M. Walton, *Western River Transportation: The Era of Early Internal Development, 1810–1860* (Baltimore: Johns Hopkins University Press, 1975), 122. The quoted passage is on page 88.

7. Laurence Joseph Malone, *Opening the West: Federal Internal Improvements before 1860* (New York: Greenwood Press, 1998), 28–30. More recently, John Joseph Wallis and Barry R. Weingast have noted that state governments far outspent the federal government in financing improvements to transportation infrastructure (though Wallis and Weingast emphasize roads and canal projects) during the antebellum period. They argue that the relatively small role played by the federal government in that endeavor was due to the fact that it was hobbled by

the Constitution with respect to means of financing improvements and stalemated by partisan and sectional political maneuvering in Congress. See John Joseph Wallis and Barry R. Weingast, "Equilibrium Impotence: Why the States and Not the Federal Government Financed Economic Development in the Antebellum Era," NBER Working Paper Series (Cambridge, MA: National Bureau of Economic Research, 2005), 1–3, 38–40.

1. TROUBLED WATERS

1. During the 1840s, five different Mississippi River panoramas were available for viewing. See John Francis McDermott, *The Lost Panoramas of the Mississippi* (Chicago: University of Chicago Press, 1958), 17. River panoramas were also popular in Europe. One such display was "Bayne's Panorama of a voyage to Europe, together with the most interesting portion of the celebrated river Rhine. . . . The truthful illustrations of what we have often read, seems [sic] to impress the memory stronger than even books can do, for the most attentive student." *Scientific American* 5, no. 28 (March 31, 1850), 218; Cornell University's "Making of America" Web site, http://moa.cit.cornell.edu/.

2. McDermott, *The Lost Panoramas of the Mississippi*, 139. The May 18, 1849, *Charleston (SC) Mercury* carried a notice that the "Panorama of the Mississippi River" being shown in that city's Hibernian Hall would close a few days later before leaving for its next city of call. Six days later, under the heading "The Panorama," the paper reported that "[b]y request of many who have been unable to obtain eligible seats during the past week, from the crowded state of the Hall, this magnificent illustration of the 'Great Father of Waters' will remain open for three nights more" (*Charleston (SC) Mercury*, May 24, 1849).

3. McDermott, *The Lost Panoramas of the Mississippi*, appendix A.

4. Ibid., 17. A detailed account of Henry Lewis's panorama is given in Joseph Earl Arrington, "Henry Lewis' Moving Panorama of the Mississippi River," *Louisiana History* 6 (Summer 1965): 239–72. See also John W. Reps, *Views and Viewmakers of Urban America: Lithographs of Towns and Cities in the United States and Canada, Notes on the Artists and Publishers, and a Union Catalog of Their Work, 1825–1925* (Columbus: University of Missouri Press, 1984), 190.

5. George Conclin, *A Book for All Travelers. Conclin's New River Guide, or A Gazetteer of All the Towns on the Western Waters* (Cincinnati: H. S. and J. Applegate, 1849), 116.

6. Conclin, *Conclin's New River Guide*, 88. For a thorough account of the quake and the damage it caused, see Jay Feldman, *When the Mississippi Ran Backwards: Empire, Intrigue, Murder, and the New Madrid Earthquake* (New York: Free Press, 2005), esp. 150–59, 168–71.

7. Mid-nineteenth-century accounts of the consequences of the New Madrid earthquake refer to the "sunk country." See, for example: "Sixteenth Meeting of the British Association for the Advancement of Science . . . Wednesday, Sept. 16 [1846]," *Living Age* 11, no. 130 (November 7, 1846): 353; and Charles Lyell, F.R.S., "Lyell's Second Visit to America," *North American Review* 6 (October 1849): 346. The phrase "sunk ground" is used in "An Earthquake or Two," *Harper's New Monthly Magazine* 11 (November 1855): 799. The bluff on which the town of New Madrid sat is described as having "sunk down to the level of the river" and been "afterward

submerged" in T. B. Thorpe, "Remembrances of the Mississippi," *Harper's New Monthly Magazine* 12, no. 67 (December 1855): 33. All of the foregoing are available on the Cornell University "Making of America" Web site, http://cdl.library.cornell.edu/cgi-bin/moa.

8. These observations are from Dickens's *American Notes*, 143 and 145, quoted in John W. Reps, *Cities of the Mississippi: Nineteenth-Century Images of Urban Development* (Columbia: University of Missouri Press, 1994), 162.

9. James T. Lloyd, *Lloyd's Steamboat Directory, and Disasters on the Western Waters*. . . . (Cincinnati: James T. Lloyd, 1856), 51.

10. See "Flood at Cairo," *Living Age* 58 (17 July 1858), 200; Cornell University "Making of America" Web site, http://cdl.library.cornell.edu/cgi-bin/moa.

11. *Memorial of the Mayor and City Council of the City of St. Louis, Praying an Appropriation for the Improvement of the Harbor of That City*, 30th Cong., 1st sess., 1848, S. Misc. Doc. 84, serial 511, 1.

12. Cincinnati's population in 1850 was 115,436; the same year, the population of New Orleans was 116,375. Population figures for Cincinnati, New Orleans, and St. Louis in 1850 are from J. D. B. De Bow, Superintendent of the United States Census, *Statistical View of the United States. . . Being a Compendium of the Seventh Census* (Washington: Beverly Tucker, Senate Printer, 1854), table 241.

13. "Boat Building at St. Louis," article 8 in *Western Journal and Civilian* 1, no. 1 (January 1848): 46.

14. Conclin, *Conclin's New River Guide*, 81.

15. Louis C. Hunter, *Steamboats on the Western Rivers: An Economic and Technological History* (Cambridge: Harvard University Press, 1949; reprint, Mineola, NY: Dover, 1993), 49. The corresponding figures that year for other major river cities were 2,977 steamboat arrivals in New Orleans, 2,885 in Pittsburgh, and 4,007 in Cincinnati (ibid., table 2 [appendix]).

16. *Memorial of a Number of Citizens of St. Louis, Missouri, Praying an Appropriation for the Removal of Obstructions in the Western Rivers, and for the Improvement of the Harbor of That City*, 28th Cong., 1st sess., 1844, S. Doc. No. 185, serial 434, 6.

17. Ibid., 7.

18. *Memorial of the Mayor and City Council of the City of St. Louis*, March 21, 1848, serial 511, 1.

19. Ibid., 1–2.

20. For a good specialized account of St. Louis and its steamboats in this period, see William E. Lass, "The Fate of Steamboats: A Case Study of the 1848 St. Louis Fleet," *Missouri Historical Review* 96 (October 2001): 2–15.

21. Accounts of the sequence of events are plentiful, though not in all cases reliable. The standard versions of the origin, spread, and aftermath of the great St. Louis fire of 1849 are those printed in the newspapers of the day, and most subsequent accounts are derivative of them. See, for example, the account in the *Western Journal and Civilian* 2, no. 5 (May, June, July, August, 1849): 347–48. For later accounts, see: J. Thomas Scharf, *History of Saint Louis City and County, From the Earliest Periods to the Present Day: Including Biographical Sketches of Representa-*

tive Men, 2 vols. (Philadelphia: Louis H. Everts, 1883), 1: 820; and John V. Morris, *Fires and Firefighters* (Boston: Little, Brown, 1955), 152–53. Morris erroneously gives "the evening of May 19, 1849," as the date of the fire's beginning (152).

22. Scharf, *History of Saint Louis City and County*, 1: 819–20.

23. An account of the St. Louis fire appeared in the *New York Herald* on May 21, 1848. In this story, the third run by the newspaper in four days, Targee was identified as an auctioneer.

24. Charles van Ravenswaay, *Saint Louis: An Informal History of the City and Its People, 1764–1865*, ed. Candace O'Connor (St. Louis: Missouri Historical Society Press, 1991), 385.

25. The federal arsenal at St. Louis sent shipments of war materiel, including ammunition, artillery, muskets, rifles, pistols, swords, and powder, to the army in Mexico during the war. See "Statement of the Principal Articles of Ordnance and Ordnance Stores Manufactured and Issued to the Army from the St. Louis Arsenal, Missouri, since the Commencement of the War with Mexico, to This Date [February 14, 1848]," in *Memorial of the Mayor and City Council of the City of St. Louis*, 1848, serial 511, 5–6.

26. van Ravenswaay, *Saint Louis*, 385.

27. Scharf, *History of Saint Louis City and County*, 1: 820. The account of the fire in the May 21, 1848, edition of the *New York Herald* had a St. Louis dateline of Friday, May 18, and identified the drug store in which Targee was killed as that of Doenich & Valloux.

28. *New York Herald*, May 21, 1848.

29. See *Western Journal and Civilian* 2, no. 5 (May, June, July, August, 1849): 347–48; and Morris, *Fires and Firefighters*, 152–53. The number of steamboats lost in the fire may, in fact, have been a bit smaller or larger than twenty-three. In a brief discussion of the fire and its impact on St. Louis's steamboats, William E. Lass states that twenty-three steamboats were lost to the flames. He relies for that figure on secondary sources (see Lass, "The Fate of Steamboats," 7). Using the Berman database and the list in Lytle's *Merchant Steam Vessels of the United States, 1807–1868*, I was able to determine that twenty-two steamboats had perished at St. Louis on May 17, 1849. The account printed in the *Western Journal and Civilian*, published in St. Louis, listed the names of twenty-three steamboats as having been lost in the "Great Fire in St. Louis." There are, however, errors of commission and omission in that account. First, two side-wheelers—the *Timour* and the *Prairie*—are listed among those vessels lost. Neither boat, however, was there on the fateful night. Second, although the account in the *Journal and Civilian* includes a steamboat with the name *American Eagle*, there were, in fact, two steamboats of that name lost in the fire. Both boats were side-wheelers, but one was of 295 tons displacement and had been built seven years earlier in Louisville, and the other was a considerably smaller and newer craft of 216 tons that had been built in 1847 in Freedom, Pennsylvania. See Lytle, *Merchant Steam Vessels of the United States, 1807–1868*, 8. Morris states that twenty-six steamboats were set afire on May 17 in the St. Louis fire but does not say how many, if any, of this number escaped destruction (152). That there is uncertainty as to the number of steamboats lost in the fire is hardly surprising. The figure of twenty-three steamboats is the one given in secondary accounts for the number of boats lost at St. Louis in the great fire. That does not, of course, make the figure accurate; it has simply acquired the status of a fact by dint of frequent and sequential repetition.

30. This characterization does not, of course, embrace the cost in human life associated with the *Sultana* tragedy of April 1865, when that steamer, filled to overflowing with returning Union army veterans, many former prisoners of war, exploded and sank, killing more than two thousand of those aboard. For a detailed account of the calamity, see Gene Eric Salecker, *Disaster on the Mississippi: The* Sultana *Explosion, April 27, 1865* (Annapolis: Naval Institute Press, 1996).

31. The possibility of arson is mentioned in Scharf, *History of Saint Louis City and County*, 820.The figures on buildings and lives lost are from Morris, *Fires and Firefighters*, 153.

32. *Western Journal and Civilian* 2, no. 5 (May, June, July, August, 1849): 347.

33. Ibid., 348.

34. Ibid. Property losses, both on the water and ashore, were undoubtedly far higher. Scharf, citing the city assessor's figures, placed the total monetary loss at $6,102,290: $5 million of goods ashore; $600,000 in steamboats and goods aboard; and $502,290 in buildings (*History of Saint Louis City and County*, 820).

35. *Western Journal and Civilian* 2, no. 5 (May, June, July, August, 1849): 348.

36. For a discussion of town boosterism in the West of this period, see Daniel Boorstin, *The Americans: The National Experience* (New York: Random House, 1965), 161–68.

37. See "Nothing Later from St. Louis," dateline Philadelphia, May 20, 1849, in the *New York Herald*, May 21, 1849.

38. One of the nation's leading newspapers, the *New York Herald*, ran four reports of the St. Louis fire over the course of five days—two of the reports, one below the other, in the same issue (*New York Herald*, May 19, 21, 22, 1848). News reports of the fire also ran in the *Charleston (SC) Mercury* on May 19, 21, and 22. Even a periodical with a relatively narrow focus, the weekly *Scientific American*, took notice of the burning of St. Louis and the steamboats at its levee and ran a short report of the fire in its May 26, 1849, issue (vol. 4, no. 36, 282; http://cdl.library.cornell.edu/cgi-bin/moa). The magazine ran another squib in its August 25, 1849, issue (vol. 5, no. 49, 386; http://cdl.library. cornell.edu/cgi-bin/moa).

39. Great fires were an inescapable fact of urban life during the nineteenth century, especially in the decades before the Civil War, because of a combination of densely populated neighborhoods of wooden buildings, narrow streets, and rudimentary arrangements for firefighting. In one large city after another, fire claimed vast areas, large amounts of property, and many lives, as in: New York City in 1835 and 1845; Pittsburgh in 1845; Philadelphia in 1839 and 1850; and San Francisco in 1851. See Morris, *Fires and Firefighters*, 380–81; see also Rev. E. J. Goodspeed, D.D., *History of the Great Fires in Chicago and the West. A Proud Career Arrested by Sudden and Awful Calamity....* (New York: H. S. Goodspeed, 1871), 665–66.

40. McDermott, *Lost Panoramas of the Mississippi*, 13.

41. *St. Louis Weekly Reveille*, August 6, 1849, 2184, quoted in McDermott, *Lost Panoramas of the Mississippi*, 153, original italics.

42. Henry Lewis to his brother, George F. Lewis, Cincinnati, July 29, 1849, quoted in McDermott, *Lost Panoramas of the Mississippi*, 133.

43. *Western Journal and Civilian* 3, no. 1 (October, 1849): 70.

44. Predictably, the drawing power of the touring panoramas waned as their novelty wore

off and as each one found itself in competition with a host of others for the same urban markets. While Lewis worked on his panorama, there were five other river panoramas before the public, and as many as another twenty panoramas on various other subjects. See Arrington, "Henry Lewis' Moving Panorama of the Mississippi River," 266.

45. R.B.J. Twyman, publisher of a Memphis, Tennessee, business directory, attributed the phrase "the American Nile" to "Gen. [Andrew] Jackson, writing in the 'Port Folio,' published in Philadelphia, in 1820." The context for the phrase is the following passage: "The general advantages of Memphis, are owing to its being founded on the Mississippi, one of the largest and most important rivers on the globe, and the high road for all the commerce of the vast and fertile valley through which it flows. This noble river, which may, with propriety, be denominated the *American Nile*, is about two thousand five hundred and eighty miles from its head to its mouth, and with its branches, waters two-thirds of the territory of the United States." See *Twyman's Memphis Directory and General Business Advertiser, for 1850, with a Brief History of Memphis Annexed* (Memphis: R.B.J. Twyman, 1849), 103–4, original italics. A similar comparison was made in the course of a discussion of the swamps on the west bank of the Mississippi River in *The Southern Business Directory*. It is asserted there that when that swampland was drained and protected from subsequent flooding "it requires no prophet to foresee that this great valley will be "The Nile of the United States" (quoted in John P. Campbell, *The Southern Business Directory and General Commercial Advertiser* [Charleston, SC: Steam Power Press of Walker and James, 1854], 157).

46. *Scientific American* 5, issue 25 (March 9, 1850), 194; http://cdl.library.cornell.edu/cgi-bin/moa.

47. Erik F. Haites, James Mak, and Gary M. Walton, *Western River Transportation: The Era of Early Internal Development, 1810–1860* (Baltimore: Johns Hopkins University Press, 1975), 108.

48. The best account of the inspection effort is presented in Hunter, *Steamboats on the Western Rivers*, chap. 13. The Steamboat Law of 1852 stipulated in some detail the qualifications, responsibilities, and liabilities of pilots licensed under the act. Section 9, clause 10, of the act stated: "It shall be unlawful for any person to serve as a Pilot on Steam Vessels carrying passengers, who is not licensed by the Inspectors, under a penalty of One Hundred Dollars for each offense." Section 38 of the act assured passengers that "[a]ny person sustaining loss or injury through the carelessness, negligence, or wilful [sic] misconduct of any Pilot . . . may sue such Pilot and recover damages for any such injury caused as aforesaid by any such Pilot" (see "Duties and Liabilities of Pilots Licensed under the Steamboat Law of August 30, 1852," Norman Collection, Group VII: Loose Manuscripts, Box 6 of 6, 0:68, Special Collections, Hill Memorial Library, Louisiana State University Libraries, Baton Rouge).

49. Hunter, *Steamboats on the Western Rivers*, 541–42; Haites, Mak, and Walton, *Western River Transportation*, 109–10.

50. See Haites, Mak, and Walton, *Western River Transportation*, 109–10. The rate of loss, due to fire and explosion, is calculated in two ways for each of the periods. One of these ways uses the average number of vessels in service, derived from figures given in Haites, Mak, and Walton, *Western River Transportation*, appendix B. The ratio of losses from fire and explosion to average number of vessels in service is 135:650, or 0.21, for the 1847–51 period and 153:733,

or 0.21, for the 1853–57 period. The other, admittedly less realistic method of calculating the rate of loss uses, as the denominator, the cumulative sum of the number of vessels in service in each year of the five-year periods 1847–51 and 1853–57. The values of the ratio for each period are 135:3,250, or 0.042, and 153:3,664, or 0.042, respectively.

51. Statement by W. W. Guthrie, local inspector, originally published in the *Cincinnati Gazette*, and reprinted in *De Bow's Review* 19 (October 1855): 466.

52. Paul F. Paskoff, Steamboat Database, abstracted and compiled by the author from Bruce D. Berman, *Encyclopedia of American Shipwrecks* (Boston: Mariners Press, 1972).

53. There is considerable uncertainty about the number of fatalities arising from steamboat boiler explosions on the western rivers. As Louis Hunter notes, "the number of casualties in accidents before 1853 were based on newspaper reports which the investigations of the supervising inspectors themselves showed frequently to have exaggerated the loss of life" (*Steamboats on the Western Rivers*, 542). Figures from the 1861 and 1863 reports of the Board of Supervising Inspectors of Steamboats indicate a significant reduction in the number of steamboat boiler explosions—from fifty on just the western rivers during the five-year period 1848–52 (that is, the period before the Act of 1852) to twenty on both eastern and western rivers during the five-year period 1854–58. The reduction in the loss of life due to those incidents from one period to another was more significant—from 1,155 fatalities on just the western rivers during the 1848–52 period to 224 on both eastern and western rivers during the 1854–58 period (ibid., 541). Assuming, for the sake of discussion, the accuracy of the number of explosions and fatalities on the western rivers during the 1848–52 period, it is not unreasonable to speculate that the sharp drop in the number of fatalities recorded after the Act of 1852 went into effect was in no small way due to the legislation's requirement that steamboats be equipped with "[f]ire engines and hose as well as life preservers and lifeboats" (ibid., 537).

54. This is essentially also Hunter's conclusion, though he suggests that the resurgence of steamboat losses during the late 1850s was due to the fact that inspectors lacked the "authority to compel" compliance with the Act of 1852 and also to the fact that "the difficulties with which they had to contend were too numerous, complex, and too little understood to be resolved in a few years" (ibid., 542).

55. Daniel Webster, "Public Dinner at Philadelphia," December 2, 1846, in Daniel Webster, *Works of Daniel Webster*, 6th ed. (Boston: Little, Brown, 1853), 2: 339.

56. Quoted in *Niles National Register* 74, no. 1925 (December 20, 1848): 394.

57. "On the Democratic Tendencies of Science," *American Journal of Education* 1, no. 2 (January 1856): 167.

58. Frederick Way Jr., comp., *Way's Packet Directory, 1848–1994: Passenger Steamboats of the Mississippi River System since the Advent of Photography in Mid-Continent America*, rev. ed. (Athens: Ohio University Press, 1994), 69.

59. *Memorial of the Citizens of Cincinnati, to the Congress of the United States, Relative to the Navigation of the Ohio and Mississippi Rivers* (Cincinnati: L'Hommedieu, 1843), 17. The memorial, or petition, was the product of a meeting held on November 4, 1842 (ibid., 3).

60. A study of the states of origin of steamboats in service on Mississippi's Yazoo River before the Civil War finds the same sort of geographical concentration of steamboat building: of

172 documented steamboats on the Yazoo in that period, 129—or 75 percent—came from four Ohio River states and about another 10 percent from smaller boatyards along the Ohio; all told, then, approximately 85 percent of the steamboats on the Yazoo River came from Ohio River boatyards. See Harry P. Owens, *Steamboats and the Cotton Economy: River Trade in the Yazoo-Mississippi Delta* (Jackson: University Press of Mississippi, 1990), 10. A comparable but somewhat greater degree of geographical concentration was evident in the origin of the steamboats that plied Cypress Bayou and the lakes west of Shreveport, Louisiana, during the antebellum period. Four Ohio River states—Pennsylvania, Ohio, Kentucky, and Indiana—accounted for just over 90 percent of 232 steamboats (the states of origin of another eight steamboats are unknown). See Jacques D. Bagur, *A History of Navigation on Cypress Bayou and the Lakes* (Denton: University of North Texas Press, 2001), 721. Almost all steamboats were built in one boatyard, but, of the 671 steamboat packets for which more or less complete data are available in *Way's Packet Directory*, six, or less than one-tenth of 1 percent, were built in two different places. In such instances, the hull of the vessel was built at one site, and the vessel was then completed at another. The construction of three of these six packets was conducted at two sites within the same state; the other three were completed in states other than the ones in which the hulls were built.

61. The figure of $15,000 as the cost of a steamboat on the western rivers is a very conservative estimate. In 1843, a figure of $15,478 was computed as the average value for the steamboats, of whatever vintage and tonnage, "running on the Western rivers" that year. According to the same source, twenty-six steamboats were built in 1843. Using the individual dollar cost assigned to each of those vessels, a calculation of the average value of the boats built that year yields a figure of $14,846. The foregoing figures are taken or computed from "Steamboats on the Western Rivers," 1843, in Leonard V. Huber, comp., *Advertisements of Lower Mississippi River Steamboats, 1812–1920: A Scrapbook with Introduction and Index of Vessels and Lines* (West Barrington, RI: Steamship Historical Society of America, 1959), 97. Substantially higher average values of $27,900 and $36,000 for steamboats built during the 1840s and 1850s, respectively, for the New Orleans–Louisville trade on the Mississippi and Ohio rivers are given in Haites, Mak, and Walton, *Western River Transportation*, table C–1 (appendix).

62. Cincinnati's boatyards built seven steamboat packets in 1850. Using the conservative figure of $15,000 as the estimated average value of a steamboat, the seven boats represented $105,000 of manufactured product, roughly one-half of 1 percent of the more than $20 million in manufactured output produced throughout the city's county of Hamilton. Similarly, the four packets constructed in Louisville, valued here at $60,000, were one-half of 1 percent of the $11 million in manufactured goods produced in that city's county of Jefferson. The source of the values of manufactured product in Hamilton County, Ohio, and Jefferson County, Kentucky, in 1850 is the *Statistical View of the United States . . . Being a Compendium of the Seventh Census*, 295 and 247, respectively.

63. For Washington County's manufacturing output, see ibid., 301.

64. Figures are derived from Paul F. Paskoff, Steamboat Packet Database, itself abstracted from Way, *Way's Packet Directory*, passim.

65. Ibid.

66. Ibid.

67. *William's Cincinnati Almanac, Business Guide and Annual Advertiser, 1850*. First issue (Cincinnati: S. Williams, 1850), 164–65, 184.

68. The figure for 1856 is derived from the business listings in J. M. Paschall and C. B. Riggs, comps, *Paschall & Riggs' First Annual Memphis City Directory and General Business Advertiser, for 1856–7* (Memphis: J. M. Paschall and C. B. Riggs, 1856), 218; the figure for 1859 is derived from the listings in Tanner, Halpin & Co., *Memphis City Directory, for 1859. Being a Complete General and Business Directory for the Entire City* (Memphis: Hutton and Clark, 1859), 221; the figure for 1860 is similarly derived from the listings in *Williams' Memphis Directory, City Guide, and Business Mirror*, vol. 1, 1860 (Memphis: Cleaves and Vaden, 1860), 355–67.

69. *Twyman's Memphis Directory and General Business Advertiser, for 1850, with a Brief History of Memphis Annexed* (Memphis: R.B.J. Twyman, 1849), 112.

70. Ibid.

71. E. R. Marlett, M.D., and W. H. Rainey, comps. and publishers, *W. H. Rainey & Co.'s Memphis City Directory, and General Business Advertiser, for 1855 & '6. Also: A Business Directory* (Memphis: D. O. Dooley & Co., Whig Book and Job Office, 1855), 244.

72. See, for example, entries and advertisements in the following city business directories: for Nashville, Tennessee, Campbell, *The Southern Business Directory and General Commercial Advertiser*, vol. 1, pt. 2, 143–44; for Memphis, Tennessee, *Williams' Memphis Directory, City Guide, and Business Mirror*, vol. 1, 1860, 72, 74, including the advertisement for Illinois ice on 72; for Vicksburg, Mississippi, H. C. Clarke, comp., *A General Directory for the City of Vicksburg* (Vicksburg: H. C. Clarke, 1860), appendix, passim.

73. See Carl A. Brasseaux and Keith P. Fontenot, *Steamboats on Louisiana's Bayous: A History and Directory* (Baton Rouge: Louisiana State University Press, 2004), table 3.1.

74. See Hunter, *Steamboats on the Western Rivers*, 320–22. Also see Way, *Way's Packet Directory*, passim.

75. See Haites, Mak, and Walton, *Western River Transportation*, 72.

76. Hunter, *Steamboats on the Western Rivers*, 236; Floyd M. Clay, *History of Navigation on the Lower Mississippi*, National Waterways Study, U.S. Army Engineer Water Resources Support Center, Institute for Water Resources (Washington, D.C.: Superintendent of Documents, U.S. Government Printing Office, distributor, 1983), 11–12.

77. Paskoff, Steamboat Database, derived from sources described in appendix A.

78. Lloyd, *Lloyd's Steamboat Directory*.

79. With some rhetorical excess, a request from citizens of Cincinnati for federal aid to remove hazards, especially snags, from the western rivers, characterized a snagging of a steamboat as a "scene . . . terrific beyond description" (*Memorial of the Citizens of Cincinnati*, 35).

80. Lloyd, *Lloyd's Steamboat Directory*, 163; Hunter, *Steamboats on the Western Rivers*, 274.

81. The number of passengers aboard the *Shepherdess* is given as "between sixty and seventy" in Lloyd, *Lloyd's Steamboat Directory*, 163. Hunter puts the number of passengers aboard the *Shepherdess* at "two hundred or more" (*Steamboats on the Western Rivers*, 275). As Hunter notes, the "Graveyard" was located between St. Louis and Cairo (ibid., 235). More specifically, the notoriously snag-infested stretch of river lay between Ste. Genevieve, Missouri, which was

about 60 miles below St. Louis, and Cape Girardeau, Missouri, which was about 140 miles below St. Louis, or about 55 miles above Cairo.

82. Lloyd, *Lloyd's Steamboat Directory*, 165; other details of the later events of the snagging of the *Shepherdess* are from Hunter, *Steamboats on the Western Rivers*, 274.

83. The figures of forty and "not less than seventy" dead are from Lloyd, *Lloyd's Steamboat Directory*, 169; Hunter noted that an account submitted to Congress and reprinted in the newspaper *National Intelligencer* put the number of fatalities at "between sixty and one hundred" (*Steamboats on the Western Rivers*, 275).

84. Numerous sources describe the presence of snags in the western rivers. Among the most detailed accounts are Lloyd, *Lloyd's Steamboat Directory*, 200; Hunter, *Steamboats on the Western Rivers*, 192–200; "The Commerce and Navigation of the Mississippi River and Its Tributaries, Considered with Reference to their Improvement by the Federal Government," *Western Journal and Civilian* 1, no. 3 (March 1848): 160, 162, and passim; and *Report of the Secretary of War, Made in Compliance with a Resolution of the Senate in Relation to Work Done under the Appropriations of 1852 for the Improvement of Western Rivers and Harbors*, 33rd Cong., 1st sess., 1854, S. Exec. Doc. 51, serial 698, passim.

85. Emmeline Stuart-Wortley, *Travels in the United States, etc., during 1849 and 1850. By the Lady Emmeline Stuart Wortley* (New York: Harper and Brothers, 1851), 114; University of Michigan "Making of America" Web site, http://moa.umdl.umich.edu/moa_search.html. The phrase "floating steam-palaces" was a fairly common metaphor used to describe the larger packets of the Lower Mississippi River. Stuart-Wortley said that the *Autocrat* was "about four hundred feet long, and gorgeous as an enchanted castle inside" (ibid.). The displacement of the steamboat is given in Way, *Way's Packet Directory*, entry number 0399, 34.

86. Stuart-Wortley, *Travels in the United States*, 116.

87. Ibid.

88. Ibid., 117.

89. Charles Dickens, *American Notes*, with an introduction by Christopher Lasch (Gloucester, MA: Peter Smith, 1968), 198.

90. Ibid.

91. Frances Trollope, *Domestic Manners of the Americans*, edited, with a history of Mrs. Trollope's adventures in America, by Donald Smalley (New York: Knopf, 1949), 15. Frances Trollope was the mother of novelist Anthony Trollope. The displacement of the *Belvidere* is from William M. Lytle, *Merchant Steam Vessels of the United States 1807–1868*. Publication no. 6 (Mystic, CT: Steamship Historical Society of America, 1952), 18.

92. Trollope, *Domestic Manners of the Americans*, 23.

93. Dickens, *American Notes*, 197.

94. Ibid., 198.

95. Ibid. Dickens denied any animus for the United States and its people, insisting that, if he had any bias in the matter, it was "in favor of the United States" (ibid., xiv). That, at least, was what he said in print. Privately, as Christopher Lasch notes in his introduction to *American Notes*, Dickens was "heartily sick of America" (ibid., x).

96. The literature on involvement by southern states in and support for intrastate inter-

NOTES TO PAGES 34–36

nal improvements such as turnpikes, canals, and railroads is fairly extensive. See, for example, Charles Clinton Weaver, *Internal Improvements in North Carolina Previous to 1860*, Johns Hopkins University Studies in Historical and Political Science, series 21, nos. 3–4 (Baltimore: Johns Hopkins Press, 1903), 95; Carter Goodrich, "The Revulsion against Internal Improvements," *Journal of Economic History* 10 (November 1950): 147–48; George Rogers Taylor, *The Transportation Revolution, 1815–1860*, vol. 4 of *The Economic History of the United States* (New York: Holt, Rinehart and Winston, 1951), 25–26, 32–52, and 88–94; Albert Fishlow, *American Railroads and the Transformation of the Ante-Bellum Economy* (Cambridge: Harvard University Press, 1965), 190; Haites, Mak, and Walton, *Western River Transportation*, 106; Harry L. Watson, *Jacksonian Politics and Community Conflict: The Emergence of the Second American Party System in Cumberland County, North Carolina* (Baton Rouge: Louisiana State University Press, 1981), passim, esp. 151–97; Paul H. Bergeron, *Antebellum Politics in Tennessee* (Lexington: University of Kentucky Press, 1982); Marc W. Kruman, *Parties and Politics in North Carolina, 1836–1865* (Baton Rouge: Louisiana State University Press, 1983), passim, esp. 55–85; Peter Wallenstein, *From Slave South to New South: Public Policy in Nineteenth-Century Georgia* (Chapel Hill: University of North Carolina Press, 1987), 33, 121; and John Lauritz Larson, *Internal Improvement: National Public Works and the Promise of Popular Government in the Early United States* (Chapel Hill: University of North Carolina Press, 2001), 91–105.

97. On North Carolina's river improvement efforts, see Weaver, *Internal Improvements in North Carolina*, 49–65 passim; Watson, *Jacksonian Politics and Community Conflict*, 154–63; and Kruman, *Parties and Politics in North Carolina*, 67. Attempts made by private and state authorities in east Texas and northwest Louisiana along Cypress Bayou are discussed in Bagur, *A History of Navigation on Cypress Bayou and the Lakes*, 325–45. For an account of efforts by local, parish, and state governments to improve navigation on the bayous of south Louisiana, see Brasseaux and Fontenot, *Steamboats on Louisiana's Bayous*, 28–34, 64–65.

98. John H. Krenkel, *Illinois Internal Improvements, 1818–1848* (Cedar Rapids, IA: Torch Press, 1958), 13.

99. David Herbert Donald, *Lincoln* (New York: Simon and Schuster, 1996), 61; Larson, *Internal Improvement*, 218.

100. Larson, *Internal Improvement*, 218–19; Donald, *Lincoln*, 61; and Haites, Mak, and Walton, *Western River Transportation*, 104, 104n.

101. Stanley John Folmsbee, *Sectionalism and Internal Improvements in Tennessee, 1796–1845*. Special Studies in Tennessee History, no. 1 (Knoxville: East Tennessee Historical Society, 1939), iii.

102. Ibid., iii, 55; Charles S. Sydnor, *The Development of Southern Sectionalism, 1819–1848* (Baton Rouge: Louisiana State University Press, 1948), 86.

103. Krenkel, *Illinois Internal Improvements*, 13; Larson, *Internal Improvement*, 219.

104. Krenkel, *Illinois Internal Improvements*, 14.

105. *Resolutions of the General Assembly of Indiana, Asking an Appropriation to Improve the Navigation of the Wabash River*, 27th Cong., 3rd sess., 1843, S. Doc. 119, serial 415, 1.

106. *Report of the Secretary of War Made in Compliance with a Resolution of the Senate in Relation to the Work Done under the Appropriations of 1852 for the Improvement of Western Rivers and*

Harbors, 33rd Cong., 1st sess., 1854, S. Exec. Doc. 51, serial 698, 2. The formation of snags as a result of bank undercutting as a result of floods was described in the *Memorial of the Citizens of Cincinnati*, 32.

107. *Report of the Secretary of War*, 1854, serial 698, 3.

108. For numbers and tonnage of steamboats in operation each year, see Haites, Mak, and Walton, *Western River Transportation*, table B-1 (appendix). Numbers and tonnage of steamboats lost to natural hazards are from the Steamboat Database of the present work (see appendix A).

109. Reliable precipitation data for the Mississippi Valley region are available only for St. Louis, beginning in 1837. See U.S. Bureau of the Census, *Historical Statistics of the United States, Colonial Times to 1957* (Washington, D.C.: Department of Commerce, 1960), Series J 246–65, "Long-Record City Stations—Annual Mean Temperature and Annual Total Precipitation: 1780 to 1957," 254–55.

110. The characterization of water levels in the Lower Mississippi Valley covers the period from 1812 through 1859 and is from Clarke, *A General Directory for the City of Vicksburg*, 68.

111. The flood of 1851 was one "of the greatest floods of the nineteenth century." See George S. Pabis, "Delaying the Deluge: The Engineering Debate over Flood Control on the Lower Mississippi River, 1846–1861," *Journal of Southern History* 64 (August 1998): 439.

2. POLITICS BEFORE POLK

1. E. C. Nelson, "Presidential Influence on the Policy of Internal Improvements," *Iowa Journal of History and Politics* 4 (January 1906): appendix A, 53–54.

2. Ibid., 55.

3. Ibid., 21–23; John Niven, *John C. Calhoun and the Price of Union: A Biography* (Baton Rouge: Louisiana State University Press, 1988), 55–56.

4. Charles Sellers, *The Market Revolution: Jacksonian America, 1815–1846* (New York: Oxford University Press, 1991), 76; George Rogers Taylor, *The Transportation Revolution, 1815–1860*, vol. 4 of *The Economic History of the United States* (New York: Holt, Rinehart and Winston, 1951), 19; Frank Bourgin, *The Great Challenge: The Myth of Laissez-Faire in the Early Republic* (New York: Harper and Row, 1990), 167.

5. Niven, *John C. Calhoun and the Price of Union*, 57; see also John Lauritz Larson, "'Bind the Republic Together': The National Union and the Struggle for a System of Internal Improvements," *Journal of American History* 74 (September 1987): 363.

6. John Lauritz Larson, *Internal Improvement: National Public Works and the Promise of Popular Government in the Early United States* (Chapel Hill: University of North Carolina Press, 2001), 139–40; Sellers, *The Market Revolution*, 149–50.

7. Larson, *Internal Improvement*, 139–40; Sellers, *The Market Revolution*, 149–51; Feller, *The Public Lands in Jacksonian Politics*, 57 (Madison: University of Wisconsin Press).

8. Derived from data presented in Nelson, "Presidential Influence on the Policy of Internal Improvements," appendix A. Charles Sellers asserts that Monroe's improvements policy

"placed the federal government fully at the service of the republic's capitalist destiny" (*The Market Revolution*, 151).

9. Sellers, *The Market Revolution*, 151–52; Nelson, "Presidential Influence on the Policy of Internal Improvements," appendix A.

10. U.S. Bureau of the Census, *Historical Statistics of the United States, Colonial Times to 1957* (Washington, D.C.: Department of Commerce, 1960), series A 123–180, 13.

11. Ibid., series Y 155–204, 693.

12. See table D.17.

13. Ibid.

14. Ibid.

15. Nelson, "Presidential Influence on the Policy of Internal Improvements," appendix A.

16. Letter to the Reverend Charles W. Upham, February 2, 1837, in John Adams and John Quincy Adams, *The Selected Writings of John and John Quincy Adams*, ed. Adrienne Koch and William Peden (New York: Knopf, 1946), 389. For a thought-provoking speculation on some of the profound long-term consequences of Jackson's movement away from the basis and thrust of Adams's internal improvements program, see Richard R. John, *Spreading the News: The American Postal System from Franklin to Morse* (Cambridge: Harvard University Press, 1995), 256.

17. *Gales and Seaton's Register of Debates in Congress, Comprising the Leading Debates and Incidents of the Second Session of the Twentieth Congress*, 20th Cong., 2nd sess., vol. 5, December 16, 1828 (Washington: Gales and Seaton, 1830), 106.

18. Sellers, *The Market Revolution*, 192–94.

19. James D. Richardson, *A Compilation of the Messages and Papers of the Presidents* (New York: Bureau of National Literature, 1897), 3: 1050–51. Helpful explanations of the various considerations that figured in Jackson's Maysville Road veto are given in Feller, *The Public Lands in Jacksonian Politics*, 137–42, and in Larson, *Internal Improvement*, 183–84.

20. Richardson, *A Compilation of the Messages and Papers of the Presidents*, 3: 1015; Sellers, *The Market Revolution*, 312.

21. Richard Ellis poses a similar question, asking "How is this to be explained?" in *The Union At Risk: Jacksonian Democracy, States' Rights and the Nullification Crisis* (New York: Oxford University Press, 1989), 24.

22. *Congressional Globe*, 28th Cong., 1st sess., House, April 23, 1844, 529; for Holmes's ties to Calhoun, see Niven, *John C. Calhoun and the Price of Union*, 289. Richard Ellis concludes that the reason that Jackson's internal improvements policy was not "absolutely consistent" was that "[t]he issue was simply too vague and complex and too political for that to be possible" (*The Union at Risk*, 25). A more likely explanation is that Jackson was being calculatingly inconsistent, just as he had been with respect to the tariff question during the election campaign of 1824. See Harry L. Watson, *Liberty and Power: The Politics of Jacksonian America* (New York: Hill and Wang, 1990), 80–81. For a discussion of the role of Jackson's supporters in muddying his position on the tariff during the 1824 campaign, see Harry L. Watson, *Andrew Jackson vs. Henry Clay: Democracy and Development in Antebellum America* (Boston: Bedford/St. Martin's, 1998), 70.

23. See Harry L. Watson, *Liberty and Power: The Politics of Jacksonian America* (New York: Hill and Wang, 1990), 87–88, 93–95.

24. Sellers, *The Market Revolution*, 303, 312–13, 316.

25. *The Development of Southern Sectionalism, 1819–1848* (Baton Rouge: Louisiana State University Press, 1948), 180.

26. Ibid., 180–81.

27. U.S. Bureau of the Census, *Historical Statistics of the United States, Colonial Times to 1957*, Series Y 155–204, 693.

28. Sydnor, *The Development of Southern Sectionalism*, 181.

29. See table D.7.

30. *Gales and Seaton's Register*, 21st Cong., 2nd sess., vol. 8, House, February 19, 1831, 756. Subsequent quotations from this debate are from this source and will be cited parenthetically by page number in the main text.

31. *Gales and Seaton's Register*, 22nd Cong., 1st sess., vol. 8, House, June 4, 1832, 3248, pt. 3. Subsequent quotations from this debate are from this source and will be cited parenthetically by page number in the main text.

32. Niven, *John C. Calhoun and the Price of Union*, 20–23; see also Drew Gilpin Faust, *James Henry Hammond and the Old South: A Design for Mastery* (Baton Rouge: Louisiana State University Press, 1982), 148.

33. *Gales and Seaton's Register*, 22nd Cong., 1st sess., vol. 8, House, June 4, 1832, 3256, pt. 3.

34. Nelson, "Presidential Influence on the Policy of Internal Improvements," appendix A, 62.

35. Ibid., appendix A, 60

36. Ibid., appendix B, 67

37. Ibid., 62.

38. *Message from the President of the United States, Returning the Bill, Entitled "An Act for the Improvement of Certain Harbors, and the Navigation of Certain Rivers"; with His Objections to the Same*, 22nd Cong., 2nd sess., 1832, H. Exec. Doc. 17, serial 233, 1.

39. *Biographical Directory of the American Congress, 1774–1949: The Continental Congress . . . and the Congress of the United States. . . .*, 81st Cong., 2nd sess., H. Doc. 607 (Washington, D.C.: U.S. Government Printing Office, 1950), 771.

40. *Gales & Seaton's Register*, 22nd Cong., 1st sess., vol. 8, House, June 4, 1832, 3252, pt. 3.

41. Ibid. This interpretation is somewhat at odds with that of Mathew Crenson, who, quoting Jackson, notes his objections to "[a] liberal internal improvements policy [which] could inflame petty jealousies, give encouragement to 'personal ambition and self-aggrandizement,' and 'sap the foundations of public virtue.'" See Crenson, *The Federal Machine: Beginnings of Bureaucracy in Jacksonian America* (Baltimore: Johns Hopkins University Press, 1975), 173. See also Larson, *Internal Improvement*, 184–85. Still, when considering improvements measures, particularly river improvements measures, what Jackson objected to was federal funding of projects that were only local, that is intrastate, in scope; federal funding of truly national projects almost invariably received his support.

42. Nelson, "Presidential Influence on the Policy of Internal Improvements," appendix A, 62.

43. Ibid.; *Statement of Appropriations and Expenditures for Public Buildings, Rivers and Harbors, Forts, Arsenals, Armories, and Other Public Works from March 4, 1789 to June 30, 1882*, 47th Cong., 1st sess., 1881, S. Exec. Doc. 196, serial 1992, passim.

44. See Paul Studenski and Herman E. Krooss, *Financial History of the United States*, 2nd ed. (New York: McGraw-Hill, 1963), 112–13; Robert Seager II, *and Tyler too: A Biography of John and Julia Gardiner Tyler* (New York: McGraw-Hill 1963), 160–61; and Sellers, *The Market Revolution*, 412.

45. *Congressional Globe and Appendix*, 27th Cong., 2nd sess., vol. 11, no. 1, December 16, 1841, 1.

46. Ibid., January 24, 1842, 168.

47. Ibid.; Seager, *and Tyler too*, 152, 155; Michael F. Holt, *The Rise and Fall of the Whig Party in America: Jacksonian Politics and the Onset of the Civil War* (New York: Oxford University Press, 1999), 168. See also William W. Freehling, *The Road to Disunion*, vol. 1, *Secessionists at Bay, 1776–1854* (New York: Oxford University Press, 1990), 350–51.

48. *Congressional Globe and Appendix*, 27th Cong., 2nd sess., vol. 11, no. 1, January 25, 1842, 170.

49. Ibid., 215.

50. Sellers, *The Market Revolution*, 403–4.

51. Nelson, "Presidential Influence on the Policy of Internal Improvements," appendix A, 63–64, 67, and appendix B, 67.

52. Richardson, *Compilation of the Messages and Papers of the Presidents*, 5: 2185.

53. Ibid., 2184.

54. Ibid., 2186.

3. POLITICS: POLK AND POST-POLK

1. *Gales & Seaton's Register of Debates in Congress*, 22nd Cong., 1st sess., vol. 8, June 4, 1832, 3255, pt. 3

2. Ibid.

3. Glyndon G. Van Deusen, *The Jacksonian Era, 1828–1848*, New American Nation Series (New York: Harper Torchbooks, Harper and Row, 1959), 206–7. For a more recent and sympathetic appraisal of Polk's veto of the improvements bill of 1846, see Charles Sellers, *The Market Revolution: Jacksonian America, 1815–1846* (New York: Oxford University Press, 1991), 425.

4. For an incisive assessment of the bearing of Polk's veto on the principles and future of the Democratic Party, see John Lauritz Larson, *Internal Improvement: National Public Works and the Promise of Popular Government in the Early United States* (Chapel Hill: University of North Carolina Press, 2001), 240.

5. Sellers, *The Market Revolution*, 473–74. John C. Calhoun to Thomas G. Clemson, Brussels, August 8, 1846, in John C. Calhoun, *The Papers of John C. Calhoun*, ed. Clyde N. Wilson (Columbia: University of South Carolina Press, 1995), 23: 399.

6. See Don E. Fehrenbacher, *Chicago Giant: A Biography of "Long John" Wentworth* (Madison, WI: American History Research Center, 1957), 66.

7. *Appendix to the Congressional Globe*, 29th Cong., 1st sess., House, February 10, 1846, 455. Wentworth later became a Republican. *Biographical Directory of the American Congress*, 1993.

8. *Appendix to the Congressional Globe*, 29th Cong., 1st sess., House, March 9, 1846, 480.

9. Charles Sellers, *James K. Polk, Continentalist, 1843–1846* (Princeton: Princeton University Press, 1966), 319–20; Eric H. Walther, *The Fire-Eaters* (Baton Rouge: Louisiana State University Press, 1992), 49, 55–56. The determination of southern congressmen and senators to resist extensions of federal power into the states, including federally financed river improvements, likely often owed its vigor to a concern that such projects, if allowed, would serve as precedent for future extensions of federal authority to limit or abolish slavery. An insistence upon a strict construction of the Constitution was the means to avert such intrusions, and southern politicians attempted to apply that standard to a number of disparate questions before Congress, as in their opposition to a federal prohibition of mail deliveries on Sunday. On the latter question, see Gaines M. Foster, *Moral Reconstruction: Christian Lobbyists and the Federal Legislation of Morality, 1865–1920* (Chapel Hill: University of North Carolina Press, 2002), 18. For a discussion of the more general point, first articulated publicly by North Carolinian Nathaniel Macon, that a federal government that assumed the power to undertake internal improvements could next undertake to abolish slavery, see Larson, *Internal Improvement*, 105, 126.

10. *Appendix to the Congressional Globe*, 29th Cong., 1st sess., House, March 10, 1846, 357.

11. Sellers, *James K. Polk*, 445, 468–70, 486.

12. *Appendix to the Congressional Globe*, 29th Cong., 1st sess., House, March 10, 1846, 357.

13. Ibid., May 28, 1846, 688, original italics.

14. Ibid., March 11, 1846, 401.

15. Ibid., March 10, 1846, 357.

16. *Gales & Seaton's Register of Debates in Congress*, 21st Cong., 2nd sess., vol. 7, February 19, 1831, 757–58.

17. *Memorial of the Citizens of Cincinnati, to the Congress of the United States, Relative to the Navigation of the Ohio and Mississippi Rivers* (Cincinnati: L'Hommedieu, 1843), 16.

18. Vicki Vaughn Johnson, *The Men and the Vision of the Southern Commercial Conventions, 1845–1871* (Columbia: University of Missouri Press, 1992), 15.

19. John C. Calhoun, "Address on Taking the Chair of the Southwestern Convention" (dateline "Memphis, Nov. 13th, 1845"), in *Papers of John C. Calhoun*, 22: 281.

20. "The Louisville Canal," in Daniel Webster, *Works of Daniel Webster*, 6th ed. (Boston: Little, Brown, 1853), 4: 248.

21. James Gadsden to John C. Calhoun, Charleston, SC, October 21, 1845, in *Papers of John C. Calhoun*, 22: 236.

22. James Gadsden to John C. Calhoun, Charleston, SC, October 24, 1845, ibid., 22: 238. Gadsden was hardly alone in this view of the relationship between the West and the South. For expressions of similar sentiments, see the letter to Calhoun from John P. King, Augusta, GA,

October 27, 1845, ibid., 22: 243; and the letter to Calhoun from William C. Anderson, St. Louis, November 9, 1845, ibid., 22: 259.

23. R. B. Way, "The Mississippi Valley and Internal Improvements, 1825–1840," Mississippi Valley Historical Association, *Proceedings* 4 (1910–11): 178–79; Niven, *John C. Calhoun and the Price of Union*, 287, 293–94.

24. Johnson, *The Men and the Vision of the Southern Commercial Conventions*, 15.

25. Niven, *John C. Calhoun and the Price of Union*, 281.

26. Ibid., 273–89.

27. Ibid., 287.

28. Ibid., 293–94.

29. Ibid., 295–96.

30. *The Year of Decision: 1846* (Boston: Little, Brown, 1943), 103.

31. Paul F. Paskoff and Daniel J. Wilson, eds., "J. D. B. De Bow and the *Commercial Review*," in *The Cause of the South: Selections from De Bow's Review, 1846–1867* (Baton Rouge: Louisiana State University Press, 1982), 2, 8. In July 1850, about three months after Calhoun's death, the *Review* carried an editorial by De Bow, "The Cause of the South," which called for a program of southern economic development, even to the point of self-sufficiency, and the proposed drive for economic independence from the North displaced the journal's original call for an interregional economic alliance between the South and the West. Such an alliance, De Bow said, was not practicable; only through self-sufficiency could the South hope to preserve slavery.

32. "Report on the Memphis Memorial," in Senate of the United States, June 26, 1846, *Papers of John C. Calhoun*, 22: 210.

33. Ibid., 22: 222.

34. John C. Calhoun to James Edward Colhoun, July 2, 1846, ibid., 22: 261

35. *Appendix to the Congressional Globe*, 29th Cong., 1st sess., House, February 10, 1846, 454.

36. *Memorial of Sundry Delegates from Certain Counties and Towns in Pennsylvania, for an Appropriation to Improve the Navigation of the Monongahela River from Pittsburg to the Cumberland Road at Brownsville, &c.*, 23rd Cong., 2nd sess., 1835, S. Doc 98, serial 268, 1.

37. Ibid., 2.

38. Ibid., 2–3.

39. *Memorial of the Legislature of Arkansas, Asking for an Appropriation for Completing the Removal of the Raft in Red River*, 29th Congress, 1st sess., 1845, S. Doc 31, serial 472, 2.

40. *Memorial of John B. Sterigere and Others, Praying an Appropriation for the Improvement of the Alleghany River*, 25th Cong., 2nd sess., 1838, S. Doc. 174, serial 316, 1.

41. Ibid., 2.

42. Ibid.

43. *Memorial of a Number of Citizens of Bond County, Illinois, Praying Appropriations for the Improvement of the Navigation of the Western Rivers and Lakes, and for the Completion of the Cumberland Road.* 28th Cong., 1st sess., 1844, S. Doc. 216, 1.

44. Ibid., 2.

45. Ibid., 4.

46. Support for railroads by state governments took a variety of forms, including offers of generous terms in the charters of incorporation, construction of rights of way and laying track, and subsidies of railroad construction with credits and cash. See George Rogers Taylor, *The Transportation Revolution, 1815–1860*, vol. 4 of *The Economic History of the United States* (New York: Holt, Rinehart and Winston, 1951), 88–96. On the subject of support for railroads by state governments and by the federal government, see also Colleen A. Dunlavy, *Politics and Industrialization: Early Railroads in the United States and Prussia* (Princeton: Princeton University Press, 1994), 51–52, 56–63, and 100–114 passim. Railroad enthusiasm is discussed in greater detail in chapter 4 of the present work.

47. *Memorial of the Chicago Convention, in Favor of the Improvement of Harbors and Rivers by the General Government*, 30th Cong., 1st sess., 1848, S. Doc. 146, serial 511, 1.

48. Ibid., 25.

49. Ibid., 27.

50. All information about congressional service is from the *Biographical Directory of the American Congress*.

51. Silbey examined this sort of behavior by analyzing a number of roll-call votes taken on these issues. See Silbey, *The Shrine of Party: Congressional Voting Behavior, 1841–1852* (Pittsburgh: University of Pittsburgh Press, 1967). With respect to internal improvements, Silbey concluded that "local and regional pressures were important" (116).

52. Democratic congressmen representing districts along Lake Erie broke ranks with their party over the improvements question by supporting improvements bills in the decades before the Civil War. See Stanley Norman Pearce, "Constituency or Party? Democratic Congressmen and Internal Improvements on Lake Erie, 1825–1860" (Ph.D. diss., Southern Connecticut State University, 1994).

53. Harold W. Stanley and Richard G. Niemi, *Vital Statistics on American Politics, 1999–2000* (Washington, D.C.: CQ Press, 2000), 35.

54. Ibid.; the definitive account of the implosion of the Whig Party is given in Michael F. Holt, *The Rise and Fall of the American Whig Party: Jacksonian Politics and the Onset of the Civil War* (New York: Oxford University Press, 1999), chap. 20 passim and, in particular, table 31. For the connection between the Whig Party's decline in the South and the issue of slavery, see William J. Cooper Jr., *The South and the Politics of Slavery, 1828–1856* (Baton Rouge: Louisiana State University Press, 1978), chap. 9, esp. 322–41. A useful study of the rise and fall of the Whig Party in one southern state is John M. Sacher, *A Perfect War of Politics: Parties, Politicians and Democracy in Louisiana, 1824–161* (Baton Rouge: Louisiana State University Press, 2003).

55. *Statement of Appropriations and Expenditures*, serial 1992, 227.

56. *Congressional Globe*, 32nd Cong., 1st sess., House, July 29, 1852, 1999; Cartter represented Ohio's Eighteenth Congressional District, which lay to the south of Akron. It was not a district that would have directly benefited from the river improvements called for by Ward.

57. Ibid., 2000.

58. Ibid.

59. Horace Greeley, *A Political Text-book for 1860: Comprising a Brief View of Presidential*

Nominations and Elections. . . . (New York: Tribune Association, 1860), 23; "Making of America" Web site, http://moa.umdl.umich.edu.

60. James Pinkney Hambleton, *A Biographical Sketch of Henry A. Wise, with a History of the Political Campaign in Virginia in 1855* (Richmond, VA: J. W. Randolph, 1856), 491; "Making of America" Web site, http://moa.umdl.umich.edu.

61. Appropriations for improvements listed under the heading "Miscellaneous" typically involved projects on a river that flowed along or through two or more states or territories. As such, the projects were not listed under the names of the individual states affected by them. For a detailed explanation of the method used here to apportion miscellaneous appropriations, see appendix B.

62. The figure of $167,500 is a conservative one. Not included in it are another $85,000 appropriated for specific projects in Mississippi Valley states: $20,000 for Indiana and $65,000 for Illinois. These amounts are excluded because, although both Indiana and Illinois are Mississippi Valley/Ohio Valley states, they are also Great Lakes states and are therefore presumably implicated in Campbell's alleged conspiracy of interests. Also excluded are $35,000 for Pennsylvania, in this case because, although western Pennsylvania is in the watershed of the Ohio River valley (and therefore also in that of the Mississippi River valley), almost all of the money appropriated to the state for internal improvements before the Civil War went to the eastern half of the state or to its harbor at Erie on Lake Erie. See *Statement of Appropriations and Expenditures*, serial 1992, passim.

63. *Congressional Globe*, 28th Cong., 1st sess., vol. 13, House, April 22, 1844, 528, original italics.

64. *Appendix to Congressional Globe*, 29th Cong., 1st. sess., March 11, 1846, 360.

65. Ibid., 361.

66. Ibid.

67. Ibid., 363.

68. *Memorial of the Chicago Convention*, 20.

69. Ibid., 21.

70. Ibid., 21–22.

71. *Congressional Globe*, 32nd Cong., 1st. sess., Senate, August 23, 1852, 1138.

72. Stephen A. Douglas to Joel A. Matteson, January 2, 1854, in Stephen A. Douglas, *The Letters of Stephen A. Douglas*, ed. Robert W. Johannsen (Urbana: University of Illinois Press, 1961), 272–82.

73. *Congressional Globe*, 32nd Cong., 1st. sess., Senate, August 23, 1852, 1137.

74. Douglas to Matteson, January 2, 1854, in *Letters of Stephen A. Douglas*, 279.

75. Ibid., 275.

76. Ibid., 276.

77. *Appendix to the Congressional Globe*, 32nd Cong., 1st sess., Senate, August 23, 1852, 1133. Subsequent quotations from this debate are from this source and will be cited parenthetically by page number in the main text.

78. Capitalization in the original.

79. Polk's message of December 15, 1847, cited several instances when Congress had ap-

proved the levying of tonnage duties by states, including Georgia in 1787 and again in 1804 to clear Savannah Harbor of wrecks from the Revolutionary War. See James K. Polk, "Veto Message (Pocket Veto) to the House of Representatives," in James D. Richardson, *A Compilation of the Messages and Papers of the Presidents* (New York: Bureau of National Literature, 1897), 5: 2466–67.

80. For Soulé's political orientation, see Roger W. Shugg, *Origins of Class Struggle in Louisiana: A Social History of White Farmers and Laborers during Slavery and After, 1840–1875* (University: Louisiana State University Press, 1939), 158.

81. For Weller's military service, see the entry on Weller in the *Biographical Directory of the American Congress, 1990–91*.

82. *Appendix to the Congressional Globe*, 29th Cong., 1st sess., House, March 16, 1846, 437.

83. Ibid., March 10, 1846, 358.

84. Ibid., March 16, 1846, 437. For an incisive discussion of Jefferson Davis's strict constructionist view of the Constitution and its specific application to the question of whether the federal government could legitimately fund river improvements, see William J. Cooper Jr., *Jefferson Davis, American* (New York: Knopf, 2000), 118, 169.

85. *Appendix to the Congressional Globe*, 29th Cong., 1st sess., House, March 16, 1846, 437.

86. The reference is to Berlin in the years before 1914. See Barbara Tuchman, *The Proud Tower: A Portrait of the World before the War, 1890–1914* (New York: Bantam Books, 1971), 378.

87. On the latter stages of the disintegration of the Whig Party, see Holt, *The Rise and Fall of the American Whig Party*, 954–58; on the consonance of Pierce's views with those of leading southern Democrats, see Cooper, *The South and the Politics of Slavery*, 333.

88. "A Pierce-t," *Vanity Fair*, May 12, 1860, 312; http://moa.umdl.umich.edu/moa_html.

89. Roy Franklin Nichols, *Franklin Pierce: Young Hickory of the Granite Hills* (Philadelphia: University of Pennsylvania Press, 1958), 98.

90. Ibid., 104.

91. Stephen A. Douglas to Charles H. Lamphier, 11 November 1853, in *Letters of Stephen A. Douglas*, 268.

92. For total numbers of vetoes, see Stanley and Niemi, *Vital Statistics on American Politics*, tables 6–9.

93. *Message of the President of the United States, Returning to the House of Representatives a Bill Entitled "An Act Making Appropriations for the Repair, Preservation, and Completion of Certain Public Works, Heretofore Commenced under Authority of Law," with His Objections*, 33rd Cong., 2nd sess., 1855, S. Exec. Doc. 17, serial 751, 12.

94. Nichols, *Franklin Pierce*, 354–55.

95. *Veto Message. Message from the President of the United States, Returning the Bill of the House No. 392, Making Certain Appropriations, &c.*, 33rd Cong. 2nd sess., 1854, H. Exec. Doc. 2, serial 780, 2.

96. Nichols, *Franklin Pierce*, 355.

97. *Statement of Appropriations and Expenditures*, serial 1992, passim.; Nelson, "Presidential Influence on the Policy of Internal Improvements," appendix A, 66.

98. Nichols, *Franklin Pierce*, 466–68.

99. Quoted in the *Warsaw (IL) Bulletin*, June 12, 1856, quoted in Allan Nevins, *Ordeal of the Union*, vol. 2, *A House Dividing, 1852–1857* (New York: Scribners,' 1947), 502.

100. *Message of the President of the United States, Assigning His Reasons for Not Approving a Bill, Entitled "An Act Making an Appropriation for Deepening the Channel over the St. Clair Flats, in the State of Michigan,"* 36th Cong., 1st sess., 1860, S. Exec. Doc. 6, serial 1027, 1.

101. Ibid., 2–5.

102. *Joint Resolutions of the Legislature of the State of Michigan, Relative to the State of the Union*, 36th Cong., 2d sess., 1861, H. Misc. Doc. 38, serial 1103, 1.

4. WAYS AND MEANS

1. The growth in the number of the federal government's civilian workers is consistent with a more general trend noted by John Joseph Wallis, who points out that federal spending per capita increased, though not consistently, from one decade to the next during the period 1800 to 1860. See John Joseph Wallis, "American Government Finance in the Long Run: 1790 to 1990," *Journal of Economic Perspectives* 14 (Winter 2000), table 1.

2. For an insightful analysis of the development of the postal system in the United States from the beginning of the American Revolution until the business success of Morse's telegraph, see Richard R. John, *Spreading the News: The American Postal System from Franklin to Morse* (Cambridge: Harvard University Press, 1995). John is concerned with the institutional growth of the Post Office and also with the "communications revolution" that it began and with that revolution's social, cultural, political, and economic consequences. Also useful, but more narrowly focused on the rise of a federal bureaucracy during the Jacksonian era, is Matthew A. Crenson, *The Federal Machine: Beginnings of Bureaucracy in Jacksonian America* (Baltimore: Johns Hopkins University Press, 1975). Crenson notes that the two largest and, for the public, most directly felt agencies of the federal government were the Post Office and the Land Office (xi). For an account of the transformation of the United States government into the juggernaut that it had become by the end of the Civil War, see Heather Cox Richardson, *The Greatest Nation of the Earth: Republican Economic Policies during the Civil War* (Cambridge: Harvard University Press, 1997). In pursuit of her goal of demonstrating the extent and depth of this transformation, Richardson tends to understate the antebellum national government's capacities and accomplishments.

3. All dollar amounts, except for the army, are derived from data contained in the *Statement of Appropriations and Expenditures for Public Buildings, Rivers and Harbors, Forts, Arsenals, Armories, and Other Public Works from March 4, 1789 to June 30, 1882*, 47th Cong., 1st sess., 1881, S. Exec. Doc. 196, serial 1992. The figure for the United States Army is the sum of the individual annual entries listed for the War Department in U.S. Bureau of the Census, *Historical Statistics of the United States, Colonial Times to 1970*, electronic ed., ed. Susan Carter et al. [machine-readable data file] (Cambridge University Press, 1997), chapter Y: Government Series Y 457–465, "Outlays of the Federal Government: 1789–1970."

4. *Statement of Appropriations and Expenditures*, serial 1992, passim.

5. Ibid.

6. Erik F. Haites, James Mak, and Gary M. Walton, *Western River Transportation: The Era of Early Internal Development, 1810–1860* (Baltimore: Johns Hopkins University Press, 1975), 95–96.

7. Ibid., 100 and 100 n. 43.

8. They performed a regression, reproduced here, of appropriations in a given year, A_t, on expected total revenue the preceding year, TR_{t-1}, and derived the equation, $A_t = -121{,}790 + .00843 TR_{-1}$, for which the Adjusted $R^2 = .31$; the F-statistic = 8.1665; Prob (F-statistic) = .0120; and the Durban-Watson statistic, a measure of serial correlation, = 1.5358 (generally, a value of less than 2.0 indicates positive serial correlation). These results pertain to the period 1824–40 and are significant. The same basic regression performed for 1841–60 did not yield results that were either statistically significant or reliable, essentially because, for several years in this latter period, there were no appropriations for western river improvements. See ibid., 100 and 100 n. 43.

9. The results of a regression of appropriations for western improvements in a given year, A_t, on the preceding year's appropriation, A_{t-1}, and the preceding year's budgetary balance, B_{t-1}, are as follows: Adjusted $R^2 = .4930$; F-statistic = 16.5596; Prob (F-statistic) = .0010; and the Durban-Watson statistic = 2.0420. The equation is: $A_t = .665307 A_{t-1} + .005734 B_{t-1} + \in$, where $\in$ is an error term.

10. In periods of economic distress, statements in congressional documents to the effect that the straitened circumstances of the federal Treasury necessitated reduced appropriations for and expenditures on improvements were not uncommon. See, for example, the report of the House of Representatives Committee on Commerce titled *Improvement of Bars, Harbors, and Rivers in the Territory of Florida*, 27th Cong., 2nd sess., 1842, H. Rep. 886, serial 410. For the quotation in the text, see House of Representatives Committee on Commerce, *Harbors and Rivers (to Accompany Bill H. R. no. 483)*, 35th Cong., 1st sess., April 15, 1858, H. Rep. 251, serial 965, 2.

11. Although not always included by historians in accounts of the antebellum internal improvements program, contemporaries often considered light stations, beacons, and marker buoys to be part of the program as, apparently, did Congress in its report of 1881 on public works. See *Statement of Appropriations and Expenditures*, serial 1992. If this amount is included in the total of congressional appropriations for the program of internal improvements to natural waterways, harbors, and coastal waters, then the federal government's investment in the program came to $29,592,958 by the end of 1860.

12. House of Representatives Committee on Commerce, *Light-Houses, &c., and Surveys for 1838 (to Accompany Bill H. R. no. 712)*, 25th Cong., 2nd sess., 1838, H. Rep. 752, serial 335, 1.

13. Ambrose Bierce, *The Devil's Dictionary* (1911; New York: Dell, 1991), 102.

14. See *Statement of Appropriations and Expenditures*, serial 1992, for appropriations for light stations, marker buoys, and beacons. For coastal tonnage, see Douglass C. North, *The Economic Growth of the United States, 1790–1860* (Englewood Cliffs, NJ: Prentice-Hall, 1961), appendix 2, table G–IX. For net tonnage clearing American ports in the foreign trade, see Susan B. Carter et al., eds., *Historical Statistics of the United States: Earliest Times to the Present*, Millennial ed., vol. 4, pt. D, Economic Sectors (New York: Cambridge University Press, 2006),

table Df594–605, "U.S. and Foreign Vessels Entered and Cleared—Net Tonnage, by Type of Port, 1789–1995." The correlation coefficients among the three variables—coastal tonnage, tonnage clearing American ports in the foreign trade, and appropriations for light stations, marker buoys, and beacons—during the forty-year period 1821–60 are very high and are significant at better than the .001 level of confidence:

	Appropriations	Coastal tonnage	Clearing for foreign ports
Appropriations	1.0000	.8097	.7578
Coastal tonnage		1.0000	.9648
Clearing for foreign ports			1.0000

15. "Report of the Chief, Topographical Engineers, Bureau of Topographical Engineers, Washington, November 17, 1848," in *Report of the Chief Engineer to the Secretary of War, at the Opening of the First Session of the Thirtieth Congress* (Washington: Wendell and Van Benthuysen, 1848), 1.

16. The history of President James Buchanan's decision to send the army into Utah to reestablish federal authority there is well known. A good summary of the Mormon War is given in Kenneth M. Stampp's *America in 1857: A Nation on the Brink* (New York: Oxford University Press, 1990), 196–208. Stampp notes that, notwithstanding the absence of battles, the army sent to Utah "suffered heavy losses from sickness and the bitter cold of a mountain winter" (205).

17. For what is still the best succinct characterization of the suspicions about Polk's motives and actions in taking the United States to war, see David M. Potter, *The Impending Crisis, 1848–1861*, completed and edited by Don E. Fehrenbacher (New York: Harper and Row, 1976), 4.

18. *Congressional Globe*, n.s., 29th Cong., 2nd sess., House, February 18, 1847, 451.

19. Ibid.

20. The Supreme Court's decision in favor of Pennsylvania and against the bridge company was not unanimous: Chief Justice Roger B. Taney and Associate Justice Peter V. Daniel dissented. A brief synopsis of the majority decision and Taney's and Daniel's minority opinions is given in Clifford M. Lewis, S.J., "The Wheeling Suspension Bridge," *West Virginia History* 33, no. 3 (1972): 215–20. An indication of the importance attached to the controversy and its associated early legal proceedings is the coverage accorded them in such a specialized publication as *Scientific American*. See *Scientific American* 4, no. 52 (September 15, 1849), 414; Cornell University's "Making of America" Web site, http://moa.cit.cornell.edu, 414. For the economic implications of the decision for the steamboat industry, see Louis C. Hunter, *Steamboats on the Western Rivers: An Economic and Technological History* (Cambridge.: Harvard University Press, 1949; reprint, Mineola, NY: Dover, 1993), 489. An interesting perspective on the action taken by Congress to protect the bridge is offered in a contemporary account that was unabashedly biased in favor of the bridge interests. See Eli Bowen, *Rambles in the Path of the Steam-Horse . . . Embracing*

a *General Historical and Descriptive View of the Scenery, Agricultural and Mineral Resources, and Prominent Features of the Travelled Route from Baltimore to Harper's Ferry, Cumberland, Wheeling, Cincinnati, and Louisville* (Philadelphia: W. Brownell and W. W. Smith, 1855), 359; University of Michigan's "Making of America" Web site, http://moa.umdl.umich.edu.

21. George Rogers Taylor, *The Transportation Revolution, 1815–1860*, vol. 4 of *The Economic History of the United States* (New York: Holt, Rinehart and Winston, 1951), 25, 30.

22. Ibid., 102.

23. Contemporaries were emphatic about the linkage between river improvements and the sale of public lands and the subsequent settlement and economic development of that land. Assertions along these lines were a staple of the memorials and petitions sent to Congress to urge appropriations for particular river improvements, as well as of the reports by the Bureau of Topographical Engineers. See, for example, *Letter from the Secretary of War [Lewis Cass], Transmitting the Information Required by a Resolution of the House of Representatives of the 2d Instant, in Relation to Certain Works of Internal Improvement*, 24th Cong., 1st sess., 1836, H. Doc. 212, serial 290, 1–2; *Petition of a Number of Citizens of Missouri, Praying an Appropriation for the Improvement of the Current River*, 26th Cong., 1st sess., 1840, S. Doc. 302, serial 359, 1; *Memorial of Numerous Citizens of Illinois, Praying an Appropriation of Public Lands for the Improvement of Rock River*, ibid., 1840, S. Doc. 492, serial 360, 1–3; *Memorial of the Legislature of Arkansas, Asking for an Appropriation for Completing the Removal of the Raft in Red River*, 29th Cong., 1st sess., 1845, S. Doc. 31, serial 472, 1; *Memorial of the Legislature of Mississippi, Asking the Appropriation to That State of Alternate Sections on the Leaf, Pascagoula, and Chickasahay Rivers, for the Improvement of Those Rivers*, ibid., 1846, S. Doc. 224, serial 474, 1; *Report from the Bureau of Topographical Engineers, on the Subject of Rivers and Harbors*, 31st Cong., 1st sess., 1850, H. Misc. Doc. 54, serial 582, 16; and *Memorial of the Legislature of the State of Minnesota, Asking an Appropriation for the Improvement of St. Croix River*, 36th Cong., 2nd sess., 1861, H. Misc. Doc. 37, serial 1103, 1. The connection between internal improvements in the states of the Old West and the subsequent increases in that region's supply of agricultural products available to the East and South is described in North, *The Economic Growth of the United States, 1790–1860*, 143.

24. The number of towns formed in Indiana, Mississippi, Missouri, and Ohio during the years 1835–56 is strongly correlated with interest rates in Indiana (correlation coefficient, $R = 0.4757$, for 22 cases; and $R^2 = 0.2263$). In this case, the interest rates in Indiana are a proxy for Lower Ohio River valley, Lower Missouri River valley, and Mississippi River valley regional interest rates. The equation describing the relationship is TOWNS = $-65.566 + 29.947(\text{RATE}) - 1.773(\text{RATE}^2)$. This equation is a polynomial of the second degree, which is consistent with how we would expect interest rates to affect town formation: at low to moderate rates of interest, an increase in the rate is accompanied by an increase in the number of towns formed; once, however, interest rates exceed the range of 6 to 8 percent, the number of towns formed decreases. The numbers of towns formed are derived from the following sources: for Indiana, Ronald L. Baker and Marvin Carmony, *Indiana Place Names* (Bloomington: Indiana University Press, 1975); for Mississippi, but only through 1848, see A. Hutchinson, *Code of Mississippi: Being an Analytical Compilation of the Public and General Statutes of the Territory and State with Tabular References to the Local and Private Acts, from 1798 to 1848*. . . . (Jackson, MS: Price

and Fall, State Printers, 1848); for Missouri, see Robert L. Ramsay, *Our Storehouse of Missouri Place Names*, Missouri Handbook No. 2, *The University of Missouri Bulletin* (Columbia: University of Missouri, 1952); and for Ohio, William D. Overman, *Ohio Town Names: The Origin of the Names of over 500 Ohio Cities, Towns and Villages* (Akron, OH: Atlantic Press, 1958) and David Lindsey, *Ohio's Western Reserve: The Story of Its Place Names* (Cleveland: Press of Western Reserve University and the Western Reserve Historical Society, 1955). For interest rates in Indiana, see Howard Bodenborn and Hugh Rockoff, "Regional Interest Rates in Antebellum America," in Claudia Goldin and Hugh Rockoff, eds., *Strategic Factors in Nineteenth Century American Economic History: A Volume to Honor Robert W. Fogel*, National Bureau of Economic Research Conference Report (Chicago: University of Chicago Press, 1992), table 5.2. The interest rates used here are, strictly speaking, "interest rate proxies." For a detailed discussion of the calculation of the latter, see the source, 164–66.

25. Most sales of public lands were made to individuals. See Benjamin Horace Hibbard, *A History of the Public Land Policies*, foreword by Paul W. Gates (Madison: University of Wisconsin Press, 1965), 101.

26. A recent exploration of the role of internal improvements, primarily roads, in spurring the populating and economic development of the Old Northwest is Laurence Joseph Malone, *Opening the West: Federal Internal Improvements before 1860* (New York: Greenwood Press, 1998). This book is essentially Malone's 1991 doctoral dissertation, done at the New School for Social Research. It is a contribution to a slowly growing body of literature that, by accretion, has begun to make a convincing case for the view that nineteenth-century federal policy measures promoted economic growth and development. In attempting to demonstrate the nature of the connection between federal funding of improvements and western development, Malone brings to bear a large body of quantitative data. Although his argument and evidence suggest that there was such a connection, the case can be made more conclusively and rigorously, as is done herein.

27. The pattern of railroad-induced town formation is a familiar one, and evidence of it in several states is plentiful. For Indiana, see Baker and Carmony, *Indiana Place Names*. For Ohio, see Overman, *Ohio Town Names*, and Lindsey, *Ohio's Western Reserve: The Story of Its Place Names*. Similar information is available for other western river states. For Kentucky, see Robert M. Rennick, *Kentucky Place Names* (Lexington: University Press of Kentucky, 1984); for Missouri, see Ramsay, *Our Storehouse of Missouri Place Names*; for Michigan, see Walter Rummage, *Michigan Place Names: The History of the Founding and the Naming of More Than Five Thousand Past and Present Michigan Communities* (Detroit: Wayne State University Press, 1986).

28. Town formation data for five states (Kentucky is included) are charted in figure 17, as opposed to data for four states in figure 14, because Kentucky land sales data were not available.

29. The cost of a dredger and scows is from *Report of the Colonel [J. J. Abert, Commanding Corps Topographical Engineers] of Corps of Topographical Engineers, for the Year 1855*, appendix B, 51; hourly dredging rate is from J. J. Abert, Col. Corps of Topographical Engineers, Note of July 15, 1846, in *Report of the Secretary of War, in Compliance with a Resolution of the Senate, Showing the Aggregate Appropriations for Improving Rivers and Harbors since July 1, 1836, and the*

Sum Expended in Dredging Operations, with the Cost per Cubic Yard of Removing Obstructions, 29th Cong., 1st sess., 1846, S. Exec. Doc. 451, serial 478, 11.

30. As is often the case with respect to many facets of the history of nineteenth-century steamboats on the western rivers, the best account of Shreve's invention and its application is Hunter, *Steamboats on the Western Rivers*, 192–200. A less rigorous account is to be found in Edith McCall, *Conquering the Rivers: Henry Miller Shreve and the Navigation of America's Inland Waterways* (Baton Rouge: Louisiana State University Press, 1984), 180–211. For accounts of the single-hulled snag boat, see, for example, *Report of the Secretary of War, Made in Compliance with a Resolution of the Senate in Relation to the Work Done under the Appropriations of 1852 for the Improvement of Western Rivers and Harbors*, 33rd Cong., 1st sess., 1854, S. Exec. Doc. 51, serial 698, 10, which concerns snag removal on the Arkansas River. The report states, in part: "The channel of the rivers, especially in low water, is too narrow and crooked, and the current too rapid to admit of the successful operation of a twin snag-boat of the usual size, although in an elevated stage of the river such a boat may operate to great advantage." Also see *Report of the Colonel of Corps of Topographical Engineers, for the Year 1855*, 33.

31. *Report of the Colonel of Corps of Topographical Engineers, for the Year 1855*, appendix B, "Annual Report on Western River Improvements, September 1, 1855, by Lieutenant Colonel S. H. Long, Superintendent Western River Improvements," Office Western River Improvements, Louisville, September 1, 1855. "Supplementary Observations, Comprising Sundry Desultory Views and Suggestions in Reference to the Improvement of the Western Rivers," 32.

32. Ibid., 33.

33. Ibid., report by Chas. A. Fuller, "U.S. Agent and Engineer O.R. Improvement, &c, &c. ('Office of the Ohio River Improvements')" to Lieut. Col. S. H. Long, "Superintendent W. R. [Western River] Improvements, Louisville, Kentucky," Louisville, September 1, 1853, 21.

34. Ibid., 21–22.

35. Ibid., 8.

36. Ibid., 10.

37. Ibid., 11.

38. *Report of the Colonel of Corps of Topographical Engineers, for the Year 1855*, appendix—Doc. 13, "Annual Report on Improvement of Ohio and Red Rivers," Office Red River Improvements, Louisville, September 1, 1855, 52.

39. Ibid., appendix B, "Annual Report on Western River Improvements," September 1, 1855, 28.

40. Ibid.

41. Ibid., appendix—Doc. 4, "Tour for Funds, Inspection, and Payment," by Thos. R. Sinton, Clerk, Office Western River Improvements, Louisville, April 30, 1855, 40.

42. Although called a "light-draught" snag boat, *Terror* was a "twin" snag boat, that is, one with twin hulls. The cost of constructing and fully equipping such a vessel was estimated at $30,000; the approximate cost of a single-hull, or "light-draught," snag boat, was $15,000. See ibid., appendix B, "Annual Report on Western River Improvements," September 1, 1855, 33.

43. Ibid., 40; ibid., appendix B—Doc. 14, "Inspection Report of June 20, 1855," Louisville, June 20, 1855, 61.

5. FACTORS OF DESTRUCTION

44. Ibid., appendix B—Doc. 7, "Annual Report on Improvement of Mississippi Rapids," Keokuk, August 25, 1855, 43.

1. See Louis C. Hunter, *Steamboats on the Western Rivers: An Economic and Technological History* (Cambridge: Harvard University Press, 1949; reprint, Mineola, NY: Dover, 1993), 65–76, esp. 69–70, and Adam I. Kane, *The Western River Steamboat* (College Station: Texas A&M University Press, 2004), chaps. 3 and 4. William Thompson, *A Tradesman's Travels, in the United States and Canada, in the Years 1840, 41 & 42* (Edinburgh, 1842), 54–55, quoted in Charles S. Sydnor, *The Development of Southern Sectionalism, 1819–1848* (Baton Rouge: Louisiana State University Press, 1948), 257.

2. Unless otherwise indicated, the data and findings presented here pertain specifically to the steam packets that plied the major western rivers. The reason for relying on data about packets is simply that they are the vessels for which the most complete information about hull sizes and engines exists. Packets were among the largest steamboats, and, while they often carried freight, their speed and plush accommodations were designed to attract passengers. Although packets were generally larger and equipped with more powerful engines than the common run of steamboats, any conclusions drawn about the operation of packets, especially with respect to vulnerability to river hazards, would apply with even more force to the more numerous ordinary steamboats. I have relied primarily on the unsurpassed account in Hunter, *Steamboats on the Western Rivers*, and on material in Kane, *The Western River Steamboat*. Erik F. Haites, James Mak, and Gary M. Walton, *Western River Transportation: The Era of Early Internal Development, 1810–1860* (Baltimore: Johns Hopkins University Press, 1975), presents useful information of a more general nature concerning the architecture, machinery, and construction of steamboats.

3. This statement is based on an analysis of data in the Steamboat Packet Database, and supported by Hunter, *Steamboats on the Western Rivers*, 75 and table 5.

4. Hunter, *Steamboats on the Western Rivers*, 76. See also U.S. Congress, House, Department of the Treasury, *Report on the Internal Commerce of the United States*, William F. Switzler, pt. 2 of *Report on the Commerce and Navigation of the U.S.*, 50th Cong., 1st sess., 1888, H. Exec. Doc. 6, serial 2552, 206.

5. Figures 18 and 19 are derived from Hunter, *Steamboats on the Western Rivers*, tables 7 and 8. Hunter's tables 7 and 8 are themselves derived from his table 4, "Average Hull Dimensions of Western Steamboats by Selected Tonnage Groups, 1818–1880." In his account of its sources at the foot of table 4, he notes: "The number of vessels from which the averages were computed ranged, with minor exceptions, from 10 to 25, with 20 the most common." Although Hunter's imprecision with respect to the number of steamboats used to compute the average values does not undermine his main point, the reader should bear in mind the fact that the impression given in figures 18 and 19 of a pronounced convergence of values is probably somewhat overdrawn.

6. The source for figures 20 and 21 is the Steamboat Packet Database, abstracted from in-

formation given in Frederick Way Jr., comp., *Way's Packet Directory, 1848–1994: Passenger Steamboats of the Mississippi River System since the Advent of Photography in Mid-Continent America*, rev. ed. (Athens: Ohio University Press, 1994). While figures 20 and 21 pertain to a smaller period than do figures 18 and 19, derived from Hunter's data, they are also more detailed, charting changes in the values of hull dimension ratios on an annual basis over a nineteen-year period, rather than in the selected years used by Hunter. The two sets of figures, however, are mutually reinforcing and together establish the important point made by Hunter and elaborated upon here.

7. *Memorial of the Citizens of Cincinnati, to the Congress of the United States, Relative to the Navigation of the Ohio and Mississippi Rivers* (Cincinnati: L'Hommedieu, 1843), 24. The explanation of the phenomenon given by Louis C. Hunter is both succinct and accurate: "The resistance offered by the water to the passage of the boat is caused by the combined effect of friction of the wetted surface and the energy required to produce waves. With increased speed the wave-making resistance mounts more rapidly than frictional resistance and is relatively less for a long vessel than for a short one" (*Steamboats on the Western Rivers*, 87).

8. Thomas A. McMahon and John Tyler Bonner, *On Size and Life* (New York: Scientific American Books, 1983), 47. They analyze the phenomenon with respect to the design of rowing shells.

9. Hunter, *Steamboats on the Western Rivers*, 88.

10. Sixty-four percent of steam packets produced in Ohio and 67 percent of those turned out in Kentucky were 183 feet or shorter. For the basic data and sources on which these percentages are based, see chap. 1, table 1.

11. Derived from: Louis C. Hunter, *A History of Industrial Power in the United States, 1780–1930*, vol. 2, *Steam Power* (Charlottesville: Published for the Hagley Museum and Library by the University of Virginia Press, 1985), table 28; and *Subject-Matter Index of Patents for Inventions Issued by the United States Patent Office from 1790 to 1873, Inclusive*, vols. 1–3 (New York: Arno Press, 1976), passim.

12. A striking set of figures on the increases in western river steamboat speeds—an increase of 308 percent during the two decades prior to about 1840 and an increase by another 90 percent from 1840 to the Civil War—is given in: *Report on the Internal Commerce of the United States*, Department of the Treasury, 50th Cong., 1st sess., 1888, H. Exec. Doc. 6, serial 2552, 207, pt. 2. For the fastest steamboats on the western rivers—the packets—upstream and downstream speeds averaged fifteen miles per hour; for the trade as a whole, in average boats, typical speeds did not hit ten miles per hour. See George Rogers Taylor, *The Transportation Revolution, 1815–1860*, vol. 4 of *The Economic History of the United States* (New York: Holt, Rinehart and Winston, 1951), 138. Haites, Mak, and Walton, in *Western River Transportation* (69–72), offer an alternative set of measures of speed. If, instead of tendering steamboat speed in terms of miles per hour, one measures it in terms of the number of days per round trip or the number of round trips per year, then other factors emerge that were at least as important as improvements in hull design and engine power to increasing average speed. These were the development of routinized procedures for the collection and handling of cargo and the increase in the number of months during which steam navigation of the rivers was possible. The latter devel-

opment resulted primarily from changes in hull designs, improvements made to the river channels, and "an increase in the skills and know-how of pilots and captains" (72).

13. I am indebted to Dr. Daniel S. Allen for having called this measure of relative power to my attention. He used it to analyze changes in the technology of nineteenth-century British oceangoing shipping. See Daniel S. Allen, "The Impact of Technological Change on the Economic Viability of Individual Production Centers: The Case of the 1840–1880 British Ocean-Going Iron and Steam Shipbuilding Industry" (Ph.D. diss., Louisiana State University, 1997). Measurements of steam engine pressure, indicated as pounds per square inch, or psi, are available for thirty-two steamboats, recorded in 1850. The vessels were built between 1842 and 1850, and the average pressure of their engines was about 131 psi. Although there was a pronounced fluctuation from one year to the next in the average engine psi of steamboats built in a particular year, there was no clear pattern. Moreover, there was considerable variability in psi among steamboats built in a given year. The foregoing results are derived from a detailed set of data presented in Kane, *The Western River Steamboat*, appendix 2, which is itself drawn from R. Hyde Walworth, *Order of Reference of the Supreme Court of the United States in the Case of the State of Pennsylvania, Complainant, against the Wheeling & Belmont Bridge Company and Others, Defendants* (Washington, D.C., 1850).

14. Data pertaining to steam-powered vessels for 1819 and 1829 are for one vessel each year and are from Hunter, *Steamboats on the Western Rivers*, appendix table 14. Data pertaining to steam vessels for 1838, 1849, 1859, 1869, 1879, 1889, 1899, and 1899 fall into two groups, each of which is from a different source: figures on aggregate horsepower are from Hunter, *A History of Industrial Power in the United States*, vol. 2, *Steam Power*, table 30; aggregate gross steam vessel tonnage figures are from the U.S. Bureau of the Census, *Historical Statistics of the United States, Colonial Times to 1970*, electronic ed., ed. Susan Carter et al. [machine-readable data file] (Cambridge: Cambridge University Press, 1997), chapter Q: Transportation Series Q 417–32. The computed values of horsepower per ton for steam vessels are derived from these two sets of data. Data pertaining to steam locomotives are for representative locomotives and are from John H. White Jr., *American Locomotive: An Engineering History, 1830–1880*, rev. ed. (Baltimore: Johns Hopkins University Press, 1997), figure 36. Data pertaining to gasoline tractors are from C. H. Wendel, *Nebraska Tractor Tests since 1920* (Osceola, WI: Motorbooks International, 1993), 543–46 passim.

15. This finding is at odds with that reported by Hunter in *Steamboats on the Western Rivers*, 143. Hunter is correct that, prior to 1840, average length of the stroke of a steamboat engine's piston doubled from "four to eight feet." He is wrong, however, when he goes on to say that following "1840 engine cylinders underwent little further development." He cites five of the largest steamboats, all packets, that had strokes of eight feet, another that had a stroke of ten feet, and still another with an eleven-foot stroke. The source of Hunter's error probably lies in the small number of steamboats on which he based his generalizations. He relies on 33 steamboats for engine data. By contrast, the findings reported in this study are based on engine data for 267 steam packets. For the two post-1840 decades, this study relies on data for 264 packets. That Hunter's findings and those reported here should differ is therefore not surprising. Contrary to Hunter's conclusion that, after 1840, stroke lengths remained high, data for the 264 packets

indicate that mean stroke length in 1840 was 8.0 feet and peaked shortly thereafter, in 1845, at 8.2 feet. In 1850, mean stroke length was 7.5 feet; five years later, it was 6.5 feet; and, in 1860, it was 6.4 feet.

16. Hunter, *Steamboats on the Western Rivers*, 130; Kane, *The Western River Steamboat*, 70.

17. Kane, *The Western River Steamboat*, 70.

18. The points made in this paragraph come primarily from the account in Hunter, *Steamboats on the Western Rivers*, 129–33.

19. See Hunter, *Steamboats on the Western Rivers*, 142–46, esp. 142–43. The data for 264 steam packets, referred to in note 15, above, indicate that Hunter was wrong about the increase in engine size.

20. This observation is based on analysis of the data abstracted from Way, *Way's Packet Directory, 1848–1994,* passim. A corollary to this observation is that the larger values for average relative engine size in part reflected the inclusion of a number of low-pressure engines in the calculation, the low-pressure engines usually being considerably larger than the high-pressure design.

21. Hunter, *Steamboats on the Western Rivers*, 111–12.

22. The source of figure 30 is Howard Bodenborn and Hugh Rockoff, "Regional Interest Rates in Antebellum America," in Claudia Goldin and Hugh Rockoff, eds., *Strategic Factors in Nineteenth Century American Economic History: A Volume to Honor Robert W. Fogel,* National Bureau of Economic Research Conference Report (Chicago: University of Chicago Press, 1992), table 5.2. The interest rates used here are, strictly speaking, "interest rate proxies." For a detailed discussion of their calculation of the latter, see the source, 164–66. The interest rates pertain to Pennsylvania, outside Philadelphia, and are therefore "country rates." The reason for using them here is that western Pennsylvania's eighteen boatyards accounted for about 31 percent of the total number of yards in which steam packets were built between 1840 and 1860; those eighteen boatyards built 34 percent of all the packets constructed during the same period (see chap. 1, table 1.1).

23. The reasoning here is that engine construction, which, as a substantial capital undertaking, took considerable time, did not respond immediately to interest-rate fluctuations. Rather, the response is assumed to have been lagged one year on interest movements. This assumption is a restrained one. During the 1820s and 1830s, the output of stationary steam engines by that industry's major manufacturers was between three and five engines per year. See Hunter, *A History of Industrial Power in the United States, 1780–1930,* vol. 2, *Steam Power,* 232–33. Fragmentary returns from the 1850 and 1860 federal censuses for Kentucky, a center of steamboat building, indicate that eight firms that built machinery turned out a total of 124 steam engines, stationary and marine, alike, an annual average output of about 16 steam engines per firm. These figures are derived from ibid, appendix 2 and table 2.

24. *Conclin's New River Guide, or a Gazetteer of All the Towns on the Western Rivers . . . Compiled from the Latest and Best Authority. With Forty-four Maps* (Cincinnati: H. S. and J. Applegate, 1849), 89.

25. *Report of the Secretary of War, Made in Compliance with a Resolution of the Senate in Relation to the Work Done under the Appropriations of 1852 for the Improvement of Western Rivers and*

Harbors, 33rd Cong., 1st sess., 1854, S. Exec. Doc. 51, serial 698, table, "Positions and Distances from Mouth of the Missouri (downward) to Natchez," p. 4.

26. The best accounts of the rise of the railroad industry and its role in the economic history of the United States remain Taylor, *The Transportation Revolution, 1815–1860;* Douglass C. North, *The Economic Growth of the United States, 1790–1860* (Englewood Cliffs, NJ: Prentice-Hall, 1961); Robert William Fogel, *Railroads and American Economic Growth: Essays in Econometric History* (Baltimore: Johns Hopkins Press, 1964); Albert Fishlow, *American Railroads and the Transformation of the Ante-Bellum Economy* (Cambridge: Harvard University Press, 1965); and Alfred D. Chandler Jr., *The Visible Hand: The Managerial Revolution in American Business* (Cambridge: Belknap Press of Harvard University Press, 1977), esp. chap. 3.

27. "Railroads vs. Steamboats," *Scientific American*, March 2, 1850, 185, http://moa.cit.cornell.edu.

28. "Reception at Savannah," May 26, 1847, reprinted "From the *Savannah Republican*," June 3, 1847, in Daniel Webster, ed., *The Works of Daniel Webster*, 6th ed. (Boston: Little, Brown, 1853), 2: 404.

29. Ibid., 2: 405. Webster made similar remarks in the "Opening of the Northern Railroad to Grafton, N.H.," August 28, 1847, ibid., 2: 411.

30. U.S. Bureau of the Census, *Historical Statistics of the United States, Colonial Times to 1970*, electronic ed., ed. Susan Carter et al. [machine-readable data file] (Cambridge: Cambridge University Press, 1997), chapter J: Land and Water Utilization Series, J 21–2, "Public Land Grants by United States to Aid in Construction of Railroads, Wagon Roads, Canals, etc.: 1823 to 1871."

31. John Lauritz Larson, *Internal Improvement: National Public Works and the Promise of Popular Government in the Early United States* (Chapel Hill: University of North Carolina Press, 2001), 260. On page 6, he notes that: "After midcentury, railroad developers and other innovative entrepreneurs, while they never strayed far from government subsidies and protection, trumpeted with rising conviction the superiority of strictly private enterprise over public works." Larson points out, however, that early railroads "demanded and received public investments in one form or another . . . including land grants" (226).

32. The value of the public land grants in 1853 and 1856 were computed by multiplying the number of acres of the grants in each of the two years by the valuation of at least $2.50 per acre, which was the figure used in making the Illinois Central Railroad land grant of 1850. See Benjamin Horace Hibbard, *A History of the Public Land Policies*, foreword by Paul W. Gates (Madison: University of Wisconsin Press, 1965) 245. The figures for the railroad industry's gross capital formation in 1853 and 1856 are from Fishlow, *American Railroads and the Transformation of the Ante-Bellum Economy*, table 52 (appendix). These capital formation figures are for *all* railroads in the United States, not just those receiving grants of public lands.

33. See Fishlow, *American Railroads and the Transformation of the Ante-Bellum Economy*, 190–191. See also Carter Goodrich, "The Revulsion against Internal Improvements," *Journal of Economic History* 10 (November 1950): 167.

34. See Taylor, *The Transportation Revolution*, 86–90; Colleen A. Dunlavy, *Politics and In-*

dustrialization: Early Railroads in the United States and Prussia (Princeton: Princeton University Press, 1994), 137.

35. Peter Wallenstein, *From Slave South to New South: Public Policy in Nineteenth-Century Georgia* (Chapel Hill: University of North Carolina Press, 1987), 59.

36. On the subject of Midwestern railroad enthusiasm, see Fishlow, *American Railroads and the Transformation of the Ante-Bellum Economy,* 191–96; and Taylor, *The Transportation Revolution,* 92–94. Douglass North tabulates the railroad mileage extant in each of seven midwestern states—every one of them lying along the Ohio, Missouri, or Mississippi rivers or along the Great Lakes—in eight specific years, mainly in the 1850s, from 1845 through 1861. From these figures, one can deduce the number of miles constructed from one year to the next, and it is clear that the transportation system of the region became increasingly railroad-oriented during the 1850s. See North, *The Economic Growth of the United States,* table 10. A similar, but more detailed, listing of states, including those of New England, the Middle Atlantic and South Atlantic regions, and the Gulf South, is provided by Taylor, *The Transportation Revolution,* 79. Taylor's table provides a decennial list of canal and rail mileage in place; thus it covers 1830, 1840, 1850, and 1860.

37. *Williams' Memphis Directory, City Guide, and Business Mirror,* vol. 1, *1860* (Memphis: Cleaves and Vaden, 1860), 16. An overstated account of the reasons for railroad enthusiasm in Memphis and Louisville is given in Maury Klein, *History of the Louisville & Nashville Railroad,* in Thomas B. Brewer, gen. ed., *Railroads of America* (New York: Macmillan, 1972), 18. Klein states that "by the late 1840s the Mississippi and Ohio rivers above Memphis had become dangerously unreliable. . . . By the 1850s it was considered foolhardy to risk a valuable cargo on the river, where *losses reached nearly a ship a day* and the shells of grounded steamers littered the shoreline [my emphasis]." During the decade of the 1850s, the average number of steamboats lost each year on *all* of the western rivers was 48. By Klein's account, the number should have been closer to at least two hundred, allowing for times of the year when low water levels or ice precluded a high volume of steamboat traffic.

6. THE SUCCESS OF PUBLIC POLICY

1. For instances when advocates of improvements made this sort of argument, see *Memorial of the Citizens of Wilmington, North Carolina, for a Further Appropriation for Improving the Navigation of Cape Fear River,* 24th Cong., 1st sess., 1836, S. Doc. 370, serial 283, 1; *Memorial of the Town Council and Citizens of Erie, Pa., Praying an Appropriation for the Improvement of Presque Isle Harbor,* 27th Cong., 2nd sess., 1841, S. Doc. 23, serial 396, 1; and *Memorial of Citizens of Tennessee, Praying an Appropriation to Repair the Breach in the Dam at the Head of Cumberland Island, in the Ohio River,* 30th Cong., 1st sess., 1848, S. Misc. Doc. 99, serial 511, 1.

2. Erik F. Haites, James Mak, and Gary M. Walton stressed the impact of constitutional and economic considerations on limiting the effectiveness of river improvements, especially after 1838, in *Western River Transportation: The Era of Early Internal Improvement, 1810–1860* (Baltimore: Johns Hopkins University Press, 1975), chap. 7, esp. 95–100. A somewhat more positive appraisal of the effectiveness of the river improvements effort may be found in the

course of a discussion of snag removal by Louis Hunter in his classic study, *Steamboats on the Western Rivers: An Economic and Technological History* (Cambridge: Harvard University Press, 1949; reprint, Mineola, NY: Dover, 1993), chap. 4. The extent to which political considerations limited the effectiveness of the U.S. Army engineers' river hazard–removal projects is treated in Forest G. Hill, *Roads, Rails & Waterways: The Army Engineers and Early Transportation* (Norman: University of Oklahoma Press, 1957), chap. 7. Todd Shallat offers a more critical view of the army's engineers and their work in his study, *Structures in the Stream: Water, Science, and the Rise of the U.S. Army Corps of Engineers* (Austin: University of Texas Press, 1994). More recently, Ari Kelman has explored how the army engineers' efforts to clear snags from the Mississippi were often frustrated by bank undercutting and the subsequent creation of snags, both of which were consequences of the deforestation along the river to provide its steamboats with fuel. See Ari Kelman, "Forests and Other River Perils," in *Transforming New Orleans and Its Environs: Centuries of Change*, ed. Craig E. Colten (Pittsburgh: University of Pittsburgh Press, 2000), 45–63 passim.

3. The nature and sources of the quantitative data that are the basis of the work presented here are discussed at length in appendix A.

4. The figures for losses in 1843 are from Leonard V. Huber, comp., *Advertisements of Lower Mississippi River Steamboats, 1812–1920: A Scrapbook with Introduction and Index of Vessels and Lines* (West Barrington, RI: Steamship Historical Society of America, 1959), 97; the report of losses for 1848 was reprinted in *Niles' National Register* 74, no. 1925 (December 20, 1848): 394; the figures for 1855 were reprinted in *De Bow's Review* 19, no. 4 (October 1855), 466; losses for the first half of 1860 were reported by the editor of the *Louisville Journal* and reprinted in *Hunt's Merchants' Magazine* 43, no. 4 (October 1860): 486. The highly suspect character of these published figures on steamboat losses becomes clear when one considers them in conjunction with the number of steamboats in operation during each of the years in question. As the following table illustrates, a comparison of reported losses, relative to the number of vessels in service, results in improbably high rates of loss (the figure reported for 1855 is excluded from these calculations because it almost certainly pertained to the Ohio River, alone):

Year	Published Losses	Steamboats Operating	Rate of Loss (%)
1843	51 (19)	449	11.4 (4.2)
1848	251 (50)	666	37.7 (7.5)
1860	125 (66)	817	15.3 (8.1)

The two figures in parentheses on each line above pertain, respectively, to the number of losses, due to all causes, determined to have occurred in the indicated years, as documented in this study, and the loss rate calculated by using them. The number of steamboats in operation is from Haites, Mak, and Walton, *Western River Transportation*, table B-1 (appendix).

5. One obstacle to arriving at an estimate of the value of vessel and cargo losses is the fact that vessels were seldom lost during their first, or even second, year of operation but rather in subsequent years. Thus, depreciating vessels, from what can only be rough estimates of their

initial capitalization, is an exercise that is fraught with error. Even less reliable—in fact, almost completely speculative—is any estimate of total cargo losses. This is not to say, however, that proponents of the river improvements program were shy about offering such figures.

6. *Memorial of the Citizens of Cincinnati, to the Congress of the United States, Relative to the Navigation of the Ohio and Mississippi Rivers* (Cincinnati: L'Hommedieu, 1843), 35.

7. "Digging the Dirt at Public Expense: Governance in the Building of the Erie Canal and Other Public Works," in *Corruption and Reform: Lessons from America's Economic History*, ed. Edward L. Glaeser and Claudia Goldin, National Bureau of Research Conference Report (Chicago: University of Chicago Press, 2006): 104.

8. The appropriations figure for each state is the accumulated total of all measures from 1821 through 1860, inclusive. The population figures for the states, from which their respective shares of the national population of 23,191,876 are computed, come from the 1850 census. The use of the states' population proportions in 1850 is intended mainly as a heuristic device to illustrate a point. The national population reported in the 1820 census was 9,638,453; that reported in the 1860 census was 31,443,321.

9. A simple linear regression of *state-specific* appropriations for river and harbor improvements [APP] to each state on its proportion of the national population in 1850 [POP%] confirms what one would have expected, that is, that there was a fairly strong relationship between the two: APP = 7,629,699 * POP% + 70,834 (R^2 = 0.170; P-Value = 0.0166 for 33 cases, each of which is a state). There may also have been a relationship between the size of state-specific appropriations made to each state and each state's number of rivers or number of navigable miles of rivers, though that relationship is not likely to have been especially robust because successful appropriations bills did not fund improvements of intrastate rivers.

10. The matter of "miscellaneous appropriations" is less straightforward than that of state-specific appropriations. When speaking of the latter, there is no doubt as to which state received a particular appropriation. In the case of the miscellaneous appropriations, Congress earmarked funds for a river, such as the Ohio River, or part of a river system, such as the Ohio between Louisville, Kentucky, and Cairo, Illinois, continuing down the Mississippi River to New Orleans. In such instances, the appropriation clearly affected two or more states, and one is presented with the problem of determining the best method of apportioning the appropriation among the states along or through which the river or system of rivers flowed. The method used here is the one suggested by the language of the appropriations measures themselves and by the reports of the army's engineers who oversaw the removal of snags, the clearing of trees from riverbanks, and the dredging of river channels. For a discussion of this method, see appendix B.

11. In figure 34, the oblique line traces the general relationship among the states between the incidence of wrecks due to natural causes, especially snags, and total federal appropriations for river improvements. A regression of total appropriations for river improvements (APP) on the number of wrecks due to natural causes (WRECK) gives the equation of this line and indicates a strong relationship between the two variables: APP = 5,069.33*WRECK + 107,926.5 (R^2 = .514; P-Value < 0.0001 for 31 cases, each of which is a state). Not included among the states plotted in figure 34 or in the regression are Delaware and New York, which,

as is explained in the text, received the largest appropriations from Congress, despite low vessel loss rates, probably because of the importance of their respective harbors.

12. Total steamboat losses due to fire, explosion, and human error, as well as to natural hazards, on Louisiana's rivers, lakes, and coastal waters were 268. The state with the next-highest losses due to natural and all causes was Missouri, with 140 and 233, respectively. All figures on losses are derived from the databases described in appendix A.

13. During the forty years from 1821 through 1860, the port of New York City accounted for one-third of the nation's exports and 60 percent of its imports. See George Rogers Taylor, *The Transportation Revolution, 1815–1860*, vol. 4 of *The Economic History of the United States* (New York: Holt, Rinehart and Winston, 1951), 196. Although Philadelphia's shares of exports and imports were far smaller—smaller even than they had been during the first two decades of the century—domestic and foreign commerce on Delaware Bay and on the Delaware River, including that of Pennsylvania, New Jersey, and Delaware, made the safety of navigation on the bay a matter of national importance. For the share of total federal revenues contributed by customs revenues, see table D.17.

14. As was suggested in chapter 1, the surge in losses due to snags in 1841 and 1842, evident in figures 35 and 36, and also in figures 1 and 2, may well have been due to a combination of higher than normal precipitation in the watersheds of the Mississippi River system during the period 1839–42 and low water levels on the Lower Mississippi River. The greater precipitation contributed to some flooding and bank undercutting along the rivers, both of which uprooted and washed larger numbers of trees into the river channels. This, along with lower water levels downriver, probably aggravated the problem of snags. Early reliable precipitation data for the Mississippi Valley region are available only for St. Louis, beginning in 1837. See U.S. Bureau of the Census, *Historical Statistics of the United States, Colonial Times to 1957* (Washington, D.C.: U.S. Government Printing Office, 1960), Series J 246–265, "Long-Record City Stations—Annual Mean Temperature and Annual Total Precipitation: 1780 to 1957," 254–55. Water levels in the Lower Mississippi River basin are from: H. C. Clarke, comp., *A General Directory for the City of Vicksburg* (Vicksburg, MS: H. C. Clarke, 1860), appendix, "High and Low Water Years."

15. Snags claimed 68 percent of the twenty-two documented steamboats lost on the bayous of southeastern Louisiana between 1831 and 1861 and 73 percent of the eleven documented steamboats wrecked on Cypress Bayou and a formation of lakes of northwestern Louisiana west of Shreveport. The loss-rate on the southeastern Louisiana bayous is derived from Carl A. Brasseaux and Keith P. Fontentot, *Steamboats on Louisiana's Bayous: A History and Directory* (Baton Rouge: Louisiana State University Press, 2004), table B.1 (appendix). The loss-rate on the Cypress Bayou and lakes of northwestern Louisiana is derived from Jacques D. Bagur, *A History of Navigation on Cypress Bayou and the Lakes* (Denton: University of North Texas Press, 2001), table 11–1.

16. Hunter, *Steamboats on the Western Rivers*, 84–85.

17. Ibid., 171–72.

18. Of these 841 steamboats, 20 percent were stern-wheelers. Of 211 steamboats operating on northwest Louisiana's Cypress Bayou and the lakes during the period 1841–60, 20 percent were stern-wheelers. The latter percentage is derived from Bagur, *A History of Navigation on*

Cypress Bayou and the Lakes, table A–2 (appendix). The relative importance of stern-wheelers could vary, depending upon the waterway. Thus, on the Yazoo River in Mississippi from 1831 through 1860, stern-wheelers accounted for only 10 percent of the steamboats operating there. This figure is derived from Harry P. Owens, *Steamboats and the Cotton Economy* (Jackson: University Press of Mississippi, 1990), appendix A.

19. *Western River Transportation*, 59–73, esp. 72.

20. Ibid., 122.

21. Specifically, the relationship was negatively exponential in character.

22. Haites, Mak and Walton, *Western River Transportation*, 72–73.

23. The problem of autocorrelation does not arise with respect to the rate of vessel loss due to snags because the tonnage unit of measurement is eliminated by calculating the value of that ratio, that is, the number of tons lost to snags divided by the number of tons in operation.

24. Haites, Mak, and Walton, *Western River Transportation*, 60, 61n, 119–21; several memorials from state legislatures to Congress expressed the view of contemporaries that river improvements reduced the cost of transportation. See, for example, *Memorial of the Galena [Illinois] Chamber of Commerce, Praying an Appropriation for the Improvement of the Mississippi River, at the Des Moines and Rock River Rapids*, 26th Cong., 1st sess., 1840, S. 150, serial 357, 1; *Report: The Committee on Commerce, to Whom Were Referred Sundry Memorials Asking Congress to Make an Appropriation to Improve the Navigation of the Mississippi River and Its Principal Tributaries.* . . . , 27th Cong., 3rd sess., 1843, S. Doc. 137, serial 415, 5; *Memorial of a Number of Citizens of St. Louis, Missouri, Praying an Appropriation for the Removal of Obstructions in the Western Rivers, and for the Improvement of the Harbor of that City*, 28th Cong., 1st sess., 1844, S. Doc. 185, serial 434, 12; and *Memorial of a Number of Citizens of Bond County, Illinois, Praying Appropriations for the Improvement of the Navigation of the Western Rivers and Lakes, and for the Completion of the Cumberland Road*, ibid., 1844, S. Doc. 216, serial 434, 2.

25. This is borne out by a linear regression of the annual ratio of tons lost by snags on the western rivers to total tons in operation there (%SNAG) on the logarithm of annual net federal expenditures for river improvements (logNETEX). The regression yields the following equation and measurements: %SNAG = .096- 0.14 * logNETEX (R^2 = 0.241; P-Value = 0.0032 for 40 cases, each case representing a year during the period 1821–60, inclusive).

26. *Report of the Secretary of War, Made in Compliance with a Resolution of the Senate in Relation to the Work Done under the Appropriations of 1852 for the Improvement of Western Rivers and Harbors*, 33rd Cong., 1st sess., 1854, S. Exec. Doc. 51, serial 698, 9.

27. This result is not the equivalent of the "indirect [social] savings" attributable to western river steamboats, estimated by Erik Haites and James Mak. See their article "Social Savings Due to Western River Steamboats," in *Research in Economic History: An Annual Compilation of Research* 3 (Greenwich, CT: JAI Press, 1978), 263–304, esp. 294–98.

28. The point is worth emphasizing that it is within the national arena that the contribution of the federal government's river improvements program to economic development is considered here. Of course, the role of the federal government's river improvements program in promoting development was not the extent of the public sector's involvement in the ante-

bellum economy. In fact, as John Joseph Wallis and Barry R. Weingast note, the aggregate expenditures for internal improvements, broadly defined, by the several state governments far exceeded the appropriations and expenditures of the national government for similar purposes. See Wallis and Weingast, "Equilibrium Impotence: Why the States and Not the Federal Government Financed Economic Development in the Antebellum Era," NBER Working Paper Series (Cambridge, MA: National Bureau of Economic Research, 2005), 1. Wallis and Weingast put total federal expenditures on "transportation improvements" from 1790 to 1860 at about $60 million and note that that amount was far overshadowed by the $450 million estimated by Carter Goodrich in his *Government Promotion of American Canals and Railroads* (New York: Columbia University Press, 1960) as the total of state government spending for similar purposes during the same period.

EPILOGUE

1. Albert Stein, "Mississippi Valley: Remarks on the Improvement of the River Mississippi," *De Bow's Review* 9 (December 1850): 594–601.

2. Herman Haupt, *The Problem of the Mississippi*, reprint from the *Gulf Ports Marine Journal* (1897): 8.

3. Ellet's chief antagonist was U.S. Army engineer Andrew A. Humphreys, who, along with other members of the army's Corp of Topographical Engineers, viewed Ellet with a jaundiced eye and considered him to be ill-informed and, consequently, dangerous. For a careful and judicious treatment of the dispute between Ellet and his critics, especially Humphreys, see George S. Pabis, "Delaying the Deluge: The Engineering Debate over Flood Control on the Lower Mississippi River, 1846–1861," *Journal of Southern History* 64 (August 1998): 430–45.

4. *The Problem of the Mississippi*, 8.

5. Ibid.

6. U.S. Department of Commerce, Bureau of the Census, *Statistical Abstract of the United States, 1990*, 110th ed. (Washington, D.C.: U.S. Government Printing Office, 1990), table 546, "Estimates of Total Dollar Costs of American Wars."

7. Quoted in Allan Nevins, *The Organized War to Victory, 1864–1865*, vol. 4 of *The War for the Union*, in *The Ordeal of the Union* (New York: Scribner's, 1971), 373.

8. *Congressional Globe*, 42nd Cong., 2nd sess., Senate, June 6, 1872, 4286.

9. Ibid.

10. Ibid., 4291.

11. Ibid., 4297.

12. Ibid.

13. Ibid., 4298.

14. *Statement of Appropriations and Expenditures for Public Buildings, Rivers and Harbors, Forts, Arsenals, Armories, and Other Public Works from March 4, 1789 to June 30, 1882*, 47th Cong., 1st sess., 1881, S. Exec. Doc.196, serial 1992, 224–70, 286.

15. Isaac Lippincott, "A History of River Improvement," *Journal of Political Economy* 22 (1914): 652.

16. The St. Louis Congressional Convention was one of several that were held in various river cities after the war. See Lippincott, "A History of River Improvement," 651–52.

17. *Proceedings of the Congressional Convention Held in the City of St. Louis, on 13th, 14th and 15th Days of May, 1873* (St. Louis: Woodward, Tiernan and Hale, 1873), vii.

18. Ibid., 7. Notwithstanding Governor Woodson's disclaimer, the convention had an obvious political cast. Congressmen and governors made up the delegations of the twenty or more states present. See Lippincott, "A History of River Improvement," 652.

19. Samuel S. Cox, "Speeches. I. Finance, Tariffs, Etc. Comparison of Expenses between 1858 and 1864, and of the Taxation of England and America in 1858," excerpted from his speech of June 12, 1858, in *Eight Years in Congress, from 1857 to 1865. Memoir and Speeches* (New York: Appleton, 1865) University of Michigan's "Making of America" Web site, http://moa.umdl.umich.edu, 31–35; *Congressional Record*, 46th Cong., 2nd sess. (May 17, 1880): 3438.

20. Cox, *Eight Years in Congress*, 332.

21. Ibid., 6.

22. *Congressional Record*, 46th Cong., 2nd sess. (May 17, 1880): H 3439.

23. *Biographical Directory of the American Congress, 1774–1949*, 1334.

24. *Congressional Record*, 46th Cong., 2d sess. (May 17, 1880): H 3439.

25. Horace Greeley, *The American Conflict; A History of the Great Rebellion in the United States of America, 1860–'65: Its Causes, Incidents, and Results. . . .* (Hartford, CT: O. D. Case, 1866); University of Michigan's "Making of America" Web site, http://moa.umdl.umich.edu, 282.

26. *Congressional Record*, 46th Cong., 2d sess. (May 17, 1880): H 3440.

27. Ibid. For Reagan's role on the Commerce Committee with respect to rivers and harbors legislation, see Terry L. Seip, *The South Returns to Congress: Men, Economic Measures, and Intersectional Relationships, 1868–1879* (Baton Rouge: Louisiana State University Press, 1983), 222, 232–33.

28. *Congressional Record*, 46th Cong., 2d sess. (May 17, 1880): H 3441.

29. *Biographical Directory of the American Congress, 1774–1949*, 1723.

30. *Congressional Record*, 46th Cong., 2d sess. (May 17, 1880): H 3438.

31. Ibid., 3441.

32. Ibid.

33. See Lippincott, "A History of River Improvement," 655. In 1884, Congress created a Missouri River Commission, which went out of existence eighteen years later because traffic volume on the Missouri had fallen to the point where the activities of the commission no longer made economic sense (ibid., 656). The most comprehensive account of the events and political maneuvering leading up to the formation of the Mississippi River Commission is given in George S. Pabis, "Restraining the Muddy Waters: Engineers and Mississippi River Flood Control, 1846–1881" (Ph.D. diss., University of Illinois, Chicago Circle, 1997). See also Pabis, "Delaying the Deluge," 452–53; Martin Reuss, *Designing the Bayous: The Control of Water in the Atchafalaya Basin, 1800–1995* (Alexandria, VA: Office of History, U.S. Army Corps of Engineers, 1998), 356; and Louis C. Hunter, *Steamboats on the Western Rivers: An Economic*

and Technological History (Cambridge: Harvard University Press, 1949; reprint, Mineola, NY: Dover, 1993), 213–15.

34. *Statement of Appropriations and Expenditures*, serial 1992, 286.

35. During the five-year period 1881–85, an average of 370,103 tons of vessels operated on the western rivers. Of this tonnage, 6,409 tons were lost due to all known causes, including snags, which claimed only 148 tons. (An additional 10,155 tons were lost due to unknown causes and are excluded here because, almost without exception, vessels lost to unknown causes met their fate in the Gulf of Mexico, and not on the western rivers.) The loss rate calculated from these figures is 1.7 percent. An alternative loss rate of 2.7 percent results from using the figure of 6,409 tons lost due to all causes for the period 1881–85, and that of 246,184 steamboat tonnage enrolled in ports on the western rivers during the period 1880–84. That the two periods used in the calculation of this alternative loss rate differ in their respective beginning and ending years is due to the fact that the source of the tonnage in operation used the 1880–84 periodization. In any event, the calculation of the alternative loss rate is probably not significantly affected as a result of this disparity. Sources of the foregoing figures are: U.S. Bureau of the Census, *Historical Statistics of the United States, Colonial Times to 1957*, Series Q 169 174, "Documented Merchant Vessels, by Geographic Region, 1816–1957," 446, for the number of tons in operation during the 1881–85 period; the Shipwreck Database, an explanation of which is given in appendix A, for the number of tons lost due to all known causes and the number of tons lost due to unknown causes; and Hunter, *Western River Transportation*, table 18, for the number of steamboat tons enrolled in western river ports during the 1880–84 period.

APPENDIX B

1. *Statement of Appropriations and Expenditures for Public Buildings, Rivers and Harbors, Forts, Arsenals, Armories, and Other Public Works from March 4, 1789 to June 30, 1882*, 47th Cong., 1st sess., 1881. S. Exec. Doc. 196. serial 1992, 249.

2. Ibid., 255 (emphasis added).

3. *Report of the Secretary of War, Made in Compliance with a Resolution of the Senate in Relation to the Work Done under the Appropriations of 1852 for the Improvement of Western Rivers and Harbors*, 33rd Cong., 1st sess., 1854, S. Exec. Doc. 51, serial 698, table, "Positions and Distances from Mouth of the Missouri (downward) to Natchez," p. 4.

APPENDIX C

1. These particular indices are used here because each is readily available in the literature and, in the cases of the indices for Cincinnati and New Orleans, reflected price movements in major commercial centers on the Ohio and Mississippi rivers, respectively. Those two cities likely were the sources of supplies for the hazard-removal efforts.

2. The figures are from Benjamin Horace Hibbard, *A History of the Public Land Policies*

(New York: Macmillan, 1924; reprint, Madison: University of Wisconsin Press, 1965), table X; different figures—a significantly higher value for receipts in 1853 and somewhat lower values for receipts in 1854 and 1855—are given in Arthur H. Cole, "Cyclical and Sectional Variations in the Sale of Public Lands, 1816–60," *Review of Economic Statistics* (1927), table 1.

WORKS CITED

PRIMARY SOURCES

Manuscript and Printed Sources

Adams, John, and John Quincy Adams. *The Selected Writings of John and John Quincy Adams*. Edited by Adrienne Koch and William Peden. New York: Knopf, 1946.

Biographical Directory of the American Congress, 1774–1949: The Continental Congress . . . and the Congress of the United States. . . . 81st Cong., 2nd sess. H. Doc. 607. Washington, D.C.: U.S. Government Printing Office, 1950.

Bowen, Eli. *Rambles in the Path of the Steam-Horse. . . . A General Historical and Descriptive View of the Scenery, Agricultural and Mineral Resources, and Prominent Features of the Travelled Route from Baltimore to Harper's Ferry, Cumberland, Wheeling, Cincinnati, and Louisville*. Philadelphia: W. Bromwell and W. W. Smith, 1855.

Calhoun, John C. *The Papers of John C. Calhoun*. Vol. 22, *1845–1846*, edited by Clyde N. Wilson. Columbia: University of South Carolina Press, 1995.

———. *The Papers of John C. Calhoun*. Vol. 23, *1846*, edited by Clyde N. Wilson and Shirley Bright Cook. Columbia: University of South Carolina Press, 1996.

Campbell, John P. *The Southern Business Directory and General Commercial Advertiser*. Vol. 1. Charleston, SC: Steam Power Press of Walker and James, 1854.

Clarke, H. C., comp. *A General Directory for the City of Vicksburg, Containing the Name and Address of Every Professional and Business Man and Resident of the City, Historical Sketches of the State of Mississippi and the City of Vicksburg*. . . . Vicksburg, MS: H. C. Clarke, 1860.

Clayton, Frank M., comp. *Landings on All the Western and Southern Rivers and Bayous Showing Location, Post-Offices, Distances, &c. Also, Tariff of Premiums on Insurance to all Points*. St. Louis: Woodward, Tiernan and Hale, 1881.

Conclin, George. *A Book for All Travelers. Conclin's New River Guide, or A Gazetteer of All the Towns on the Western Waters: Containing Sketches of the Cities, Towns, and Countries Bordering on the Ohio and Mississippi Rivers, and their Principal Tributaries; Together with their Population, Products, Commerce, &c. in 1848; and Many Interesting Events of History Connected with Them. Compiled from the Latest and Best Authority. With Forty-four Maps*. Cincinnati: H. S. and J. Applegate, 1849.

Congressional Globe and Appendix.

Cox, Samuel S. *Eight Years in Congress, from 1857 to 1865. Memoir and Speeches.* New York: D. Appleton, 1865. University of Michigan's "Making of America" Web site. http://www.umdl.umich.edu/moa.

De Bow, J.D.B., comp. Superintendent of the United States Census, *Statistical View of the United States . . . Being a Compendium of the Seventh Census.* Washington, D.C.: Beverly Tucker, Senate Printer, 1854.

Dickens, Charles. *American Notes.* With an introduction by Christopher Lasch. Gloucester, MA: Peter Smith, 1968.

Douglas, Stephen. *The Letters of Stephen A. Douglas.* Edited by Robert W. Johannsen. Urbana: University of Illinois Press, 1961.

Ellet, Charles, Jr. *The Mississippi and Ohio Rivers: Containing Plans for the Protection of the Delta from Inundation; and Investigations of the Practicability and Cost of Improving the Navigation of the Ohio and Other Rivers by Means of Reservoirs. . . .* Philadelphia: Lippincott, Grambo, 1853.

———. "Of the Physical Geography of the Mississippi Valley, with Suggestions for the Improvement of the Navigation of the Ohio and Other Rivers." In *Contributions to the Physical Geography of the United States,* pt 1. Smithsonian Contributions to Knowledge (December 1849). Washington, D.C.: Smithsonian Institution, 1851.

Gales and Seaton's Register of Debates in Congress, Comprising the Leading Debates and Incidents of the Second Session of the Twentieth Congress. Vol. 5. Washington, D.C.: Gales and Seaton, 1830.

Goodspeed, Rev. E. J., D.D. *History of the Great Fires in Chicago and the West. A Proud Career Arrested by Sudden and Awful Calamity; Towns and Counties Laid Waste by the Devastating Element. Scenes and Incidents, Losses and Sufferings . . . to Which is Appended a Record of the Great Fires in the Past.* New York: H. S. Goodspeed, 1871.

Greeley, Horace. *The American Conflict: A History of the Great Rebellion in the United States of America, 1860–'65: Its Causes, Incidents, and Results. . . .* Hartford: O. D. Case, 1866. University of Michigan's "Making of America" Web site. http://moa.umdl.umich.edu.

———. *A Political Text-book for 1860: Comprising a Brief View of Presidential Nominations and Elections. . . .* New York: Tribune Association, 1860. University of Michigan's "Making of America" Web site. http://moa.umdl.umich.edu.

Hambleton, James Pinkney. *A Biographical Sketch of Henry A. Wise, with a History of the Political Campaign in Virginia in 1855.* Richmond, VA: J. W. Randolph, 1856. University of Michigan's "Making of America" Web site. http://moa.umdl.umich.edu.

Howland, S. A. *Steamboat Disasters and Railroad Accidents in the United States.* Worcester, MA: Dorr, Howland, 1846.

Huber, Leonard V., comp. *Advertisements of Lower Mississippi River Steamboats, 1812–*

1920: A Scrapbook with Introduction and Index of Vessels and Lines. West Barrington, RI: Steamship Historical Society of America, 1959.

Lewis, Henry. *The Valley of the Mississippi: Illustrated*. Edited by Bertha L. Heilbron. Translated by A. Hermina Poatgieter. St. Paul: Minnesota Historical Society, 1967.

Lloyd, James T. *Lloyd's Steamboat Directory, and Disasters on the Western Waters. . . .* Cincinnati: James T. Lloyd, 1856.

Lyell, Sir Charles. "Lyell's Second Visit to America." *North American Review* 6 (October 1849); Cornell University's "Making of America Web site." http://library5.library.cornell.edu/moa.

Marlett, E. R., M.D., and W. H. Rainey, comps. *W. H. Rainey & Co.'s Memphis City Directory, and General Business Advertiser, for 1855 & '6. Also: A Business Directory*. Memphis: D. O. Dooley & Co., Whig Book and Job Office, 1855.

Memorial of the Citizens of Cincinnati, to the Congress of the United States, Relative to Navigation of the Ohio and Mississippi Rivers. Cincinnati: L'Hommedieu, 1843.

New Orleans Chamber of Commerce. "Documents Relating to the Improvement of the Navigation of the Mississippi River. New Orleans: Chamber of Commerce, 1837.

Norman Collection. Special Collections. Hill Memorial Library, Louisiana State University Libraries. Baton Rouge.

Olmsted, Denison, L.L.D. "On the Democratic Tendencies of Science." *American Journal of Education*. 1 (January 1856): 164–73.

Paschall, J. M., and C. B. Riggs, comps. *Paschall and Riggs' First Annual Memphis City Directory and General Business Advertiser, for 1856-7*. Memphis: J. M. Paschall and C. B. Riggs, 1856.

Proceedings of the Congressional Convention Held in the City of St. Louis, on 13th, 14th and 15th Days of May, 1873. St. Louis: Woodward, Tiernan and Hale, 1873.

Professional and Business Directory of the City of Jackson, Miss. Jackson: J. L. Power, 1860.

"Report of the Chief, Topographical Engineers, Bureau of Topographical Engineers, Washington, November 17, 1848." In *Report of the Chief Engineer to the Secretary of War, at the Opening of the First Session of the Thirtieth Congress*. Washington, D.C.: Wendell and Van Benthuysen, 1848.

Richardson, James D., comp. *A Compilation of the Messages and Papers of the Presidents, 1789–1907*. Vols. 2–5. New York: Bureau of National Literature and Art, 1908.

Stein, Albert. "Mississippi Valley: Remarks on the Improvement of the Mississippi River." *De Bow's Review* 9 (December 1850): 594–601.

Stuart-Wortley, Emmeline. *Travels in the United States, etc. during 1849 and 1850. By the Lady Emmeline Stuart Wortley*. New York: Harper and Brothers, 1851.

Tanner, Halpin, and Co., comp. *Memphis City Directory, for 1859. Being a Complete General and Business Directory of the Entire City*. Memphis: Hutton and Clark, 1859.

Thomas, E. S. *Reminiscences of the Last Sixty-five Years, Commencing with the Battle of Lexington. Also, Sketches of His Own Life and Times.* 2 vols. Hartford, CT: Printed by Case, Tiffany and Burnham, for the author, 1840.

Thorpe, T. B. "Remembrances of the Mississippi." *Harper's New Monthly Magazine* 12, no. 67 (December 1855). Cornell University's "Making of America" Web site. http://library5.library.cornell.edu/moa.

Trollope, Frances. *Domestic Manners of the Americans.* Edited, with a history of Mrs. Trollope's adventures in America, by Donald Smalley. New York: Knopf, 1949.

Twyman, R. B. J. *Twyman's Memphis Directory and General Business Advertiser, for 1850, with a Brief History of Memphis Annexed.* Memphis: R.B.J. Twyman, 1849.

Webster, Daniel. *The Works of Daniel Webster.* 6th ed. 6 vols. Boston: Little, Brown, 1853.

Williams, S. *Williams' Cincinnati Almanac, Business Guide and Annual Advertiser, 1850.* First issue. Cincinnati: S. Williams, 1850.

Williams' Memphis Directory, City Guide, and Business Mirror. Vol. 1, 1860. Memphis: Cleaves and Vaden, 1860.

Compilations of Data

Berman, Bruce D. *Encyclopedia of American Shipwrecks.* Boston: Mariners Press, 1972.

Carter, Susan B. et al., eds. *Historical Statistics of the United States: Earliest Times to the Present.* Millennial ed. New York: Cambridge University Press, 2006.

Lytle, William M., and Forrest R. Holdcamper. *Merchant Steam Vessels of the United States, 1790–1868: The Lytle-Holdcamper List, Initially Compiled from Official Merchant Marine Documents of the United States and Other Sources.* Revised and edited by C. Bradford Mitchell, with the assistance of Kenneth R. Hall. Staten Island, NY: Steamship Historical Society of America; Baltimore: distributed by University of Baltimore Press, 1975.

Neville, Bert. *Directory of River Packets in the Mobile-Alabama-Warrior-Tombigbee Trades, 1818–1932, with Illustrations, Charts and Table of Landings.* Selma, AL: 1967.

———. *Directory of Tennessee River Steamboats (1821–1928), with Illustrations.* Selma, AL: 1963.

Stanley, Harold W., and Richard G. Niemi. *Vital Statistics on American Politics, 1999–2000: A Comprehensive Reference of over 200 Tables and Figures.* Washington, D.C.: CQ Press, 2000.

Subject-Matter Index of Patents for Inventions Issued by the United States Patent Office from 1790 to 1873, Inclusive. Vols. 1–3. New York: Arno Press, 1976.

U.S. Bureau of the Census. *Historical Statistics of the United States, Colonial Times to 1957.* Washington, D.C.: Department of Commerce, 1960.

———. *Historical Statistics of the United States, Colonial Times to 1970*. Electronic ed. [machine-readable data file]. Edited by Susan Carter et al. Cambridge: Cambridge University Press, 1997.

Way, Frederick, Jr., comp. *Way's Packet Directory, 1848–1994: Passenger Steamboats of the Mississippi River System since the Advent of Photography in Mid-Continent America*. Rev. ed. Athens: Ohio University Press, 1994.

Wendel, C. H. *Nebraska Tractor Tests since 1920*. Osceola, WI: Motorbooks International, 1993.

Periodicals and Newspapers

Charleston Mercury
De Bow's Review (Commercial Review of the South and West)
Harper's New Monthly Magazine, 1855
Hunt's Merchants' Magazine and Commercial Review
Journal of the Franklin Institute
Living Age, 1846, 1858
Memphis Daily Appeal
Natchez Courier
New Orleans Daily Picayune
New York Herald
Niles' National Register, Containing Political, Historical, Geographical, Scientific, Statistical, Economical, and Biographical Documents, Essays and Facts: Together with Notices of the Arts and Manufactures, and Record of the Events of the Times. Edited by Jeremiah Hughes. Baltimore: Printed by Hughes, 1848–49.
North American Review, 1849
Scientific American
Western Journal and Civilian, Devoted to Agriculture, Manufactures, Mechanic Arts, Internal Improvement, Commerce, Public Policy, and Polite Literature. Title varies. Vols. 1 (January 1848)–15 (April 1856).

Government Documents

U.S. Congress. House. *Letter from the Secretary of the Treasury, Transmitting, in Compliance with a Resolution of the House of Representatives of the 26th of May Last, a Statement of Disbursements Made, since the Year 1789, for Fortifications, Light-houses, Public Debt, Revolutionary and Other Pensions, and Internal Improvements, &c.* 21st Cong., 2nd sess., 1830. H. Doc. 11. Serial 206.

———. House. *Letter from the Secretary of the Treasury, Transmitting a Statement of Appropriations and Disbursements for Fortifications, Lighthouses, Public Debt, Revolu-*

tionary and Other Pensions, Internal Improvements, &c., &c., 22nd Cong., 1st sess., 1832. H. Doc. 200. Serial 220.

———. *Message from the President of the United States, Returning the Bill, Entitled "An Act for the Improvement of Certain Harbors, and the Navigation of Certain Rivers"; with His Objections to the Same.* 22nd Cong., 2nd sess., 1832. H. Exec. Doc. 17. Serial 233.

———. Senate. *Memorial of the Legislature of Missouri Praying That an Appropriation Be Made for Removing Obstructions in, and Repairing the Harbor of the City of St. Louis.* 23rd Cong., 1st sess., 1834. S. Doc. 21. Serial 238.

———. House. *Memorial and Resolutions Adopted by a Convention of Delegates from New York and Pennsylvania, Assembled at Warren, Pennsylvania, in Relation to Internal Improvements.* 23rd Cong., 1st sess., 1834. H. Doc. 84. Serial 256.

———. House. Committee on the Public Lands. *Inquiry into the Expediency and Justice of Granting to the State of Louisiana, in Aid of Internal Improvements, the Same Extent of Land Which Has Heretofore Been Granted by Congress to Other Western States, and Particularly to the State of Alabama.* . . . 23rd Cong., 1st sess., 1834. H. Rep. 159. Serial 260.

———. Senate. Committee on Roads and Canals. *Inquire into the Expediency of Making an Appropriation to Improve the Navigation of the Wabash River.* . . . 23rd Cong., 2nd sess., 1834. S. Rep. 18. Serial 267.

———. Senate. *Memorial of Sundry Delegates from Certain Counties and Towns in Pennsylvania, for an Appropriation to Improve the Navigation of the Monongahela River from Pittsburg to the Cumberland Road at Brownsville, &c.* 23rd Cong., 2nd sess., 1835. S. Doc. 98. Serial 268.

———. House. *Letter from the Secretary of War, Transmitting Reports in Relation to the Progress of Internal Improvements Carried on by the General Government in the State of North Carolina.* 23rd Cong., 2nd sess., 1835. H. Doc. 59. Serial 272.

———. House. *Report of the Engineer Appointed by the Commissioners of the Fund Appropriated by the States of Illinois and Indiana for the Improvement of the Navigation of the Wabash River, to Survey the Obstructions Therein, with the Estimate of the Probable Cost of the Proposed Improvements.* 23rd Cong., 2nd sess., 1835. H. Doc. 70. Serial 272.

———. House. *Report of the Secretary of the Treasury, of Expenditures for Fortification, Light-houses, Public Debt, Revolutionary and other Pensions, Internal Improvements, &c., from 1791 to 1833.* 23rd Cong., 2nd sess., 1835. H. Doc. 89. Serial 273.

———. Senate. *Memorial of the Citizens of Wilmington, North Carolina, for a Further Appropriation for Improving the Navigation of Cape Fear River.* 24th Cong., 1st sess., 1836. S. Doc. 370. Serial 283.

———. House. *Letter from the Secretary of War, Transmitting the Information Required by a Resolution of the House of Representatives of the 2d Instant, in Relation to Certain Works of Internal Improvement.* 24th Cong., 1st sess., 1836. House Doc. 212. Serial 290.

———. House. *Report by the Secretary of War on Improvement of Rivers and Harbors, Old Works; Amount Heretofore Appropriated, and Amount Required for Completion.* 25th Cong., 2nd sess., 1838. H. Doc. 90. Serial 325.

———. House. Committee on Commerce. *Light-Houses, &c., and Surveys for 1838 (to Accompany Bill H. R. no. 712).* 25th Cong., 2nd sess., April 6, 1838. H. Rep. 752. Serial 335.

———. House. *Report: The Committee on Commerce, to Which the Subject of Light-houses, Buoys, &c. Was Referred.* . . . 25th Cong., 2nd sess., 1838. H. Rep. 752. Serial 335.

———. Senate. *Documents in Relation to Internal Improvements in the State of Illinois.* 25th Cong., 2nd sess., 1838. S. Doc. 99. Serial 315.

———. Senate. *Documents Relating to Internal Improvements in the State of Illinois . . . Report of the Board of Public Works.* 25th Cong., 2nd sess., 1838. S. Doc. 257. Serial 316.

———. Senate. *Documents in Relation to Internal Improvements in the State of Illinois.* 25th Cong., 2nd sess., 1838. S. Doc. 259. Serial 316.

———. Senate. *Memorial of John B. Sterigere and Others, Praying an Appropriation for the Improvement of the Alleghany River.* 25th Cong., 2nd sess., 1838. S. Doc. 174. Serial 316.

———. Senate. *Resolutions of the Legislature of New York, Relative to a Survey of the Navigable Waters of the Northern and Western Lakes and Rivers, and the Improvement of the Harbors on the Same.* 25th Cong., 2nd sess., 1838. S. Doc. 237. Serial 316.

———. House. *Memorial of the Legislative Assembly of Wisconsin, Praying for an Appropriation of $200,000 for the Improvement of the Navigation of the Mississippi River, at the Des Moines River and Rock-River Rapids.* 25th Cong., 2nd sess., 1838. H. Doc. 169. Serial 327.

———. House. *Memorial of the Legislature of Indiana, in Relation to Lands Purchased of Miami Indians.* 25th Cong., 2nd sess., 1838. H. Doc. 214. Serial 328.

———. Senate. *Resolutions of the General Assembly of Maryland, to Obtain an Appropriation for the Improvement of the Harbor of Havre de Grace.* 25th Cong., 3rd sess., 1839. S. Doc. 226. Serial 340.

———. Senate. *Report from the Secretary of War, in Compliance with a Resolution of the Senate, in Reference to the Expenditure of the Appropriation for the Removal of the Sandbar at the Mouth of the Connecticut River.* 25th Cong., 3rd sess., 1839. S. Doc. 289. Serial 342.

———. House. *Letter of the Honorable J. D. Doty, Relating to the Bill to Authorize the Territory of Wisconsin to Establish a System of Internal Improvements.* 25th Cong., 3rd sess., 1839. Serial 347.

———. House. *Memorial of the Mayor and City Council of Baltimore, Praying a Continuance of the Appropriation Heretofore Made for Preserving the Harbor.* 25th Cong., 3rd sess., 1839. H. Doc. 219. Serial 348.

———. Senate. *Report from the Secretary of War, Transmitting Copies of Reports of the Topographical Bureau in Relation to Internal Improvements in the Territory of Wisconsin, in Obedience to a Resolution of the Senate of the 15th Instant.* 26th Cong., 1st sess., 1840. S. Doc. 140. Serial 357.

———. Senate. *Memorial of the Galena [Illinois] Chamber of Commerce, Praying an Appropriation for the Improvement of the Mississippi River, at the Des Moines and Rock River Rapids.* 26th Cong., 1st sess., 1840. S. Doc. 150. Serial 357.

———. Senate. *Petition of a Number of Citizens of Missouri, Praying an Appropriation for the Improvement of the Current River.* 26th Cong., 1st sess., 1840. S. Doc. 302. Serial 359.

———. Senate. *Resolution of the General Assembly of Louisiana. To Obtain an Appropriation for the Erection of a Light-house on the Bayou Bonfouca.* 26th Cong., 1st sess., 1840. S. Doc. 348. Serial 359.

———. Senate. *Relating to the Bill (S. 342) "Making an Appropriation for Continuing the Operation of Deepening the Channel at the Mouth of the Mississippi River."* 26th Cong., 1st sess., 1840. S. Doc. 463. Serial 360.

———. Senate. *Memorial of Numerous Citizens of Illinois, Praying an Appropriation of Public Lands for the Improvement of Rock River.* 26th Cong., 1st sess., 1840. S. Doc. 492. Serial 360.

———. House. *Message from the President of the United States, Transmitting a Communication from the Secretary of War in Relation to the System of Internal Improvements Carried on by the General Government, and Showing the Operations during the Past Year, &c., &c.* 26th Cong., 1st sess., 1840. Appended to H. Exec. Doc. 2. Serial 363.

———. Senate. *Memorial of a Number of Citizens of the Town of Havre-de-Grace, in the State of Maryland, Praying an Appropriation for the Removal of Obstructions at the Entrance of the Harbor at That Place.* 26th Cong., 2nd sess., 1841. S. Doc. 216. Serial 378.

———. Senate. *Resolution of the General Assembly of Mississippi, in Favor of Appropriating the Two Per Cent. Fund of that State towards the Construction of a Railroad.* 26th Cong., 2nd sess., 1841. S. Doc. 228. Serial 378.

———. House. *Report: The Committee on Roads and Canals, to Whom Was Referred Senate Bill no. 57, Making an Appropriation to Complete the Removal of the Raft of Red River, and for Other Purposes. . . .* 26th Cong., 2nd sess., 1841. H. Rep. 141, Serial 388.

———. Senate. *Preamble and Resolutions Adopted at a Meeting of the Citizens of Surry County, VA., Adverse to the Incorporation of a National Bank, the Distribution of the Proceeds of the Public Lands, the Establishment of a Protective Tariff, &c., &c.* 27th Cong., 1st sess., 1841. S. Doc. 112. Serial 390.

———. Senate. *Memorial of the Town Council and Citizens of Erie, PA., Praying an Appropriation for the Improvement of Presque Isle Harbor.* 27th Cong., 2nd sess., 1841. S. Doc. 23. Serial 396.

———. Senate. *Resolutions of the General Assembly of Indiana, to Procure Appropriations for the Improvement of the Western Rivers, and for the Purchase of the Snag-boat Invented by Henry M. Shreve.* 27th Cong., 2nd sess., 1842. S. Doc. 112. Serial 397.

———. House. Committee on Commerce. *Improvement of Bars, Harbors, and Rivers in the Territory of Florida.* 27th Cong., 2nd sess., June 24, 1842. H. Rep. 886. Serial 410.

———. Senate. *Resolutions of the General Assembly of Indiana, Asking an Appropriation to Improve the Navigation of the Wabash River.* 27th Cong., 3rd sess., 1843. S. Doc. 119. Serial 415.

———. Senate. *Report: The Committee on Commerce, to Whom Were Referred Sundry Memorials Asking Congress to Make an Appropriation to Improve the Navigation of the Mississippi River and Its Principal Tributaries. . . .* 27th Cong., 3rd sess., 1843. S. Rep. 137. Serial 415.

———. Senate. *Memorial of the General Assembly of Indiana, Praying an Appropriation to Improve the Navigation of the Mississippi and Other Western Rivers.* 27th Cong., 3rd sess., 1843. S. Doc. 155. Serial 415.

———. House. *Explanatory of the Bill Making Appropriations for Certain Harbors and Rivers, and for Continuing the Cumberland Road, from January 1, 1843, to June 30, 1844.* 27th Cong., 3rd sess., 1843. H. Doc. 104, Serial 420.

———. Senate. *Report: The Committee on Public Lands, to Whom Were Referred Numerous Memorials, from Citizens of Wisconsin, Asking an Appropriation to Improve the Navigation of the Fox and Wisconsin Rivers, and to Connect Them by a Canal. . . .* 28th Cong., 1st sess., 1844. S. Rep. 28. Serial 432.

———. Senate. *Resolutions of the General Assembly of Indiana, to Obtain an Appropriation of Money or Land to Improve the Navigation of the Mississippi, Ohio, and Wabash Rivers.* 28th Cong., 1st sess., 1844. S. Doc. 94. Serial 432.

———. Senate. *Memorial of a Number of Citizens of St. Louis, Missouri, Praying an Appropriation for the Removal of Obstructions in the Western Rivers, and for the Improvement of the Harbor of That City.* 28th Cong., 1st sess., 1844. S. Doc. 185. Serial 434.

———. Senate. *Memorial of a Number of Citizens of Bond County, Illinois, Praying Appropriations for the Improvement of the Navigation of the Western Rivers and Lakes, and* for the Completion of the Cumberland Road. 28th Cong., 1st sess., 1844. S. Doc. 216. Serial 434.

———. Senate. *Resolutions of the Legislature of the State of Maryland, in Favor of an Appropriation for Removing Obstructions to the Navigation of the Susquehanna River.* 28th Cong., 1st sess., 1844. S. Doc. 290. Serial 435.

———. House. *Report: the Committee on Naval Affairs, to Whom Were Referred Resolutions of the State of Georgia, Asking Congress to Make an "Appropriation of a Sufficient Sum of Money to Improve the Navigation of the Savannah River, below the City of Savannah," Having Had the Same under Consideration.* . . . 28th Cong., 1st sess., 1844. H. Rep. 360. Serial 446.

———. Senate. *Memorial of the Legislature of Arkansas, Asking an Appropriation for Completing the Road from Memphis to Little Rock, and for Opening a Road between Little Rock and Fort Smith.* 29th Cong., 1st sess., 1845. S. Doc. 30. Serial 472.

———. Senate. *Memorial of the Legislature of Arkansas, Asking for an Appropriation for Completing the Removal of the Raft in Red River.* 29th Cong., 1st sess., 1845. S. Doc. 31. Serial 472.

———. Senate. *Memorial of the Legislature of Mississippi, for an Appropriation by Congress of Alternate Sections of the Public Lands to Aid in the Improvement of the Navigation of Pearl River.* 29th Cong., 1st sess., 1846. S. Doc. 212. Serial 474.

———. Senate. *Memorial of the Legislature of Mississippi, Asking for an Appropriation for the Removal of the Obstructions to the Navigation between the Mississippi River and the Gulf of Mexico.* 29th Cong., 1st sess., 1846. S. Doc. 214. Serial 474.

———. Senate. *Memorial of the Legislature of Mississippi, Asking a Grant of Public Lands for the Improvement of the Homochitto River.* 29th Cong., 1st sess., 1846. S. Doc. 215. Serial 474.

———. Senate. *Memorial of the Legislature of Mississippi, Asking the Appropriation to That State of Alternate Sections on the Leaf, Pascagoula, and Chickasahay Rivers, for the Improvement of Those Rivers.* 29th Cong., 1st sess., 1846. S. Doc. 224. Serial 474.

———. Senate. *Report of the Secretary of War, in Compliance with a Resolution of the Senate, Showing the Aggregate Appropriations for Improving Rivers and Harbors since July 1, 1836, and the Sum Expended in Dredging Operations, with the Cost per Cubic Yard of Removing Obstructions.* 29th Cong., 1st sess., 1846. S. Doc. 451. Serial 478.

———. House. *Report: The Committee on Commerce, to Whom Were Referred a Petition of Citizens of Mobile, Alabama, Praying for an Appropriation to Clear out the Harbor of Mobile, and a Joint Memorial of the Legislature of Alabama, Praying the Appropriation of a Sum of Money to Deepen the Channel of Mobile Bay.* . . . 29th Cong., 1st sess., 1846. H. Rep. 539. Serial 490.

———. House. *Report: The Committee on Public Lands, to Whom Were Referred Numerous Petitions of the Citizens of Wisconsin and Other Places, Asking for Appropriations for* the Improvement of the Fox and Wisconsin Rivers. . . . 29th Cong., 1st sess., 1846. H. Rep. 551. Serial 490.

———. Senate. *Report of the Secretary of War, in Answer to a Resolution of the Senate, Calling for a Statement of Appropriations for the Construction and Repair of Roads, Fortifications, and Harbors, and for the Improvement of Rivers*. 29th Cong., 2 sess., 1847. S. Doc. 44. Serial 494.

———. Senate. *Report of the Secretary of War, in Further Answer to a Resolution of the Senate Calling for a Statement of Appropriations for the Construction and Repair of Roads, Fortifications, and Harbors, and for the Improvement of Rivers*. 29th Cong., 2nd sess., 1847. S. Doc. 79. Serial 494.

———. Senate. *Letter from the Secretary of the Treasury, Transmitting a Report from the Acting Commissioner of the General Land Office, Relative to the Public Lands Appropriated by Alternate Sections in the States of Ohio, Indiana, Illinois, and Arkansas, for Internal Improvement, &c.* 29th Cong., 2nd sess., 1847. H. Doc. 51, Serial 499.

———. Senate. *Resolutions of the Legislature of New Jersey, in Relation to Internal Improvements*. 30th Cong., 1st sess., 1848. S. Misc. Doc. 78, Serial 511.

———. Senate. *Memorial of the Mayor and City Council of the City of St. Louis, Praying an Appropriation for the Improvement of the Harbor of That City*. 30th Cong., 1st sess., 1848. S. Misc. Doc. 84. Serial 511.

———. Senate. *Memorial of Citizens of Tennessee, Praying an Appropriation to Repair the Breach in the Dam at the Head of Cumberland Island, in the Ohio River*. 30th Cong., 1st sess., 1848. S. Doc. 99. Serial 511.

———. Senate. *Memorial of the Chicago Convention, in Favor of the Improvement of Harbors and Rivers by the General Government*. 30th Cong., 1st sess., 1848. S. Misc. Doc. 146. Serial 511.

———. House. *Resolutions of the Legislature of Mississippi, Relative to the Appropriation of the Two Per Cent. Fund to the Construction of Railroad from Jackson Eastwardly in the Direction of the Alabama Line*. 30th Cong., 1st sess., 1848. H. Misc. Doc. 56. Serial 523.

———. House. *Memorial of the Legislature of Alabama, Relative to a Grant of Land to Aid in the Construction of a Railroad between the Southern Part of the State and the Interior of the West*. 30th Cong., 1st sess., 1848. H. Misc. Doc. 57. Serial 523.

———. Senate. *Message from the President of the United States, Communicating a Report from the Secretary of the Treasury, with a Statement of the Annual Amount Appropriated on Account of the Coast Survey*. 30th Cong., 2nd sess., 1848. S. Exec. Doc. 4. Serial 529.

———. Senate. *Resolution of the Legislature of Illinois, in Favor of the Improvement of the Mississippi, Ohio, and Illinois Rivers, and of the Harbors on the Northern Lakes*. 30th Cong., 2nd sess., 1849. S. Misc. Doc. 40. Serial 533.

———. Senate. *Report: The Committee on Commerce, to Whom Was Referred the Bill from the House of Representatives Entitled "An Act Making Appropriations for the Preservation and Repairs of the Public Works upon Certain Harbors and Rivers, and for the Survey of Certain Harbors...."* 30th Cong., 2nd sess., 1849. S. Rep. 284. Serial Set 535.

———. Senate. *Resolution of the Legislature of Texas, in Favor of an Appropriation for a Light-house or Light-boat at Brazos Santiago and the Mouth of the Rio Grande.* 31st Cong., 1st sess., 1850. S. Misc. Doc. 100. Serial 563.

———. Senate. *Report: The Select Committee to Whom Was Referred the Memorial of Charles Ellet, Jr., Civil Engineer, Asking for an Appropriation of Funds to be Applied to the Improvement of the Navigation of the Ohio River, by Constructing Reservoirs on Its Tributary Streams....* 31st Cong., 1st sess., 1850. S. Rep. 191. Serial 565.

———. House. *Report from the Bureau of Topographical Engineers, on the Subject of Rivers and Harbors.* 31st Cong., 1st sess., 1850. H. Misc. Doc. 54. Serial 582.

———. House. *Memorial of the Legislature of Mississippi, Asking an Appropriation to Remove a Sand Bar from the Mouth of the Pascagoula River.* 32nd Cong., 1st sess., 1852. H. Misc. Doc. 41. Serial 652.

———. Senate. *Resolution of the Legislature of Arkansas, in Favor of an Appropriation for Removing the Obstructions to Navigation in the St. Francis River.* 32nd Cong., 2nd sess., 1853. S. Misc. Doc. 24. Serial 670.

———. House. *Resolutions of the Legislature of North Carolina, Asking an Appropriation for Deepening the Channel at the Mouth of Cape Fear River.* 32nd Cong., 2nd sess., 1853. H. Misc. Doc. 11. Serial 685.

———. Senate. *Report of the Secretary of War, Made in Compliance with a Resolution of the Senate in Relation to the Work Done under the Appropriations of 1852 for the Improvement of Western Rivers and Harbors.* 33rd Cong., 1st sess., 1854. S. Exec. Doc. 51. Serial 698.

———. Senate. *Memorial of the Legislature of Alabama, Praying an Appropriation for the Improvement of Mobile Bay.* 33rd Cong., 1st sess., 1854. S. Misc. Doc. 39. Serial 705.

———. Senate. *Report of a Committee, Accompanied by Resolutions Adopted at a Meeting of Citizens of New Hanover County, North Carolina, in Favor of an Appropriation for the Removal of the Bar in Cape Fear River.* 33rd Cong., 1st sess., 1854. S. Misc. Doc. 64. Serial 705.

———. Senate. *Message of the President of the United States, Returning to the House of Representatives a Bill Entitled "An Act Making Appropriations for the Repair, Preservation, and Completion of Certain Public Works, Heretofore Commenced under Authority of Law," with his Objections.* 33rd Cong., 2nd sess., 1855, S. Exec. Doc. 17. Serial 751.

———. House. *Veto Message. Message from the President of the United States, Returning the*

WORKS CITED

Bill of the House no. 392, Making Certain Appropriations, &c. 33rd Cong. 2nd sess., 1854, H. Exec. Doc. 2. Serial 780.

———. House. *Message from the President of the United States, Communicating at Length His Reasons for Returning to the House of Representatives the River and Harbor Bill of the Last Session of the Present Congress.* 33rd Cong., 2nd sess., 1855. H. Exec. Doc. 27. Serial 783.

———. Senate. *Report of the Secretary of War, in Compliance with a Resolution of the Senate of the 28th Ultimo, Communicating Copies of the Correspondence between the War Department and J. F. Gilmer, Relative to the Expenditure of the Appropriation for the Improvement of the Savannah River, &c.* 34th Cong., 1st sess., 1856. S. Exec. Doc. 16. Serial 815.

———. Senate. *Preamble and Resolutions of the Legislature of Mississippi, in Favor of an Appropriation to Aid in the Construction of a Canal for the Reopening of the Bayou Manshac.* 34th Cong., 1st sess., 1856. S. Misc. Doc. 42. Serial 835.

———. Senate. *Resolutions of the Legislature of the State of North Carolina, in Favor of Appropriations for the Entire Closure of New Inlet, and for the Completion of the Improvements at the Mouth of the Cape Fear River, According to the Plan of the Secretary of War.* 34th Cong., 3rd sess., 1857. S. Misc. Doc. 20. Serial 890.

———. Senate. *Resolutions of the Legislature of the State of Michigan, in Favor of an Appropriation in Money to Render Safe and Secure the Canal around the Falls of St. Mary's.* 34th Cong., 3rd sess., 1857. S. Misc. Doc. 21. Serial 890.

———. Senate. Committee on Commerce. *Memorial of the Board of Trade of the City of Pittsburg, Asking for an Appropriation of Public Lands to Enable a Company Chartered or to be Chartered, by the States Bordering on the Ohio River, to Improve Its Navigation. . . .* 34th Cong., 3rd sess., 1857. S. Rep. 319. Serial 891.

———. House. *Resolutions of the General Assembly of North Carolina, Asking Further Appropriation for the Improvement of Cape Fear River.* 34th Cong., 3rd. sess., 1857. H. Misc. Doc. 37. Serial 911.

———. Senate. *Report of the Secretary of War, Communicating, in Answer to a Resolution of the Senate, Estimates for the Improvement of Rivers and Harbors in the State of New York.* 35th Cong., 1st sess., 1858. S. Exec. Doc. 16. Serial 918.

———. House. House of Representatives Committee on Commerce. *Harbors and Rivers (to Accompany Bill H. R. no. 483).* 35th Cong., 1st sess., April 15, 1858. H. Rep. 251. Serial 965.

———. House. *Resolutions of the Legislature of the State of Wisconsin, Relative to the Payment of the Five Per Cent. Found Due to Said State for Internal Improvements.* 35th Cong., 2nd sess., 1859. H. Misc. Doc. 52. Serial 1016.

———. Senate. *Message of the President of the United States, Assigning His Reasons for Not Approving a Bill, Entitled, "An Act Making an Appropriation for Deepening the Chan-*

nel over the St. Clair Flats, in the State of Michigan." 36th Cong., 1st sess., 1860. S. Exec. Doc. 6. Serial 1027.

———. House. *Memorial of the Legislature of the State of Minnesota, Asking an Appropriation for the Improvement of St. Croix River.* 36th Cong., 2nd sess., February 8, 1861. H. Misc. Doc. 37. Serial 1103.

———. House. *Joint Resolutions of the Legislature of the State of Michigan, Relative to the State of the Union.* 36th Cong., 2nd sess., 1861. H. Misc. Doc. 38. Serial 1103.

———. Senate. *Statement of Appropriations and Expenditures for Public Buildings, Rivers and Harbors, Forts, Arsenals, Armories, and Other Public Works from March 4, 1789 to June 30, 1882.* 47th Cong., 1st sess., 1881. S. Exec. Doc. 196. Serial 1992.

———. House. Department of the Interior. *Report on the Agencies of Transportation in the U.S.* Vol. 4 of *Report of the Tenth Census,* T. C. Purdy, "History of Steam Navigation in the United States." 47th Cong., 2nd sess., 1883. H. Misc. Doc. 42. Serial 2132.

———. House. Department of the Treasury. *Report on the Internal Commerce of the United States 1887.* William F. Switzler, part 2 of *Report on the Commerce and Navigation of the United States,* 50th Cong., 1st sess., 1888. H. Exec. Doc. 6. Serial 2552.

SECONDARY SOURCES

Published Works

Arrington, Joseph Earl. "Henry Lewis' Moving Panorama of the Mississippi River." *Louisiana History* 6 (Summer 1965): 239–72.

Bagur, Jacques D. *A History of Navigation on Cypress Bayou and the Lakes.* Denton, TX: University of North Texas Press, 2001.

Baker, Ronald L., and Marvin Carmony. *Indiana Place Names.* Bloomington: Indiana University Press, 1975.

Bierce, Ambrose. *The Devil's Dictionary.* 1911. New York: Dell, 1991.

Bodenhorn, Howard, and Hugh Rockoff. "Regional Interest Rates in Antebellum America." In *Strategic Factors in Nineteenth Century American Economic History: A Volume to Honor Robert W. Fogel,* edited by Claudia Goldin and Hugh Rockoff. A National Bureau of Economic Research Conference Report. Chicago: University of Chicago Press, 1992.

Boorstin, Daniel J. *The Americans: The National Experience.* New York: Random House, 1965.

Bourgin, Frank. *The Great Challenge: The Myth of Laissez-Faire in the Early Republic.* New York: Harper and Row, 1989.

Brasseaux, Carl A., and Keith P. Fontenot. *Steamboats on Louisiana's Bayous: A History and Directory.* Baton Rouge: Louisiana State University Press, 2004.

Chandler, Alfred D., Jr. *The Visible Hand: The Managerial Revolution in American Business.* Cambridge: Belknap Press of Harvard University Press, 1977.

Clay, Floyd M. *History of Navigation on the Lower Mississippi.* National Waterways Study, U.S. Army Engineer Water Resources Support Center, Institute for Water Resources. Washington, D.C.: Superintendent of Documents, U.S. Government Printing Office, distributor, 1983.

Cole, Arthur H. "Cyclical and Sectional Variations in the Sale of Public Lands, 1816–60." *Review of Economics and Statistics* 9 (January 1927): 41–53.

Cooper, William J., Jr. *Jefferson Davis, American.* New York: Knopf, 2000.

———. *The South and the Politics of Slavery, 1828–1856.* Baton Rouge: Louisiana State University Press, 1978.

Cowdrey, Albert E. *The Delta Engineers: A History of the U.S. Army Corps of Engineers in the New Orleans District.* New Orleans: New Orleans District, 1971.

Craven, Avery O. *The Growth of Southern Nationalism, 1848–1861.* Baton Rouge: Louisiana State University Press, 1953.

Crenson, Matthew A. *The Federal Machine: Beginnings of Bureaucracy in Jacksonian America.* Baltimore: Johns Hopkins University Press, 1975.

Deák, Gloria Gilda. *Picturing America: Prints, Maps, and Drawings Bearing on the New World Discoveries and on the Development of the Territory That Is Now the United States, 1497–1899.* Vol. 1, Text; vol. 2, Illustrations. Princeton: Princeton University Press, 1988.

DeVoto, Bernard. *The Year of Decision: 1846.* Boston: Little, Brown, 1943.

Dobney, Fredrick J. *River Engineers on the Middle Mississippi: A History of the St. Louis District, U.S. Army Corps of Engineers.* St. Louis: St. Louis District, 1978.

Donald, David Herbert. *Lincoln.* New York: Simon and Schuster, 1996.

Dunlavy, Colleen A. *Politics and Industrialization: Early Railroads in the United States and Prussia.* Princeton: Princeton University Press, 1994.

Ellis, Richard E. *The Union at Risk: Jacksonian Democracy, States' Rights and the Nullification Crisis.* New York: Oxford University Press, 1987.

Engerman, Stanley L., and Kenneth L. Sokoloff. "Digging the Dirt at Public Expense: Governance in the Building of the Erie Canal and Other Public Works." In *Corruption and Reform: Lessons from America's Economic History,* edited by Edward L. Glaeser and Claudia Goldin, 95–122. National Bureau of Economic Research Conference Report. Chicago: University of Chicago Press, 2006.

Evans, George Heberton, Jr. *Business Incorporations in the United States, 1800–1943.* Publications of the National Bureau of Economic Research, no. 49. New York: National Bureau of Economic Research, 1948.

Faust, Drew Gilpin. *James Henry Hammond and the Old South: A Design for Mastery.* Baton Rouge: Louisiana State University Press, 1982.

Fehrenbacher, Don E. *Chicago Giant: A Biography of "Long John" Wentworth.* Madison, WI: American History Research Center, 1957.

Feller, Daniel. *The Public Lands in Jacksonian Politics.* Madison: University of Wisconsin Press, 1984.

Fishlow, Albert. *American Railroads and the Transformation of the Ante-Bellum Economy.* Cambridge: Harvard University Press, 1965.

Fogel, Robert William. *Railroads and American Economic Growth: Essays in Econometric History.* Baltimore: Johns Hopkins Press, 1964.

Folmsbee, Stanley John. *Sectionalism and Internal Improvements in Tennessee, 1796–1845.* Special Studies in Tennessee History, no. 1. Knoxville: East Tennessee Historical Society, 1939.

Foster, Gaines M. *Moral Reconstruction: Christian Lobbyists and the Federal Legislation of Morality.* Chapel Hill: University of North Carolina Press, 2002.

Freehling, William W. *The Road to Disunion.* Vol. 1, *Secessionists at Bay, 1776–1854.* New York: Oxford University Press, 1990.

Goodrich, Carter. "National Planning of Internal Improvements." *Political Science Quarterly* 63 (March 1948): 16–44.

———. "The Revulsion against Internal Improvements." *Journal of Economic History* 10 (November 1950): 145–69.

Govan, Thomas Payne. *Nicholas Biddle: Nationalist and Public Banker, 1786–1844.* Chicago: University of Chicago Press, 1959.

Haites, Erik F., and James Mak. "Social Savings Due to Western River Steamboats." In *Research in Economic History: An Annual Compilation of Research* 3 (1978): 263–304, esp. 294–98.

———, James Mak, and Gary M. Walton. *Western River Transportation: The Era of Early Internal Development, 1810–1860.* Johns Hopkins University Studies in Historical and Political Science, 93rd ser., no. 2. Baltimore: Johns Hopkins University Press, 1975.

Haupt, Herman. *The Problem of the Mississippi.* Reprint from the *Gulf Ports Marine Journal* (1897).

Hibbard, Benjamin Horace. *A History of the Public Land Policies.* Foreword by Paul W. Gates. Madison: University of Wisconsin Press, 1965.

Hill, Forest G. *Roads, Rails & Waterways; The Army Engineers and Early Transportation.* Norman: University of Oklahoma Press, 1957.

Historic New Orleans Collection. *The Waters of America: 19th-Century American Paintings of Rivers, Streams, Lakes, and Waterfalls.* New Orleans: Historic New Orleans Collection, 1984.

Holt, Michael F. *The Rise and Fall of the American Whig Party: Jacksonian Politics and the Onset of the Civil War.* New York: Oxford University Press, 1999.

Hulbert, Archer B. *The Paths of Inland Commerce: A Chronicle of Trail, Road, and Waterway.* In The Chronicle of America series, edited by Allen Johnson. New Haven: Yale University Press, 1920.

Hunter, Louis C. *A History of Industrial Power in the United States, 1780–1930.* Vol. 2, Steam Power. Charlottesville: University Press of Virginia, 1985.

———. *Steamboats on the Western Rivers: An Economic and Technological History.* Cambridge: Harvard University Press, 1949; reprint, New York: Dover 1993.

John, Richard R. *Spreading the News: The American Postal System from Franklin to Morse.* Cambridge: Harvard University Press, 1995.

Johnson, Emory R. "Rivers and Harbor Bills." *Annals of the American Academy of Political and Social Sciences* 2 (1892): 782–812.

Johnson, Vicki Vaughn. *The Men and Vision of the Southern Commercial Conventions, 1845–1871.* Columbia: University of Missouri Press, 1992.

Kane, Adam I. *The Western River Steamboat.* College Station: Texas A&M University Press, 2004.

Krenkel, John H. *Illinois Internal Improvements, 1818–1848.* Cedar Rapids, IA: Torch Press, 1958.

Kruman, Marc W. *Parties and Politics in North Carolina, 1836–1865.* Baton Rouge: Louisiana State University Press, 1983.

Larson, John Lauritz. "'Bind the Republic Together': The National Union and the Struggle for a System of Internal Improvements." *Journal of American History* 74 (1987): 363–87.

———. *Internal Improvement: National Public Works and the Promise of Popular Government in the Early United States.* Chapel Hill: University of North Carolina Press, 2001.

Lass, William E. "The Fate of Steamboats: A Case Study of the 1848 St. Louis Fleet." *Missouri Historical Review* 96 (October 2001): 2–15.

Lewis, Clifford M., S.J. "The Wheeling Suspension Bridge." *West Virginia History* 33, no. 3 (1972): 203–33.

Lewis, Gene D. *Charles Ellett, Jr.: The Engineer as Individualist, 1810–1862.* Urbana: University of Illinois Press, 1968.

Lindsey, David. *Ohio's Western Reserve: The Story of Its Place Names.* Cleveland: Press of Western Reserve University and Western Reserve Historical Society, 1955.

Lippincott, Isaac. "A History of River Improvement." *Journal of Political Economy* 22 (1914): 630–60.

Malone, Laurence Joseph. *Opening the West: Federal Internal Improvements before 1860* New York: Greenwood Press, 1998.

Martin, William E. *Internal Improvements in Alabama.* Johns Hopkins University Studies on Historical and Political Science, no. 20 (1902): 127–208.

Martis, Kenneth C. *The Historical Atlas of Political Parties in the United States Congress, 1789–1989.* New York and London: Macmillan, 1989.

McCall, Edith. *Conquering the Rivers: Henry Miller Shreve and the Navigation of America's Inland Waterways.* Baton Rouge: Louisiana State University Press, 1984.

McDermott, John Francis. *The Lost Panoramas of the Mississippi.* Chicago: University of Chicago Press, 1958.

McMahon, Thomas A., and John Tyler Bonner. *On Size and Life.* Scientific American Library. New York: Scientific American Books, 1983.

Merrick, George B. *Old Times on the Upper Mississippi.* Cleveland: Arthur H. Clark, 1909.

Morris, John V. *Fires and Firefighters.* Boston: Little, Brown, 1955

Nelson, E. C. "Presidential Influence on the Policy of Internal Improvements." *Iowa Journal of History and Politics* 4 (1906): 3–69.

Nevins, Allen. *The Ordeal of the Union.* 8 vols. New York: Scribner's 1947.

Nichols, Roy Franklin. *Franklin Pierce: Young Hickory of the Granite Hills.* 2nd ed. Philadelphia: University of Pennsylvania Press, 1958.

Niven, John. *John C. Calhoun and the Price of Union: A Biography.* Baton Rouge: Louisiana State University Press, 1988.

North, Douglass C. *The Economic Growth of the United States, 1790 to 1860.* Englewood Cliffs, NJ: Prentice-Hall, 1961.

Overman, William D. *Ohio Town Names: The Origin of the Names of over 500 Ohio Cities, Towns and Villages.* Akron, OH: Atlantic Press, 1958.

Owens, Harry P. *Steamboats and the Cotton Economy: River Trade in the Yazoo-Mississippi Delta.* Jackson: University Press of Mississippi, 1990.

Pabis, George S. "Delaying the Deluge: The Engineering Debate over Flood Control on the Lower Mississippi River, 1846–1861." *Journal of Southern History* 64 (August 1998): 421–54.

Parsons, Stanley B., William W. Beach, and Michael J. Dubin. *United States Congressional Districts and Data, 1843–1883.* New York: Greenwood Press, 1986.

Paskoff, Paul F. "Hazard Removal on the Western Rivers as a Problem of Public Policy, 1821–1860." *Louisiana History* 40 (1999): 261–82.

———. *Industrial Evolution: Organization, Structure, and Growth of the Pennsylvania Iron Industry, 1750–1860.* Studies in Industry and Society, edited by Glenn Porter. Baltimore: Johns Hopkins University Press, 1983.

Paskoff, Paul F., and Daniel J. Wilson eds. *The Cause of the South: Selections from "De Bow's Review," 1846–1867.* Baton Rouge: Louisiana State University Press, 1982.

Potter, David M. *The Impending Crisis, 1848–1861.* Completed and edited by Don E. Fehrenbacher. New York: Harper and Row, 1976.

Ramsay, Robert L. *Our Storehouse of Missouri Place Names*, Missouri Handbook no. 2, University of Missouri Bulletin. Columbia: University of Missouri, 1952.

Rennick, Robert M. *Kentucky Place Names*. Lexington: University Press of Kentucky, 1984.

Reps, John W. With modern photographs from the air by Alex MacLean. *Cities of the Mississippi: Nineteenth-Century Images of Urban Development*. Columbia: University of Missouri Press, 1994.

———. *Views and Viewmakers of Urban America: Lithographs of Towns and Cities in the United States and Canada, Notes on the Artists and Publishers, and a Union Catalog of Their Work, 1825–1925*. Columbus: University of Missouri Press, 1984.

Reuss, Martin. *Designing the Bayous: The Control of Water in the Atchafalaya Basin, 1800–1995*. Alexandria, VA: Office of History, U.S. Army Corps of Engineers, 1998.

Richardson, Heather Cox. *The Greatest Nation of the Earth: Republican Economic Policies during the Civil War*. Cambridge: Harvard University Press, 1997.

Rummage, Walter. *Michigan Place Names: The History of the Founding and the Naming of More Than Five Thousand Past and Present Michigan Communities*. Detroit: Wayne State University Press, 1986.

Sacher, John M. *A Perfect War of Politics: Parties, Politicians and Democracy in Louisiana, 1824–1861*. Baton Rouge: Louisiana State University Press, 2003.

Salecker, Gene Eric. *Disaster on the Mississippi: the Sultana Explosion, April 27, 1865*. Annapolis: Naval Institute Press, 1996.

Scharf, J. Thomas. *History of Saint Louis City and County, from the Earliest Periods to the Present Day: Including Biographical Sketches of Representative Men*. 2 vols. Philadelphia: Louis H. Everts, 1883.

Seager, Robert II. *and Tyler too: A Biography of John and Julia Gardiner Tyler*. New York: McGraw-Hill, 1963.

Seip, Terry L. *The South Returns to Congress: Men, Economic Measures, and Intersectional Relationships, 1868–1879*. Baton Rouge: Louisiana State University Press, 1983.

Sellers, Charles. *James K. Polk, Continentalist, 1843–1846*. Princeton: Princeton University Press, 1966.

———. *The Market Revolution: Jacksonian America, 1815–1846*. New York: Oxford University Press, 1991.

Shallat, Todd. *Structures in the Stream: Water, Science, and the Rise of the U.S. Army Corps of Engineers*. Austin: University of Texas Press, 1994.

Shugg, Roger W. *Origins of Class Struggle in Louisiana: A Social History of White Farmers and Laborers during Slavery and After, 1840–1875*. University: Louisiana State University Press, 1939.

Silbey, Joel H. *The Shrine of Party: Congressional Voting Behavior, 1841–1852*. Pittsburgh: University of Pittsburgh Press, 1989.

Stampp, Kenneth M. *America in 1857: A Nation on the Brink*. New York: Oxford University Press, 1990.

Studenski, Paul, and Herman E. Krooss. *Financial History of the United States*. 2nd ed. New York: McGraw-Hill, 1963.

Sydnor, Charles S. *The Development of Southern Sectionalism, 1819–1848*. Baton Rouge: Louisiana State University Press, 1948.

Taylor, George Rogers. *The Transportation Revolution, 1815–1860*. Vol. 4 of *The Economic History of the United States*. New York: Holt, Rinehart and Winston, 1951.

Tuchman, Barbara. *The Proud Tower: A Portrait of the World before the War, 1890–1914*. New York: Bantam Books, 1971.

Twain, Mark. *Life on the Mississippi*. New York: Harper and Brothers, 1896; reprint, New York: Bantam Books, 1981.

Van Deusen, Glyndon G. *The Jacksonian Era, 1828–1848*. New American Nation Series. New York: Harper Torchbooks, Harper and Row, 1959.

van Ravenswaay, Charles. *Saint Louis: An Informal History of the City and Its People, 1764–1865*. Edited by Candace O'Connor. St. Louis: Missouri Historical Society Press, 1991.

Wallenstein, Peter. *From Slave South to New South: Public Policy in Nineteenth-Century Georgia*. Chapel Hill: University of North Carolina Press, 1987.

Wallis, John Joseph. "American Government Finance in the Long Run: 1790 to 1990." *Journal of Economic Perspectives* 14 (Winter 2000): 61–82.

Wallis, John Joseph, and Barry R. Weingast. "Equilibrium Impotence: Why the States and Not the Federal Government Financed Economic Development in the Antebellum Era." NBER Working Paper Series, June 2005, abstract, 1–49. Cambridge, MA: National Bureau of Economic Research.

Walther, Eric H. *The Fire-Eaters*. Baton Rouge: Louisiana State University Press, 1992.

Watson, Harry L. *Jacksonian Politics and Community Conflict: The Emergence of the Second American Party System in Cumberland County, North Carolina*. Baton Rouge: Louisiana State University Press, 1981.

———. *Liberty and Power: The Politics of Jacksonian America*. New York: Hill and Wang, 1990.

Way, R. B. "The Mississippi Valley and Internal Improvements, 1825–1840." *Mississippi Valley Historical Association Proceedings* 4 (1910–11): 153–80.

White, John H., Jr. *American Locomotives: An Engineering History, 1830–1880*. Rev. ed. Baltimore: Johns Hopkins University Press, 1997.

Machine-Readable Databases

Paskoff, Paul F. Steamboat Database. Abstracted and compiled by the author from Bruce D. Berman, *Encyclopedia of American Shipwrecks*. Boston: Mariners Press, 1972.

Paskoff, Paul F. Steamboat Packet Database. Abstracted and compiled by the author from Frederick Way Jr., comp., *Way's Packet Directory, 1848–1994: Passenger Steamboats of the Mississippi River System since the Advent of Photography in Mid-Continent America*. Rev. ed. Athens: Ohio University Press, 1994.

Theses and Unpublished Papers

Allen, Daniel S. "The Impact of Technological Change on the Economic Viability of Individual Production Centers: The Case of the 1840–1880 British Ocean-Going Iron and Steam Shipbuilding Industry." Ph.D. diss., Louisiana State University, 1997.

Owens, Jeffrey Alan. "Holding Back the Waters: Land Development and the Origins of the Levees on the Mississippi, 1720–1845." 3 vols. Ph.D. diss., Louisiana State University, Baton Rouge, 1999.

Pabis, George S. "Restraining the Muddy Waters: Engineers and Mississippi River Flood Control, 1846–1881." Ph.D. diss., University of Illinois, Chicago Circle, 1997.

Pearce, Stanley Norman. "Constituency or Party? Democratic Congressmen and Internal Improvements on Lake Erie, 1825–1860." Ph.D. diss., Southern Connecticut State University, 1994.

INDEX

Page references in italics refer to illustrations.

Adams, John: tariff under, 111
Adams, John Quincy: abolition petition of, 60–61; censure resolution against, 61; in congressional debates, 55; congressional memorials under, 75; on his defeat, 47–48; internal improvements under, 46–48, 49–50, 123, 192, 263n16; military spending under, 123; on Polk, 64; river improvements under, 75, 78; road improvements under, 50; and "Tariff of Abominations," 47; on threat to Union, 55. *See also* American System
Allan, Chilton, 58
Alleghany River: improvement of, 77
American National Party, 86–87; rise of, 101
American System, 46–48, 61, 192; Calhoun's opposition to, 48; Democrats' opposition to, 80; failure of, 50; Jackson's objections to, 48, 50; revenue for, 49; revival of, 60; southern opposition to, 53; tariff in, 60, 90. *See also* Adams, John Quincy
Anthony, Henry B., 193
appropriations, congressional: for army, 115; for beacons, 233, 272n22; for buoys, 233, 272n22; for canal building, 41, 178; for coasting trade, 119, 121, 122; for Cumberland Road, 44, 47; decision-making process for, 117–18; following Civil War, 191–98; for foreign commerce, 119–20; inconsistencies in, 138, 164, 165, 185; for internal improvements, 43, 44, 109; for lighthouses, 233; for Louisiana, 169; for Mexican War, 92, 191; miscellaneous, 116, 169, 205–7, 227–32; partisanship in, 189; price deflators for, 208–11; for railroad industry, 160–61, 162, 178; for river improvements, 4, 6, 40, 114–21, 128–32, 164, 165, 186; for road building, 41, 44; state-specific, 168–69, 170, 204, 205, 220–26. *See also* expenditures, federal; internal improvements; river and harbor improvements, federal
Arkansas River: improvements to, 36; snag removal on, 136, 137
Atchison, David R., 97, 99, 102
Atlantic coast: harbor dredging for, 189; shipping improvement for, 119, 121, 122
Autocrat (packet), 260n85; snagging of, 32–33

Bagur, Jacques D.: *A History of Navigation on Cypress Bayou and the Lakes*, 202
Barringer, Daniel L., 52–53
Bates, Edward, 81
Baton Rouge, Louisiana: Twain on, 12
bayous: clearing of, 34; improvements to, 261n97; steamboats on, 258n60, 285n15
beacons: appropriations for, 233, 272n11
Belfast (packet), 28
Bell, John, 97–98
Belvidere (steamboat), 260n91; grounding of, 33
Berman, Bruce D.: *Encyclopedia of American Shipwrecks*, 201
Bierce, Ambrose, 119
Blair, John, 53–54, 69
Board of Supervising Inspectors, 21, 257n53
boatyards, 24, 257n60; in Kentucky, 25, 145, 278n10; in Ohio, 25, 145, 278n10; in Pennsylvania, 25, 145, 280n22
boiler explosions, 19, 27, 255n30, 285n12; fatalities from, 257n53; of high-pressure engines, 150; numbers of, 20, 21, 256n50
boilers, steamboat: construction standards for,

INDEX

boilers, steamboat *(continued)* 20; improvements to, 146–47; legislation governing, 20. *See also* steamboat engines

Bond County, Illinois: congressional memorials from, 77

"Bonus Bill": Madison's veto of, 41, 44

Brockway, John H., 81

Brooke, Walker, 96–97

Buchanan, James: internal improvements under, 104–5; and Morman War, 273n16; public land sales under, 114; tariff under, 112

buoys: appropriations for, 233, 272n11

Butler, Andrew P., 96

Cairo, Illinois, 13; flooding at, 12–13; mechanical panoramas of, 12

Cairo Delta (newspaper), 166

Calhoun, John C.: championing of for states' rights, 53, 55, 73; and General Survey Bill, 44–45; and interest in Texas, 71; on internal improvements, 52; at Memphis commercial convention, 69, 72, 73; on Mississippi River, 73; opposition of to American System, 48; presidential candidacy of, 46, 71, 72; *South Carolina Exposition and Protest*, 47, 53, 70; on steamboats, 70; support of for western states, 69–73; on tonnage duties, 94

California, Pennsylvania: boatyards of, 25

Cambria (packet), 23

Campbell, Lewis D., 85, 269n62; in Know-Nothing Party, 86–87; on proportionality, 168

canal building: cost of, 128, 129; federal appropriations for, 41, 178

Cape Fear River: improvements to, 104

capital: federal mobilization of, 191; growth of, 187; in railroad industry, 5, 6–7, 159, 160, 161, 162, 281n32; in steamboat industry, 128, 129, 152–53

capitalism: influence of in American society, 187; neo-Hamiltonian, 106; role of in antebellum economy, 6

Carson, Samuel P., 52–53, 54, 56

Cartter, David, 85, 267n56

Cass, Lewis, 98–99

Chalmers, James R., 196–97

Chandler, Zachariah, 193, 195

Charleston, South Carolina: railroad service to, 98

Charleston Mercury (newspaper), 252n2

Charlton, Robert, 98, 99

Chesapeake and Ohio Canal Company, 47

Chicago Convention (1848): congressional memorial of, 79–80; delegates to, 81; on tonnage duties, 91–92

Cincinnati, Ohio: boatyards of, 25; memorials to Congress from, 167; population of, 253n12; steamboat industry in, 143; steamboat traffic at, 26; wholesale price index for, 208–9, 289n1

Cincinnati Gazette: on steamboats losses, 166

Civil War: foreign commerce following, 193; general improvements following, 191, 195; railroads during, 192; river improvements following, 7, 191–99; steamboat losses following, 198–99

Clay, Henry: on Fiscal Corporation of the United States, 60; and General Survey Bill, 45; presidential defeat of, 71; as secretary of state, 46; slavery compromise attempt of, 88

coasting trade: appropriations for, 119, 121, 122; tonnage of, 119, 272n14

commerce, foreign: after Civil War, 193; appropriations for, 119–20; improvements for, 119–21; tonnage in, 119, 272n14

Commercial Review of the South and West (De Bow's Review), 72–73, 166, 267n31

commodities: effect of tonnage duties on, 96; price deflators for, 208–11; price swings for, 208

communications: revolution of in United States, 271n2; steamboat transport of, 23–24

Compromise of 1850, 101

Conclin's New River Guide, 14

confederationism, Jeffersonian, 2

Congress, U.S.: coalitions in, 105; corruption in, 168; debate in on Rivers and Harbors Bill of 1846, 65–68, 100; debate in on Rivers and Harbors Bill of 1852, 84–86, 87, 89, 97; debate in on tonnage duties, 90–91, 93–99; debates in on internal improvements, 40, 52–56; debates in on Mexican War, 126–27; debates in on river improvement, 2, 65–68, 84–86, 87, 89, 97, 100, 193–98, 264n30; House Committee on Commerce, 119; John Quincy Adams in, 55; log-rolling in, 67, 68, 168, 169, 186, 193, 197, 204; maverick voters in, 82–83, 234–44, 246–47; memorials to, 74–80, 105–6, 167, 234–35,

314

274n23, 286n24; on military spending, 124; sectionalism in, 82; western coalition in, 103. *See also* appropriations, congressional
Congressional Globe & Appendix, 60
Conkling, Roscoe, 194
Constitution, U.S.: Jackson's construction of, 51; Jefferson Davis on, 270n84; Monroe's view of, 41, 44; Polk's construction of, 64. *See also* river and harbor improvements, federal: constitutionality of
Corning, Erastus, 81
Cox, Samuel, 196, 197–98
Crenson, Mathew A., 264n41, 271n2
Cumberland River, snag removal on, 84
Cumberland Road: appropriations for, 44, 47; extensions of, 49, 57, 58, 59, 102; tolls for, 41
Cumberland Road bill (1822), Monroe's veto of, 44
Currier, Nathaniel, 17
Cypress Bayou: clearing of, 34; steamboat losses on, 285n15; steamboats on, 258n60

Daniel, Peter V., 273n16
Davis, Garrett, 126
Davis, Jefferson: in congressional improvements debates, 100–101; on the Constitution, 270n84; on navigation hazards, 1; report of on river improvement, 35; on threats to Union, 100–101
De Bow, James D. B., 72
De Bow's Review, 72–73, 166, 267n31
deficit, federal, 111, 113–14, 124–25; effect of on river improvements, 118–19, 121, 128; Jackson on, 51; servicing of, 122. *See also* recessions, economic
Delaware Bay: appropriations for, 169; commerce on, 285n13
Democratic Party: anti-improvement stance of, 80, 81–82; coalition of with Whigs, 83–84, 88; convention of 1856, 104; opposition of to American System, 80; representatives for Lake Erie, 267n52
Democrats, southern: support of for tonnage duties, 96–97
Democrats, western: opposition of to general improvement system, 68
De Saussure, William, 98
DeVoto, Bernard, 72

Dickens, Charles: *American Notes*, 253n8; on Cairo, 12; and experience of snagging, 33; and his view of U.S., 260n95; on Mississippi River, 34
Douglas, Stephen A.: debating skills of, 95; and Kansas-Nebraska Act, 102; on Pierce administration, 102; on R. B. Rhett, 94–95; and Rivers and Harbors Bill of 1852, 88–89; support of for internal improvements, 34–35; support of for tonnage duties, 88–89, 92–99, 103, 165, 191; on tariff system, 96
dredge boats, 132, 134; cost of, 210, 275n29
dredging, 4, 132, 134; of harbors, 189

economy, antebellum: contractions of, 59, 78; effect of general improvements on, 187; effect of railroad industry on, 281n26; following Panic of 1837, 35; history of, 178; role of capitalism in, 6; role of federal public policy in, xv, 5–6; role of river improvements in, 2, 6, 166, 286n28; surpluses in, 124–25. *See also* Panic of 1819; Panic of 1837; Panic of 1857; recessions, economic
Edward Bates (packet): burning of, 15
elections: of 1824, 45–46; of 1840, 59; of 1848, 83; of 1852, 101; of 1856, 104
Ellett, Charles, 190
Ellis, Richard, 263nn21–22
Engerman, Stanley L., 168
engines, internal-combustion, 147. *See also* steamboat engines
Era of Good Feelings: end of, 46
Evans, Alexander, 85–86
expenditures, federal: percentage of for river improvements, 124, 126, 127, 192; by presidential administration, 249; price deflators for, 208–11; for river improvements, 116, 124, 132, 209, 212–13; for snag removal, 173
expenditures, military, 121–27; Congress on, 124; for Mexican War, 118, 122, 123, 125–27

federal government: authority of over slavery, 266n9; civilian workforce of, 109, 271n1; during Jacksonian era, 271n2; and mobilization of capital, 191; and regulation of steamboats, 20–22; scale of, 109, 110; stock subscriptions by, 47, 66. *See also* Congress, U.S.; deficit, federal; United States

315

federalism, Hamiltonian, 2
Felder, John M., 54–56
Fillmore, Millard, 81, 87; public debt under, 114; river improvements under, 84; and signing of Rivers and Harbors Bill, 101
fires: aboard steamboats, 12, 19, 21, 27, 285n12; urban, 255n38. *See also* St. Louis fire (1849)
Fiscal Corporation of the United States (proposed), 60
Fugitive Slave Act, 101
Fuller, Charles, 134
Fulton, Robert, 1, 70

Gadsden, James: on West-South unity, 70, 266n22
General Survey Bill, 44–45; passage of, 45
Georgia: railroads subsidies of, 161. *See also* Savannah River
Gibbons v. Ogden (1824), 129
Gilmer, Thomas W., 61
"Graveyard" (Mississippi River), 30, 156–57, *158*, 171
Great Lakes: improvements to, 65, 132; navigation hazards on, 3
groundings, 27, 33; frequency of, 34
Gulf of Mexico: congressional memorials on, 74; mechanical panoramas of, 11; steamboat losses on, 289n35
Guthrie, W. W., 20–21

Hall, Thomas H., 48
Hamilton, Alexander, 131
harbors: appropriations for, 119; maintenance of, 189. *See also* river and harbor improvements
Harrison, William Henry, 59–60
Haupt, Herman, 190
Hibbard, Benjamin Horace: *A History of the Public Land Policies*, 289n2
Hibbard, Harry, 104
Holmes, Isaac, 50; ties of to Calhoun, 263n22
Holt, Michael F., 267n54
Houston, George S., 66, 87, 196
Humphreys, Andrew A., 287n3
Hunter, Louis C., 20, 146, 257nn53–54; on hull dimension, 278n6; on hull resistance, 278n7; on the *Shepherdess*, 259n81; on snag removal, 283n2; on steamboat engines, 279nn14–15,

280n19; on steamboat inspection, 256n48; on steamboat passengers, 259n82; *Steamboats on the Western Rivers*, 202, 251n1, 276n30, 277nn2–3, 277n5, 279n14

Illinois: economic problems of, 35; river improvements by, 34–35
Illinois Central Railroad: land grant to, 281n32
Illinois River: improvements to, 36
Indiana: economic problems of, 35; interest rates in, 274n24; river improvements by, 35; town formation in, 274n24
infrastructure, antebellum: federal role in, 5; intensive growth of, 46. *See also* internal improvements; river and harbor improvements; road building
Inspection Act of 1852, 20–21
interest rates: in economic recessions, 153–54; in Indiana, 274n24; in Pennsylvania, 155, 157, 280n22; and steamboat engine size, 154–55, 156, 157, 280n23
internal improvements: by state governments, 34–35, 97, 287n28
internal improvements, federal: under American System, 61; under Buchanan administration, 104–5; Calhoun on, 52; congressional debates on, 40, 52–56; congressional memorials on, 74–80, 105–6, 234–35; constitutionality of, 41, 44, 48, 49, 53, 54, 57, 64, 105, 109, 266n9; Democratic stance on, 80, 81–82; effect of economic crises on, 58–59; effect of slavery on, 106; effect of on Union, 99, 100–101; federal appropriations for, 43, 44, 109; federal policy on, 3, 5; under Jackson, 47, 48–58, 64, 68, 80, 263n16, 263n22, 264n41; under John Quincy Adams, 46–48, 49–50, 123, 192, 263n16; log-rolling in, 67, 68, 168, 169, 186, 193, 197, 204; under Madison administration, 41, 49; maverick voters on, 82–83, 234–44, 246–47; under Monroe administration, 40, 41, 44, 262n8; in Old Northwest, 275n26; under Pierce administration, 103; under Polk administration, 64–69, 79; by presidential administration, 42, 43; prior to Polk administration, 40; proportionality in, 69; sectional interests in, 52–56, 68–69; South Carolina's opposition to, 51–52; southern op-

position to, 51, 66, 67–68, 69; southern support for, 260n99; tariffs for, 90–91; as threat to Union, 54–56; under Tyler administration, 62–63; under Van Buren, 58–59, 62, 90; in Virginia, 51; in western states, 51; Whig Party and, 62–63. *See also* river and harbor improvements, federal

internal improvements, general, 41, 44, 63, 68, 109; effect of on economy, 187; following Civil War, 191, 195; relationship to land sales, 129–30, 274n23

Jackson, Andrew: campaigns of against Native Americans, 80; congressional memorials under, 75; constitutional views of, 51; election of, 48; internal improvements under, 47, 48–58, 64, 68, 80, 263n16, 263n22, 264n41; and Louisville Canal, 78; on national banks, 51; opposition of to American System, 48, 50; on public debt, 51; public debt under, 113–14; regional coalitions under, 50, 51; river improvements under, 54, 57–58, 74, 79, 192; road building under, 57; and "Tariff of Abominations," 47; on tariffs, 56, 64, 263n22; tariff under, 111; and use of public land sales, 47, 80, 90, 111–12, 130; veto of, of Maysville Road Bill, 49, 57, 66–67, 263n19; and western states, 50

Jacob Strader (steam packet), 151

Jefferson, Thomas: on inland republic, 130; internal improvements program of, 40–41; and Kentucky and Virginia Resolutions, 70; road building under, 40–41; tariff under, 46, 47, 111

J. M. White (packet): speed of, 23–24

John, Richard R., 271n2

Jones, George Washington, 90–91

Kane, Adam I.: *The Western River Steamboat*, 202, 279n13

Kansas: irregular warfare in, 102; military spending in, 123

Kansas-Nebraska Act, 87, 209; passage of, 102

Kelman, Ari: "Forests and Other River Perils," 283n2

Kentucky: boatyards of, 25, 145, 278n10; road building in, 49

King, T. Butler, 81

Kirkpatrick, Littleton, 81

Klein, Maury: *History of the Louisville & Nashville Railroad*, 282n37

Know-Nothing Party, 86–87; rise of, 101

Lake Erie: Democratic congressmen representing, 267n52

Larson, John Lauritz: *Internal Improvement*, 281n31

Lasch, Christopher, 260n95

Lewis, Henry: panorama of, 11, 18, 252n4, 256n44

lighthouses: appropriations for, 233

lighthouse system, 119

light stations: appropriations for, 233, 272n11

Lincoln, Abraham, 34

Lloyd's Steamboat and Railroad Directory, 31, 260n83, 260n84

loss rates, steamboat: annual, 186, 187, 283n4; due to natural hazards, 143, 155, 186, 285n14; effect of technology on, 138–39; and engine volume, 152; following Civil War, 198; on Louisiana bayous, 285n15; reductions in, 171–73; relationship of to appropriations, 166, 285n11; in tonnage, 185, 187, 286n23, 286n25, 289n35. *See also* steamboat losses

Louisiana: bayou improvements in, 261n97; river appropriations for, 169; river improvement by, 34; steamboat losses in, 285n12, 285n15

Louisville (Kentucky): boatyards of, 25; railroad enthusiasm in, 282n37

Louisville and Portland Canal Company, 57

Louisville Canal: influence of on steamboat size, 145

Louisville Canal Company, 66; Jackson's opposition to, 78

Lytle-Holdcamper List, The, 201

Macon, Nathaniel, 266n9

Madison, James: internal improvements under, 41, 49; and Kentucky and Virginia Resolutions, 70; public debt under, 114; veto of, of "Bonus Bill," 41, 44

Malone, Laurence Joseph, 202; *Opening the West*, 275n26

Marshall, Thomas Francis, 61

Maysville Road Bill: Jackson's veto of, 49, 57, 66–67, 263n19

INDEX

McCall, Edith: *Conquering the River*, 276n30

memorials to Congress: concerning Mexican War, 76; on internal improvements, 74–80, 105–6, 234–35; during John Quincy Adams administration, 75; during Polk administration, 79; regional origins of, 245; on river improvements, 74–80, 167, 234–35, 274n23, 286n24; from western states, 74–80, 105–6

Memphis, Tennessee: and commercial convention (1845), 69, 72, 73; mechanical panoramas of, 12; railroad enthusiasm in, 163, 282n37; steamboat industry at, 26, 259n58; steamboat traffic at, 26

Merchant Steam Vessels of the United States (Lytle and Holdcamper), 201

Mexican War: appropriations for, 92, 191; congressional debates on, 126–27; congressional memorials concerning, 76; effect of on river improvements, 126–27; effect of on sectionalism, 88, 117; effect of on Whig Party, 101; expenditures for, 118, 122, 123, 125–27; Polk's support for, 125–26, 273n20; popularity of in South, 159; role of St. Louis in, 13, 14

Michigan, state legislature of: resolutions to Congress, 106

military: appropriations for, 115; federal expenditures on, 121–27. *See also* Mexican War

military roads: extensions of, 49; improvements to, 57, 58; Washington on, 3

Minneapolis, Minnesota: steamboat traffic at, 27

Mississippi River: as American Nile, 256n45; appropriations for, 87, 115; British travelers on, 32–33, 34; Calhoun on, 73; confluence with Ohio, 13; Dickens on, 34; distances from mouth of, 37–38; economic importance of, 19; flooding by, 190; and flood of 1851, 36, 262n111; "Graveyard" section of, 30, 156–57, 158, 171; improvements to, 35–36, 40, 104, 205, 269n62; levee construction along, 190–91; market towns of, 24; mechanical panoramas of, 11–13, 18–19, 252n1–2, 252n4; in Minnesota, 204; natural hazards on, 22; navigable length of, 11; oxbow lakes of, 12; in Rivers and Harbors Bill of 1852, 87; snags on, 30, 35–36, 136, 283n2; steamboat productivity on, 183; steamboats losses on, 169; steamboat traffic on, 4, 14, 26, 27, 29, 169; swampland of, 256n45; during Tyler administration, 62; water levels of, 36, 38, 285n14

Mississippi River Commission, 198

Mississippi River Valley: economy of, 3–4, 6, 22–23, 156; precipitation in, 36, 38, 285n14; water levels in, 262nn109–10, 285n14

Missouri Compromise: overthrow of, 102

Missouri River: appropriations for, 115; improvements to, 36; natural hazards on, 22; in Rivers and Harbors Bill of 1852, 87; and Smith's Bar, 136; snags on, 36, 135–36; steamboat losses on, 285n12

Missouri River Commission, 288n33

Missouri River Valley: economic development of, 156

Monongahela River: improvements to, 75–76

Monroe, James: and his view of Constitution, 41, 44; internal improvements program of, 40, 41, 44, 262n8; veto of, of Cumberland Road bill, 44

Monroe doctrine, 44

Morill tariffs, 193

Mormon War, 123, 273n16

Morse, Samuel, 271n2

motive power, 147

Nashville, Tennessee: steamboat traffic at, 26

Natchez, Mississippi: distance charts to, 37–38; mechanical panoramas of, 12

national bank: under American System, 61; Jackson on, 51; Whigs and, 60

Native Americans: Jackson's campaigns against, 80

navigation hazards: on Great Lakes, 3; and river traffic, 156; and steamboat productivity, 183–84; steamboats lost to, 39; tonnage lost to, 39

navigation hazards, natural, 1, 27–39, 206; collisions caused by, 251n2; federal improvements to, 7, 22; geographical extent of, 189; Jefferson Davis on, 1; losses due to, 143, 188, 262n108; on Ohio River, 22; reductions in, 166, 171–73, 198–99; removal of, 132; tonnage lost to, 262n108. *See also* groundings; snags

New Madrid earthquake (1811), 12, 252n6; consequences of, 252n7

New Orleans, Louisiana: mechanical panoramas of, 11–12; steamboat tonnage at, 29; steamboat

318

traffic at, 26, 27, 29, 169; wholesale price index for, 208–9, 289n1

New Orleans (steamboat): sinking of, 1, 29; tonnage of, 251n1

New York City: harbor appropriations for, 169; port commerce of, 285n13; price indexes for, 208

New York Herald: on St. Louis fire, 255n38

North, Douglass, 282n36

North Carolina: river improvement by, 34, 261n99

Northwest: internal improvements in, 275n26

Northwest Ordinance (1787), 92; tonnage duties in, 98

Ohio: boatyards of, 25, 145, 278n10; transportation corporations in, 250

Ohio River: appropriations for, 115; improvements to, 24, 35, 36, 69, 85, 205; and Louisville Canal, 26; lower, 146; miscellaneous appropriations for, 206, 207, 284n10; natural hazards on, 22; public land sales along, 130; in Rivers and Harbors Bill of 1852, 87; snags on, 30, 134–35; steamboat construction for, 145–46; steamboat losses on, 166; steamboat traffic on, 27; Webster on, 70; Wheeling bridge at, 128, 273n20

Ohio River Valley: economic development of, 156; in Rivers and Harbors bill of 1852, 269n62

Olmstead, Denison: "On the Democratic Tendencies of Science," 23

Owens, Harry P.: *Steamboats and the Cotton Economy*, 202

Pacific coast: shipping improvement for, 121

packets, steamboat: *Autocrat*, 32–33, 260n85; *Belfast*, 28; *Cambria*, 23; communications transport by, 23; construction centers for, 24, 25; *Edward Bates*, 15; *Peytona*, 23; *Jacob Strader*, 151; *J. M. White*, 23–24; *White Cloud*, 15, 16; *Yorktown*, 23

Panic of 1819, 44

Panic of 1837, 35, 59; depression following, 153; effect of on commodity prices, 209

Panic of 1857, 105; effect of on public land sales, 114; railroad industry following, 154

panoramas, mechanical, 15, 255n44; of Mississippi River, 11–13, 252nn1–2, 252n4; St. Louis fire in, 18

partisanship: in river improvements, 2, 3, 82–83, 189. *See also* sectionalism

Patapsco River: improvements to, 104

Pennsylvania: boatyards of, 25, 145, 280n22; interest rates in, 155, 157, 280n22

Peytona (packet), 23

Philadelphia, Pennsylvania: port commerce of, 285n13

Pierce, Franklin, 94; election of, 101; internal improvements under, 103; Jacksonianism of, 101, 102, 103; land sales under, 113; *Vanity Fair* on, 102; veto of Rivers and Harbors Bill, 103, 104

Polk, James K.: in congressional improvement debates, 54; constitutional views of, 64; on federal funding, 65; internal improvements under, 64–69, 79; Jacksonianism of, 64, 66; memorials under, 79; military spending under, 122–23; public debt under, 114; river improvements under, 65; support of for Mexican War, 125–26, 273n20; on tonnage duties, 269n79; veto of Rivers and Harbors Bill, 65–67, 68, 91, 94, 125, 265nn3–4

Polk, William Hawkins, 85

Pomarède, Leon: panorama of, 11, 18

postal service: development of, 109, 271n2

Prairie (steamboat): loss of, 254n29

precipitation: in Mississippi River Valley, 36, 38, 262n109, 285n14; snags due to, 38

price indices, wholesale, 210–11; for Cincinnati, 208–9, 289n1; for New Orleans, 208–9, 289n1

private enterprise: Jeffersonian, 106; river improvements by, 128

public land: access to, 129, 130; development of, 188

public land grants: to railroads, 281nn31–32; for river improvements, 160, 187; in support of transportation, 161, 162; value of, 281n32

public land sales: under Buchanan administration, 114; collapse of, 112, 114, 124, 131–32; and commodity prices, 209; following Panic of 1857, 114; to individuals, 275n25; Jackson's use of, 47, 80, 90, 111–12, 130; along Ohio River, 130; under Pierce administration, 113; for railroad construction, 160–61, 162; reduction in, 78; relationship of to general improvements, 129–30, 274n23; revenue from, 46, 56, 129, 218–19; and town formation, 131 public policy, federal: effect of slavery on, 82, 88; on internal improvements, 3, 4, 5; for river improvements, 164; role of in antebellum economy, xv; role of West in, 2–3

railroad industry: capital in, 5, 6–7, 159, 160, 161, 162, 281n32; effect of on steamboat industry, 138, 157, 159, 164; effect of on U.S. economy, 281n26; federal appropriations for, 160–61, 162, 178; following Panic of 1857, 154; government subsidies for, 281n31; incorporation of, 128, 129, 159; lobbying by, 128; public land grants to, 281nn31–32; states' assistance to, 7, 79, 161–62, 268n46

railroads: bridges for, 128; during Civil War, 192; effect of on river improvements, 159; land grant funding for, 162; map of, 10; Midwestern enthusiasm for, 282n36; novelty of, 159; role of in national unity, 159–60; role of in town formation, 131, 132, 275n27; technology of, 147; transcontinental, 192; Webster on, 159–60, 161, 281n29

Raritan River: improvements to, 197

Reagan, John H., 197, 198; on Commerce Committee, 288n27

recessions, economic, 78, 116–17, 118, 124; commodity prices during, 209; congressional memorials following, 78; effect of on river improvements, 272n10; interest rates in, 153–54; of 1840s, 59. *See also* deficit, federal

Red River: snag removal from, 137; steamboat traffic on, 27

Republican Party: economic development under, 193; organization of, 104

revenue, federal, 109–10; for American System, 49; from customs receipts, 111, 285n13; from public land sales, 46, 56, 129, 218–19; sources of, 111–14, 248; from west, 73. *See also* tariffs; tonnage duties; United States Treasury

Rhett, Robert Barnwell, 94

river and harbor improvements: by private sector, 128

river and harbor improvements, federal: coalitions for, 83–84; congressional debates on, 2, 65–68, 84–86, 87, 89, 97, 100, 193–98, 264n30; congressional voting patterns on, 82–83; constitutionality of, 2, 4, 118, 197, 252n7, 282n2; by Corps of Engineers, 36, 284n10; corruption in, 164, 168; cost-effectiveness of, 186, 286n27; cost of, 38; diminishing returns in, 185–86; economic benefits of, 2, 6, 166, 286n28; effect of deficits on, 118–19, 121, 128; effect of Mexican War on, 126–27; effect of railroads on, 159; effect of recessions on, 272n10; effect of sectionalism on, 2, 165; effect of on vessel loss, 284n11; for 1824, 205; for 1827, 205–7; for 1844, 103, 104; for 1846, 65–68, 100; for 1852, 84–86, 87, 89, 97; for 1872, 193–97; for 1880, 198; expenditures for, 116, 132, 209, 212–13; federal appropriations for, 4, 6, 40, 114–21, 128–32, 164, 165, 186; under Fillmore, 84; following the Civil War, 7, 191–99; impact of on steamboat productivity, 179, 180, 183; incomplete, 165; inconsistencies in, 138, 185; under Jackson, 54, 57–58, 74, 79, 192; under John Quincy Adams, 75, 78; land grant funding for, 160, 187; magnitude of program, 189; memorials to Congress on, 74–80, 167, 234–35, 274n23, 286n24; miscellaneous, 116, 169, 204–7, 227–32, 269n61, 284n10; to Ohio River, 24; partisanship in, 2, 3, 82–83, 189, 267n51; percentage of federal expenditures, 124, 126, 127, 192; performance criteria for, 166; policy objectives of, 164; political aspects of, 2, 4–5, 38–39, 40, 121, 165–66, 204–5; under Polk, 65; by private sector, 128; proportionality in, 86, 87–88, 165–70, 204; southern states on, 3, 52–54; state-specific, 115, 116, 119, 168–69, 170, 220–26, 284n9; strengthening of Union, 77–78; success of, 184–86, 188; tonnage duties for, 88, 89–90; under Tyler administration, 62; types of, 132, 134–37; for western states, 73–84; Whig support for, 83. *See also* internal improvements, federal

riverbanks: approaches to, 143

rivers, U.S.: map of, 10. *See also* western rivers

Rivers and Harbors Bill (1844): Pierce's veto of, 103, 104

Rivers and Harbors Bill (1846): congressional debate on, 65–68, 100; Polk's veto of, 65–67, 68, 91, 94, 125, 265nn3–4

Rivers and Harbors Bill (1852): congressional debate on, 84–86, 87, 89, 97; Douglas and, 88–89; passage of, 101; provisions of, 269n62

road building: access to public lands through, 129, 130; federal appropriations for, 41, 44; under Jackson, 57; under Jefferson, 40–41; under John Quincy Adams, 50; in Kentucky, 49; under Tyler, 62

Rockwell, John A., 126, 127

sandbars: grounding on, 27

Sault Ste. Marie Canal bill, 95

Savannah River: improvements to, 52, 104, 270n79; tonnage duties at, 89, 98

Scientific American: civilizing role of steamboats in, 19

Scott, Winfield, 101

secession crisis (1850), 88, 89, 106

secession crisis (1861), 106

sectionalism: in Congress, 82; effect of Mexican War on, 88, 117; effect of on river improvements, 2, 165; role in internal improvements, 52–56, 68–69; in Whig Party, 88. *See also* partisanship

Sellers, Charles, 262n8

Shallat, Tony: *Structure in the Stream,* 283n2

Shepherdess (steamboat): fatalities on, 260n83; passengers aboard, 259n81; snagging of, 30, 32, 260n82

Sherman, John, 193

Shreve, Henry M.: snag boat design of, 134, 135, 276n30

side-wheelers (steamboats), 175, 216; longevity of, 177

Silbey, Joel H., 82, 268n51

slavery: Calhoun on, 72; effect of on improvements issues, 106; effect of on national policy, 82, 88; federal authority over, 266n9; future of, 2; John Quincy Adams and, 60–61

Smith, Truman, 93–96, 97; Douglas on, 94–95

snag boats, 134, 276n30; cost of, 210, 276n42; decommissioning of, 137; designs for, 134, 135; Topographical Corps's, 171

snagging, 1, 27; drowning following, 30; and engine size, 152, 154, 155; frequency of, 34

snag removal, 4, 134–37, 283n2; cost of, 38; federal expenditures for, 173; hazards of, 136–37; quarter boats for, 134; by states, 34; and steamboat losses, 171–73, 174; by Topographical Corps, 171, 173, 176

snags, 1, 28–30, 260n84; on Cumberland River, 84; distance table for, 37–38; effect of on steamboat speed, 152; formation of, 262n106; and hull dimensions, 143, 144, 146; losses attributable to, 171, 172, 174, 186, 187, 285nn14–15, 286n23; "planters," 28; prevention of, 35–36; "sawyers," 28, 206; tonnage loss due to, 172, 185, 187, 286n23, 286n25

Sokoloff, Kenneth L., 168

Soulé, Pierre, 99, 270n80

South Carolina: opposition of to internal improvements, 51–52

southern states: economic development of, 267n31; opposition of to American System, 53; opposition of to internal improvements, 51, 66, 67–68, 69; on river improvement, 3, 52–54; unity of with West, 69–73, 266n22, 267n31

Specie Circular: issuing of, 124, 131

Stanton, Frederick P., 65–66, 67

state governments: assistance of to railroads, 7, 79, 161–62, 268n46; extension of federal power into, 266n9; general improvements by, 106; internal improvements by, 34–35, 97, 287n28; and levying of tonnage duties, 91, 93, 260n79; and transportation infrastructure spending, 251n7

states' rights: Calhoun's championing of, 53, 55, 73; Jefferson Davis on, 101

St. Clair River: improvements to, 104–5

Steamboat Database, 201–2, 251n2

steamboat engines, 217, 279nn14–15; cylinders of, 149, 153, 279n15; and effect of size on snagging, 152, 154, 155; high-pressure, 150, 152, 153, 280n20; efficiency of, 146–52; low-pressure, 150, 280n20; relative power of, 147, 150, 279n13; safety of, 19–24; size of, 280nn19–20; size/interest rate relationships of, 154–55, 156, 157, 280n23; size-to-tonnage ratios of, 152, 153 steamboat industry, 24–27; businesses supported by, 26–27, 259n72; capitalization of, 128, 129, 152–53; in Cincinnati, 143; effect of interest rates on, 154–55, 156, 157, 280nn22–23; effect of railroads on, 138, 157, 159, 164; incorporation in, 128; at Memphis, 26, 259n68; political influence of, 128

Steamboat Law of 1852, 256n48, 257n54

steamboat losses, 39, 138–39, 143, 170, 178; after Civil War, 198–99; annual rates of, 186; to collision, 251n2; cost of, 167, 283n5; on Cypress Bayou, 285n15; due to fire, 285n12; due to human error, 285n12; due to snags, 171, 172, 174, 186, 187, 285nn14–15, 286n23; effect of river improvements on, 284n11; on Gulf of Mexico, 289n35; in Louisiana, 285n12, 285n15; on Mississippi River, 169; on Missouri River, 285n12; on Ohio River, 166; rate of, 256n50, 282n37, 283n4; snag removal and, 171–73, 174; in St. Louis fire, 254n29; stabilization of, 186–87. *See also* boiler explosions; loss rates, steamboat; navigation hazards

Steamboat Packet Database, 202

steamboat productivity, 166, 176, 177, 178–88; effect of hazard removal on, 183–84; effect of river improvements on, 179, 180, 183; inputs in, 178–79; market-related influences on, 179–80; on Mississippi River, 183; outputs in, 179, 181, 182; private enterprise in, 177; role of government in, 178; and tonnage, 180, 182–83

steamboats: Calhoun on, 70; canal-capable, 145; construction of, 24, 25, 145, 257n60, 278n10, 280n22; cost of, 257n61; depreciation of, 283n5; design improvements for, 139–40; displacement of, 139, 143, 152; economic importance of, 22–23; engineering aesthetic of, 143; federal regulation of, 20–22; fires aboard, 12, 19, 21, 27; as floating steam-palaces, 32, 260n85; groundings of, 27, 33; hull dimensions of, 139, 140–45, 146, 277n5, 278n6, 278n10; hull resistance of, 278n7; keel design of, 139; longevity of, 166, 173–76, 177, 179, 187; machinery costs of, 153; mail transportation by, 23–24; maneuverability of, 175–76; passengers on, 259n81, 277n2; passengers' fatalities on, 167, 257n53, 260n83; practicability of, 1; regulation of, 19–21, 256n48;

role in cultural life, 23; safety on, xv, 19–24; and safety legislation, 20–21; shallow-draft, 19, 173; side-wheelers, 175, 177, 216; speed of, 143, 152, 278n12; in St. Louis fire, 15, 16, 254n29; standardization of, 140, 143, 144; stern-wheelers, 175–76, 177, 216, 286n18; superstructure space of, 152; traffic density of, 156; water plane of, 139, 143

Steamboats on Louisiana's Bayous (Brasseaux and Fontenot), 202

steamboat technology, 6, 138–52; effect of on loss rates, 138–39; effect of on productivity, 179–80; hull dimension in, 139, 140–45, 146, 277n5, 278n6, 278n10; role in economic development, 23; size-to-power relationships in, 150. *See also* boilers, steamboat; steamboat engines

steam locomotives: horsepower of, 279n14

Stein, Albert, 190

stern-wheelers (steamboats), 175–76, 216; longevity of, 177; on Yazoo River, 286n18

Stevenson, Andrew, 53

St. Louis, Missouri: civic economy of, 17; Congressional Convention in (1873), 195, 288n16, 288n18; economic importance of, 14–15, 17; federal arsenal at, 254n25; harbor improvements to, 14–15; mechanical panoramas of, 13, 14–15; river interests conventions in, 195; role of in Mexican War, 13, 14; steamboat traffic at, 14, 26, 27, 253n20

St. Louis fire (1849), 15–18, 27; arson possibility in, 255n31; depictions of, 17–18; newspaper accounts of, 17, 253n21, 254n23, 254n27, 254n29, 255n38; property losses in, 255n34; steamboats in, 15, 16, 254n29; volunteer fire company in, 15, 16

St. Louis Weekly Reveille, 18

stock subscriptions: by federal government, 47, 55

Stuart-Wortley, Emmeline, 32–33, 34, 260n85

Sultana (steamboat): explosion of, 255n30

surpluses, federal, 124–25

Sydnor, Charles, 51

Taney, Roger B., 273n16

Targee, Thomas B., 17, 254n27

"Tariff of Abominations," 47

tariffs: of 1832, 56; of 1846, 83; under American

System, 61, 90; under Buchanan administration, 111; Douglas on, 96; from inland navigation, 121; for internal improvements, 90–91; under Jackson administration, 111; Jackson on, 56, 64, 263n22; under Jefferson administration, 46, 47, 111; under John Adams administration, 111; Morill, 193; revenue from, 124, 169; under Van Buren administration, 113; Walker, 83, 125; under Washington administration, 111

Taylor, George Rogers: *The Transportation Revolution*, 281n26

Taylor, Zachary: death of, 84; election of, 83

Tennessee: economic problems of, 35; river improvements in, 64

Tennessee River: improvements to, 35, 64, 66, 67, 90; snags on, 30, 32

Terror (snag boat), 134–35, 136, 137; design of, 276n42

Texas: annexation of, 71, 76; river improvement by, 34

Thurman, Allen, 194

Timour (steamboat): loss of, 254n29

tonnage: of coasting trade, 119, 272n14; and engine size, 152, 153; in foreign commerce, 119, 272n14; and hull dimensions, 140, 144, 145; losses, 36, 289n35; losses to navigation hazards, 39, 262n108; losses to snags, 172, 185, 187, 286n23, 286n25; at New Orleans, 29; role of in productivity, 180, 182–83; southern support for, 96–97; of wrecks, 215

tonnage duties: Calhoun on, 94; Chicago Convention on, 91–92; congressional debate on, 90–91, 93–99; constitutionality of, 97–98; Douglas's support for, 88–89, 92–99, 103, 165, 191; effect of on commodities, 96; levied by states, 91, 93; in Northwest Ordinance, 98; Polk on, 269n79; for river improvements, 88, 89–90; at Savannah River, 89, 98; states' levying of, 91, 93, 260n79; southern support for, 96–97

town formation: along western rivers, 130, 133, 275n28; in Indiana, 274n24; public land sales and, 131–32; role of railroad in, 131, 132, 275n28

towns: *versus* townships, 130

trade routes: specialization in, 27

transportation: of mail, 23–24; public investment in, 178; public land grants for, 161, 162

travelers, British: on Mississippi River, 32–33, 34

Trollope, Frances, 34, 260n91; experience of grounding, 33

Trumbull, Lyman, 194

Tuchman, Barbara, 101

Turnpikes: cost of, 128–29

Twain, Mark: on Baton Rouge, 12; *Life on the Mississippi*, 199

Twyman, R.B.J., 256n45

Tyler, John, 59–60; internal improvements under, 62–63; military spending under, 122, 123; on national bank, 61; public debt under, 114; river improvements under, 62; veto of, of Fiscal Corporation of the United States, 60

Union: effect of internal improvements on, 54–56, 99, 100–101; effect of railroads on, 159–60; strengthening measures for, 76–77; threats to, 54–56, 61; western states' admission to, 45

United States: communications revolution in, 271n2; population growth in, 45. *See also* Congress, U.S.; Constitution, U.S.; federal government

United States Army Corps of Engineers: river and harbor improvements by, 36, 284n10

United States Army Topographical Corps, 34, 206; officers of, 123; snag removal by, 171, 173, 176

United States Supreme Court: and Wheeling bridge decision, 128, 273n20

United States Treasury: inflows into, 111; notes issued by, 113; Specie Circular of, 124, 131; tariff revenues in, 169. *See also* revenue, federal

United States War Department: on Mississippi Graveyard, 156

Vallandigham, Clement, 196

Van Buren, Martin: internal improvements under, 58–59, 62, 90; military spending under, 123; public debt under, 114; tariff under, 113

Vicksburg, Mississippi: mechanical panoramas of, 12; steamboat traffic at, 26

Virginia: internal improvements in, 51

Wabash River, 202; improvements to, 35

Walker Tariff, 83, 125

Wallis, John Joseph, 251n7, 271n1, 287n28
Ward, William T., 84–85, 86, 87, 268n56; on proportionality, 168
War of 1812: expenditures for, 122
Washington, George: internal improvements program of, 40; on military roads, 3; on national highways, 41; tariff under, 111
Washington County, Pennsylvania: manufacturing output of, 25, 258n63
Washington Turnpike and Road Company (Maryland), 57
Way's Packet Directory, 201, 258n60, 278n6
Webster, Daniel, 60; on Ohio River, 70; on railroads, 159–60, 161, 281n29; on steamboat commerce, 22
Weingast, Barry R., 251n7, 287n28
Weller, John B., 99; extensionism of, 100; military service of, 270n81
Wentworth, John, 65; on western revenues, 73
Western and Atlantic Railroad, 161
Western Journal and Civilian, 14, 16, 18
western rivers: lower reaches of, 146; mapping of, 4; national significance of, 7; snags on, 29–30, 32–33; steamboat construction for, 24, 25; town formation along, 130, 133, 275n28
Western River Transportation (Haites, Mak, and Walton), 116–17, 118, 177–78, 202, 272nn8–9; constitutional considerations in, 282n2; productivity index of, 178–79, 180, 183–84; steamboat speed in, 278n12
western states: admission to Union, 45; agricultural products from, 274n23; boosterism in, 255n36; Calhoun's support for, 69–73; centrality claims of, 69; congressional appropriations for, 272n9; internal improvements in, 51; Jackson and, 50; and memorials to Congress, 74–80, 105–6; regional bias against, 77; revenues from, 73; river improvements for, 73–84; role in national policy, 2–3; unity with South, 69–73, 266n22, 267n31
Wheeling (Virginia) bridge: Supreme Court decision on, 128, 273n20
Whig Party: coalition of with Democrats, 83–84, 88; deterioration of, 88, 101, 268n54, 270n87; economic development platform of, 78; effect of Mexican War on, 101; and internal improvements, 62–63; and loss of congressional seats, 83; sectionalism in, 88; support of for American System, 60; support of for river improvements, 83; and Tariff of 1846, 83
White Cloud (packet): burning of, 15, 16
Whitney, Eli, 23
Woodbridge, William, 81
Woodson, Silas, 195, 288n18
wrecks: numbers of, 214; removal of, 4, 136; tonnage of, 215. *See also* navigation hazards; steamboat losses

Yancey, William Lowndes, 1, 66–67; secession threats by, 100; on western states, 69
Yazoo River: snags on, 31; steamboats on, 257n60; stern-wheelers on, 286n18
Yorktown (packet), 23
Young, Brigham, 123